Lecture Notes in Mathematics

C.I.M.E. Foundation Subseries

Volume 2348

Editors-in-Chief
Jean-Michel Morel, City University of Hong Kong, Kowloon Tong, China
Bernard Teissier, IMJ-PRG, Paris, France

Series Editors
Karin Baur, University of Leeds, Leeds, UK
Michel Brion, UGA, Grenoble, France
Rupert Frank, LMU, Munich, Germany
Annette Huber, Albert Ludwig University, Freiburg, Germany
Davar Khoshnevisan, The University of Utah, Salt Lake City, UT, USA
Ioannis Kontoyiannis, University of Cambridge, Cambridge, UK
Angela Kunoth, University of Cologne, Cologne, Germany
Ariane Mézard, IMJ-PRG, Paris, France
Mark Podolskij, University of Luxembourg, Esch-sur-Alzette, Luxembourg
Mark Policott, Mathematics Institute, University of Warwick, Coventry, UK
László Székelyhidi, MPI for Mathematics in the Sciences, Leipzig, Germany
Gabriele Vezzosi, UniFI, Florence, Italy
Anna Wienhard, MPI for Mathematics in the Sciences, Leipzig, Germany

Fondazione C.I.M.E., Firenze

C.I.M.E. stands for *Centro Internazionale Matematico Estivo*, that is, International Mathematical Summer Centre. Conceived in the early fifties, it was born in 1954 in Florence, Italy, and welcomed by the world mathematical community: it continues successfully, year for year, to this day.

Many mathematicians from all over the world have been involved in a way or another in C.I.M.E.'s activities over the years. The main purpose and mode of functioning of the Centre may be summarised as follows: every year, during the summer, sessions on different themes from pure and applied mathematics are offered by application to mathematicians from all countries. A Session is generally based on three or four main courses given by specialists of international renown, plus a certain number of seminars, and is held in an attractive rural location in Italy.

The aim of a C.I.M.E. session is to bring to the attention of younger researchers the origins, development, and perspectives of some very active branch of mathematical research. The topics of the courses are generally of international resonance. The full immersion atmosphere of the courses and the daily exchange among participants are thus an initiation to international collaboration in mathematical research.

C.I.M.E. Director (2002 – 2014)
Pietro Zecca
Dipartimento di Energetica "S. Stecco"
Università di Firenze
Via S. Marta, 3
50139 Florence
Italy
e-mail: zecca@unifi.it

C.I.M.E. Director (2023 –)
Paolo Salani
Dipartimento di Matematica e Informatica "U.Dini"
Università di Firenze
viale G.B. Morgagni 67/A
50134 Florence
Italy
e-mail: salani@math.unifi.it

C.I.M.E. Secretary
Daniele Angella
Dipartimento di Matematica "U. Dini"
Università di Firenze
viale G.B. Morgagni 67/A
50134 Florence
Italy
e-mail: daniele.angella@unifi.it

CIME activity is carried out with the collaboration and financial support of INdAM (Istituto Nazionale di Alta Matematica)

For more information see CIME's homepage: **http://www.cime.unifi.it**

Rupert Frank • Giuseppe Mingione • Lubos Pick •
Ovidiu Savin • Jean Van Schaftingen

Geometric and Analytic Aspects of Functional Variational Principles

Cetraro, Italy 2022

Andrea Cianchi • Vladimir Maz'ya • Tobias Weth
Editors

Authors

Rupert Frank
Institute of Mathematics
LMU Munich
München, Germany

Lubos Pick
Department of Mathematical Analysis,
Faculty of Mathematics and Physics Charles
University
Prague, Czech Republic

Jean Van Schaftingen
Département de Mathématique
Université Catholique de Louvain
Louvain-la-Neuve, Belgium

Giuseppe Mingione
Dipartimento di Matematica
Universitá di Parma
Parma, Italy

Ovidiu Savin
Department of Mathematics
Columbia University
New York, NY, USA

Editors

Andrea Cianchi
Dipartimento di Matematica e Informatica
"U.Dini"
University of Florence
Florence, Italy

Tobias Weth
Institute of Mathematics
Goethe-University Frankfurt
Frankfurt am Main, Germany

Vladimir Maz'ya
Department of Mathematics
Linköping University
Linköping, Sweden

ISSN 0075-8434　　　　　　　　ISSN 1617-9692　(electronic)
Lecture Notes in Mathematics
ISSN 2946-1812　　　　　　　　ISSN 2946-1820　(electronic)
C.I.M.E. Foundation Subseries
ISBN 978-3-031-67600-0　　　　ISBN 978-3-031-67601-7　(eBook)
https://doi.org/10.1007/978-3-031-67601-7

© The Editor(s) (if applicable) and The Author(s), under exclusive license to Springer Nature Switzerland AG 2024

This work is subject to copyright. All rights are solely and exclusively licensed by the Publisher, whether the whole or part of the material is concerned, specifically the rights of translation, reprinting, reuse of illustrations, recitation, broadcasting, reproduction on microfilms or in any other physical way, and transmission or information storage and retrieval, electronic adaptation, computer software, or by similar or dissimilar methodology now known or hereafter developed.
The use of general descriptive names, registered names, trademarks, service marks, etc. in this publication does not imply, even in the absence of a specific statement, that such names are exempt from the relevant protective laws and regulations and therefore free for general use.
The publisher, the authors and the editors are safe to assume that the advice and information in this book are believed to be true and accurate at the date of publication. Neither the publisher nor the authors or the editors give a warranty, expressed or implied, with respect to the material contained herein or for any errors or omissions that may have been made. The publisher remains neutral with regard to jurisdictional claims in published maps and institutional affiliations.

This Springer imprint is published by the registered company Springer Nature Switzerland AG
The registered company address is: Gewerbestrasse 11, 6330 Cham, Switzerland

If disposing of this product, please recycle the paper.

Preface

Optimization problems of geometric-analytic nature are a key to solving many open questions in mathematics and physics. The general aim is to minimize a geometric or analytic quantity, which often has the physical meaning of an energy, among a prescribed collection of sets or functions. These problems drive the development of powerful methods in the calculus of variations, the theory of partial differential equations, and geometric analysis. In turn, analytical problems benefit from information arising from optimal functional-geometric inequalities. The courses delivered at the school emphasized several aspects of the fascinating interplay between these inequalities and problems of differential and variational types. The focus was on the following main topics that are intimately related: functional and geometric inequalities, optimal norms and sharp constants in Sobolev-type inequalities, regularity of solutions to variational problems.

Starting with the seminal works of Rodemich, Talenti, and Aubin more than 40 years ago, the sharp form of functional inequalities of Sobolev type has been attracting extensive attention in the last decades. In the first part of his course, Rupert Frank reviewed the full scale of higher order and fractional Hilbertian Sobolev inequalities in the entire space and their conformally equivalent analogues on the unit sphere. In particular, he emphasized a modern rearrangement-free approach, which allows to detect best constants and optimizers also for wider classes of functional inequalities where the symmetric decreasing rearrangement technique is not available. In the second part of his course, Frank presented classical and recent results on the characterization of sharp remainder terms giving rise to the stability of Hilbertian Sobolev inequalities. He also discussed the open problem of finding remainder term estimates with explicit constants, and he finished his course with a presentation of a recent result on optimal degenerate, higher-order stability of a Sobolev inequality on a product of round spheres.

An important part of the regularity theory of partial differential equations is concerned with gradient estimates for solutions. Calderón-Zygmund type estimates and nonlinear potential theory are central aspects of the modern developments of this theory. In his course, Giuseppe Mingione offered an overview of several recent advances in these topics. Specifically, he discussed local gradient regularity results for solutions to possibly non-uniformly elliptic variational problems. The boundedness of their gradient and its Hölder continuity were discussed in quite general settings. The results presented extend and improve those in the existing literature for both autonomous and non-autonomous functionals of the calculus of variations. Applications to special instances, such as functionals with rapid non-polynomial growth, double-phase functionals, functionals with so-called $p - q$ growth, variable exponent functionals, were exhibited. In the last part of his course, Mingione also sketched a proof of one of the results exposed. This enabled the participants to taste the main difficulties to be faced and the new ideas employed in attacking the relevant problems.

In his course, Luboš Pick discussed a variety of reduction principles for functional inequalities in Banach spaces of functions of several variables. In this context, by a reduction principle, one understands a powerful technique that enables one to reduce an inequality involving functions of several variables to an equivalent but considerably simpler inequality involving a weighted integral operator of Hardy-type acting on functions defined on an interval. Such a result, wherever available, not only essentially simplifies the research of the action of operators on function spaces but has also several important applications, one of the principal ones being the characterization of the optimal partner space with respect to a given operator and a given space. The results that were covered in the course have been obtained mostly in the recent two decades. The method, which started as a technique tailored for certain specific problems concerning limiting embeddings of Euclidean–Sobolev spaces into rearrangement-invariant spaces, has evolved over the years into an independent theory with a wide variety of important consequences. Applications were offered to Euclidean–Sobolev embeddings, Gaussian–Sobolev embeddings, higher-order embeddings, general Sobolev embeddings involving Frostman measures, trace theorems. The boundedness of classical operators of harmonic analysis, such as the maximal operator, the fractional maximal operator, singular operators, the Laplace transform, or the Riesz potential, was also analyzed.

Monge-Ampère type equations are a primary instance of the so-called fully nonlinear elliptic equations. They are linked to various questions in analysis and differential geometry. After introducing the basic notions, Ovidiu Savin focused his course on a few aspects of the classical regularity theory of their solutions. In particular, geometric properties of solutions, such as those related to sections, which are critical in the approach to the regularity of solutions, were introduced. Classical Pogorelov estimates and John's lemma were presented, as well as singular solutions that provide counterexamples to the regularity when certain assumptions are dropped. The last part of the course was devoted to the celebrated Hölder regularity theory for the gradient of solutions developed by Caffarelli. Connections of the theory of Monge-Ampère equations with the Monge-Kantorovich theory of optimal transport problems, whose modern developments were ignited by the work of Brenier, were also discussed.

The course of Jean Van Schaftingen was devoted to Sobolev estimates of Gagliardo-Nirenberg type for vector differential operators in the endpoint case $p = 1$. By a classical result of Calderón and Zygmund from 1952, the ellipticity of a homogeneous vector differential operator of order k is both necessary and sufficient to control all k-th order derivatives of a vector field in the Lp -norm for any finite exponent $p > 1$. On the other hand, as shown by Ornstein in 1962, only trivial estimates on same-order derivatives extend to the endpoint case $p = 1$, while ellipticity is a necessary but not sufficient condition for endpoint Sobolev estimates. After reviewing these results, Van Schaftingen briefly discussed a collection of known endpoint inequalities including the Korn-Sobolev inequality of Strauss and the Hodge-Sobolev inequality of Bourgain and Brezis. In the main part of his course, Van Schaftingen introduced the canceling condition to characterize the class of differential operators giving rise to endpoint Sobolev estimates. For the derivation of this characterization, he elaborated a key connection to a dual inequality for L^1-vector fields belonging to the kernel of a cocanceling differential operator. This dual inequality generalizes a result of Bourgain and Brezis for divergence-free vector fields. In the presentation of underlying methods, Van Schaftingen highlighted the combination of algebraic with geometric and harmonic analysis techniques. In the last part of his course, he also included a brief presentation of associated fractional estimates.

Florence, Italy   Andrea Cianchi
Linköping, Sweden   Vladimir Maz'ya
Frankfurt am Main, Germany   Tobias Weth

Contents

1. **The Sharp Sobolev Inequality and Its Stability: An Introduction** 1
   Rupert L. Frank

2. **Nonlinear Potential Theoretic Methods in Nonuniformly Ellliptic Problems** 65
   Giuseppe Mingione

3. **Reduction Principles** 151
   Luboš Pick

4. **The Monge-Ampère Equation** 227
   Ovidiu Savin

5. **Injective Ellipticity, Cancelling Operators, and Endpoint Gagliardo-Nirenberg-Sobolev Inequalities for Vector Fields** 259
   Jean Van Schaftingen

Chapter 1
The Sharp Sobolev Inequality and Its Stability: An Introduction

Rupert L. Frank

Abstract These notes are an extended version of a series of lectures given at the CIME Summer School in Cetraro in June 2022. The goal is to explain questions about optimal functional inequalities on the example of the sharp Sobolev inequality and its fractional generalizations. Topics covered include compactness theorems for optimizing sequences, characterization of optimizers and quantitative stability.

1.1 Introduction and Outline

The Sobolev inequality on $\mathbb{R}^d$, $d \geq 3$, states that

$$\int_{\mathbb{R}^d} |\nabla u|^2 \, dx \gtrsim \left(\int_{\mathbb{R}^d} |u|^{2d/(d-2)} \, dx \right)^{(d-2)/2}, \quad (1.1)$$

provided the function u belongs to the homogeneous Sobolev space $\dot{H}^1(\mathbb{R}^d)$, defined as the completion of $C_c^1(\mathbb{R}^d)$ with respect to the L^2-norm of the gradient. We restrict ourselves in these lectures to real-valued functions. The Sobolev inequality (1.1) is of great importance in several areas of mathematical analysis,

---

© 2023 by the author. This paper may be reproduced, in its entirety, for noncommercial purposes. Partial support through US National Science Foundation grant DMS-1954995, as well as through the Deutsche Forschungsgemeinschaft through Germany's Excellence Strategy EXC-2111-390814868 and through TRR 352—Project-ID 470903074 is acknowledged.

R. L. Frank (✉)
Mathematisches Institut, Ludwig-Maximilians Universität München, München, Germany

Munich Center for Quantum Science and Technology, München, Germany

Mathematics 253-37, Caltech, Pasadena, CA, USA
e-mail: r.frank@lmu.de

© The Author(s), under exclusive license to Springer Nature Switzerland AG 2024
A. Cianchi et al. (eds.), *Geometric and Analytic Aspects of Functional Variational Principles*, C.I.M.E. Foundation Subseries 2348,
https://doi.org/10.1007/978-3-031-67601-7_1

including the calculus of variations, the theory of PDEs, differential geometry and mathematical physics.

The $\gtrsim$-sign in (1.1) means that there is a positive constant, depending only on d, such that the inequality holds with that constant on the right side. In several applications one is interested in the optimal value of this constant, that is, in the number

$$S_d := \inf_{0 \neq u \in \dot{H}^1(\mathbb{R}^d)} \frac{\int_{\mathbb{R}^d} |\nabla u|^2 \, dx}{\left(\int_{\mathbb{R}^d} |u|^{2d/(d-2)} \, dx\right)^{(d-2)/2}}.$$

Related to this is the question whether the supremum defining S_d is attained for some function u and, if so, whether one can characterize all such functions. Again motivated by applications, once this has been carried out one would like to know whether the fact that for some function $0 \neq u \in \dot{H}^1(\mathbb{R}^d)$ the quotient between the left and right sides of (1.1) is close to the optimal value S_d already implies that u is close to a function for which the supremum is attained. This question is deliberately vague. One needs to specify in which sense the closeness between two functions is understood, and in which sense the closeness of the quotient to the optimal constant is related to the closeness between u and optimal functions. It turns out that in the context of the Sobolev inequality all these questions can be answered, and this is the topic of this series of lectures.

Let us take a step back from this concrete problem. The Sobolev inequality is just one (although a paradigmatic) example of a functional inequality and the questions outlined above can be equally asked for other such inequalities. This suggests the following research program in the field of functional inequalities:

(0) Prove the validity of the functional inequality with some constant.
(1a) Show that there are optimizing functions.
(1b) Show that optimizing sequences are relatively compact (up to symmetries).
(2a) Determine the optimal constant.
(2b) Characterize the optimizers.
(2c) Show that the Hessian around optimizers is nondegenerate (up to symmetries).
(3) Show stability of the functional inequality.

The meaning of some of these assertions might not be clear at this point, but the hope is that it will be at the end of this series of lectures. The rough plan of this course is to devote each one of the first three lectures to one of the above Steps 1, 2 and 3 and to spend the fourth lecture on a related, but different inequality, where we repeat all three steps in this new setting. Step 0, namely in our case the validity of the Sobolev inequality (1.1) with some constant, will be taken for granted. In fact, an improved version of this inequality will be proved in the first lecture.

The methods used in these steps vary widely. Those in Step 1 are probably the most robust, while those in Step 2 are probably the most specialized. In the context of the Sobolev inequality Step 3 consists of a combination of the Steps 1b and 2c. In the fourth lecture, however, we will see an example of a functional inequality where additional input is needed in this step.

In order to emphasize the general nature of this program, we consider, apart from the Sobolev inequality (1.1), also its fractional counterpart

$$\int_{\mathbb{R}^d} |(-\Delta)^{s/2} u|^2 \, dx \gtrsim \left(\int_{\mathbb{R}^d} |u|^{2d/(d-2s)} \, dx \right)^{(d-2s)/d}, \qquad (1.2)$$

where s is a real number satisfying $0 < s < d/2$. In terms of the Fourier transform

$$\widehat{u}(\xi) := (2\pi)^{-d/2} \int_{\mathbb{R}^d} e^{-i\xi \cdot x} u(x) \, dx ,$$

the left side of (1.2) is equal to

$$\int_{\mathbb{R}^d} |(-\Delta)^{s/2} u|^2 \, dx = \int_{\mathbb{R}^d} |\xi|^{2s} |\widehat{u}(\xi)|^2 \, d\xi . \qquad (1.3)$$

Inequality (1.2) is valid for function u in the homogeneous Sobolev space $\dot{H}^s(\mathbb{R}^d)$ of tempered distributions whose Fourier transform belongs to $L^1_{\text{loc}}(\mathbb{R}^d)$ and for which the right side of (1.3) is finite; see, e.g., [5, Section 1.3].

It is easy to see that for $s = 1$ this definition of $\dot{H}^1(\mathbb{R}^d)$ coincides with that given before and that

$$\int_{\mathbb{R}^d} |(-\Delta)^{1/2} u|^2 \, dx = \int_{\mathbb{R}^d} |\nabla u|^2 \, dx .$$

Therefore (1.2) is indeed a generalization of (1.1).

In Lecture 4 we will discuss a version of the Sobolev inequality (1.1) on the manifold $(\mathbb{R}/T\mathbb{Z}) \times \mathbb{S}^{d-1}$, depending on the parameter $T > 0$. A specific feature of this case is that for a certain value of T property (2c) fails. It is instructive to see a repetition of the previous steps, both for those values of T where (2c) holds and where it fails.

Finally, we note that there have been some developments concerning the stability question for the Sobolev inequality since this course took place at the CIME Summer School in Cetraro in June 2022. We have made the decision not to include those in order to keep the character of these notes more elementary and instead to refer to the recent preprints and papers [43, 54, 65–67]; see also the brief remarks at the end of the third lecture.

It is my pleasure to thank the organizers of the summer school, Andrea Cianchi, Vladimir Maz'ya and Tobias Weth, as well as Paolo Salani for their kind invitation, as well as the participants of the school for their interest in these topics. I am grateful to Jean Dolbeault for his help with references and to Tobias König and Jonas Peteranderl for many useful comments on these notes.

1.2 Lecture 1: Optimizing Sequences

In this first lecture we are interested in the optimization problem

$$S_{d,s} := \inf_{0 \neq u \in \dot{H}^s(\mathbb{R}^d)} \frac{\|(-\Delta)^{s/2} u\|_2^2}{\|u\|_q^2}, \qquad (1.4)$$

where, as always in this series of lectures,

$$0 < s < \tfrac{d}{2} \quad \text{and} \quad q = \tfrac{2d}{d-2s}.$$

More specifically, we are interested in

(a) existence of an optimizer
(b) relative compactness (up to symmetries) of optimizing sequences.

The difference between (a) and (b) is that for (a) it suffices to find *one* optimizing sequence that converges, whereas for (b) one wants to show that *any* optimizing sequence has a subsequence that converges (up to symmetries); so (b) is stronger than (a). For the arguments in Lecture 2 property (a) would be enough, but in Lecture 3 we need property (b), so this is what we will prove in this lecture.

Let us explain the main difficulty when dealing with the behavior of optimizing sequences and, at the same time, explain the expression 'up to symmetries' in (b). A basic strategy in the calculus of variations to solve an optimization problem is to show that from an optimizing sequence one can extract a convergent subsequence and that its limit is an optimizer. Typically, the extracted subsequence converges a priori not in the original sense (here strong convergence in $\dot{H}^s(\mathbb{R}^d)$), but only in a weaker sense (namely weakly in $\dot{H}^s(\mathbb{R}^d)$). At this point the (noncompact) symmetries of the variational problem enter. If $u \in \dot{H}^s(\mathbb{R}^d)$ is a given function and if $(a_n) \subset \mathbb{R}^d$ and $(\lambda_n) \subset \mathbb{R}_+ = (0, \infty)$ are sequences with $|a_n| + \lambda_n + \lambda_n^{-1} \to \infty$, then the sequence

$$\lambda_n^{-d/q} u(\lambda_n^{-1}(\cdot - a_n))$$

converges weakly to zero in $\dot{H}^s(\mathbb{R}^d)$. Moreover, for these sequences the quotient in (1.4) is independent of n, reflecting the translation and dilation invariance of the optimization problem. For instance, if u is an optimizer for (1.4), then every element of this sequence is an optimizer as well, and we have constructed an optimizing sequence that converges weakly to zero. This explains why in (b) we can hope for relative compactness at most *up to translations and dilations*. These are the symmetries in question.

We now formulate the main result of this lecture, which is due to Lions.

Theorem 1.1 *Let* $0 < s < \tfrac{d}{2}$ *and* $q := \tfrac{2d}{d-2s}$. *Let* $(u_n) \subset \dot{H}^s(\mathbb{R}^d)$ *with* $\|(-\Delta)^{s/2} u_n\|_2 = 1$ *and* $\|u_n\|_q^2 \to S_{d,s}^{-1}$. *Then there is a subsequence* (u_{n_k}), *as* well as sequences $(a_k) \subset \mathbb{R}^d$ and $(\lambda_k) \subset \mathbb{R}_+$ such that the sequence of functions

$$\lambda_k^{-d/q} u_{n_k}(\lambda_k^{-1}(\cdot - a_k))$$

converges in $\dot{H}^s(\mathbb{R}^d)$ *to an optimizer of* (1.4).

Before embarking into the details of the proof, let us give a rough outline of the strategy. We argued above that translations and dilations are a possible loss of compactness. One key step in the proof of Theorem 1.1 is to show that translations and dilations (and their combination) are the *only* possible loss of compactness: that is, after applying suitable translations and dilations one can always ensure that the weak limit of a subsequence is nonzero. The mathematical tool here is a refinement of the Sobolev inequality (see Proposition 1.2 below), which involves an extra term containing a supremum over dilation and translation parameters. This allows one to translate and dilate the elements of an optimizing sequence such that this extra term stays away from zero, which translates into the integral against a fixed function in $L^{q'}(\mathbb{R}^d)$ (here and throughout: $q' = q/(q-1)$) being bounded away from zero. This implies that the weak limit is nonzero. Thus, we have shown that 'there is something somewhere'.

The second step in the proof of Theorem 1.1 is to show that 'there is nothing else anywhere else'. To explain the argument we assume that no translation, no dilation and no subsequence is necessary and we denote by u the nonzero weak limit of (u_n). Writing $u_n = u + r_n$ one can show that both the numerator and the denominator in the quotient in (1.4) asymptotically decouple in the sense that

$$\|(-\Delta)^{s/2} u_n\|_2^2 = \|(-\Delta)^{s/2} u\|_2^2 + \|(-\Delta)^{s/2} r_n\|_2^2 + o(1) \text{ and}$$

$$\|u_n\|_q^q = \|u\|_q^q + \|r_n\|_q^q + o(1).$$

The strict subadditivity of the function $\mu \mapsto \mu^{2/q}$ (since $q > 2$) can then be used to show that it is favorable to keep all the mass together, that is, to have r_n tending to zero. This argument is due to Lieb and referred to as the *method of the missing mass*.

We now turn to the details of the proof of Theorem 1.1.

1.2.1 Step 1: There Is Something Somewhere

We begin by proving the following *refined Sobolev inequality*.

Proposition 1.2 *Let* $0 < s < \frac{d}{2}$ *and* $q := \frac{2d}{d-2s}$. *Let* $\chi \in C_c^\infty(\overline{\mathbb{R}_+})$ *with* $\chi = 1$ *near the origin. Then for all* $u \in \dot{H}^s(\mathbb{R}^d)$,

$$\|u\|_q \lesssim \left(\sup_{t>0} t^{(d-2s)/4} \|\chi(-t\Delta) u\|_\infty \right)^{1-2/q} \|(-\Delta)^{s/2} u\|_2^{2/q}. \tag{1.5}$$

The supremum in (1.5) is one of the possible, equivalent norms in the Besov space $\dot{B}^{s-d/2}_{\infty,\infty}(\mathbb{R}^d)$. Our presentation, however, is selfcontained and does not need anything from the theory of these spaces.

We call the inequality in the proposition a 'refined' Sobolev inequality since it implies the Sobolev inequality (with nonsharp constant). To see this, we note that the Fourier multiplier $\chi(-t\Delta)$ acts as convolution with the function $t^{-d/2}g(t^{-1/2}\cdot)$, where

$$g(x) := (2\pi)^{-d} \int_{\mathbb{R}^d} \chi(|\xi|^2) e^{-i\xi \cdot x} \, d\xi \,. \tag{1.6}$$

Since $g \in L^{q'}(\mathbb{R}^d)$, it follows from Hölder's inequality that

$$\|\chi(-t\Delta)u\|_\infty \leq \|t^{-d/2}g(t^{-1/2}\cdot)\|_{q'} \|u\|_q = t^{-(d-2s)/4} \|g\|_{q'} \|u\|_q \,.$$

Thus, the supremum in (1.5) is $\leq \|g\|_{q'} \|u\|_q$. Inserting this inequality into (1.5), we obtain the Sobolev inequality with nonsharp constant.

Proof Using the layer cake representation (see, e.g., [74, Theorem 1.13]), we write

$$\|u\|_q^q = q \int_0^\infty |\{|u| > \tau\}| \tau^{q-1} \, d\tau$$

and bound for each fixed $\tau > 0$ with some $t > 0$ to be specified

$$|\{|u| > \tau\}| \leq |\{|\chi(-t\Delta)u| > \tau/2\}| + |\{|(1 - \chi(-t\Delta))u| > \tau/2\}| \,.$$

In particular, choosing $t = t_\tau$ such that

$$\tau/2 = t_\tau^{-(d-2s)/4} M_u, \qquad \text{where } M_u := \sup_{t>0} t^{(d-2s)/4} \|\chi(-t\Delta)u\|_\infty,$$

we see that

$$|\{|\chi(-t_\tau \Delta)u| > \tau/2\}| = 0 \,.$$

Meanwhile, we bound

$$|\{|(1 - \chi(-t_\tau \Delta))u| > \tau/2\}| \leq (2/\tau)^2 \|(1 - \chi(-t_\tau \Delta))u\|_2^2$$

and arrive at

$$\|u\|_q^q \leq q \int_0^\infty (2/\tau)^2 \|(1 - \chi(-t_\tau \Delta))u\|_2^2 \tau^{q-1} \, d\tau$$

$$= q \int_{\mathbb{R}^d} |\widehat{u}(\xi)|^2 \int_0^\infty (2/\tau)^2 |1 - \chi(-t_\tau |\xi|^2)|^2 \tau^{q-1} \, d\tau \, d\xi \,,$$

1 The Sharp Sobolev Inequality and Its Stability: An Introduction

where $\hat{u}$ denotes the Fourier transform of u. By scaling, we find

$$q \int_0^\infty (2/\tau)^2 |1 - \chi(-t_\tau |\xi|^2)|^2 \tau^{q-1} \, d\tau = C_{d,s,\chi} |\xi|^{2s} M_u^{q-2}.$$

The constant $C_{d,s,\chi}$ is finite by the properties of χ, thus proving the proposition. □

We now apply Proposition 1.2 in the setting of Theorem 1.1. Let $(u_n) \subset \dot{H}^s(\mathbb{R}^d)$ be a sequence with

$$\|(-\Delta)^{s/2} u_n\|_2 = 1 \tag{1.7}$$

and

$$\limsup_{n \to \infty} \|u_n\|_q > 0. \tag{1.8}$$

Note that this is, in particular, satisfied for an optimizing sequence. In this step, however, we will use the weaker property (1.8) rather than the optimizing property.

Our goal is to show that after a translation and a dilation, (u_n) has a subsequence with nonzero weak limit.

Inserting (1.7) and (1.8) into (1.5), we deduce that

$$\limsup_{n \to \infty} \sup_{t > 0, \, a \in \mathbb{R}^d} t^{(d-2s)/4} |(\chi(-t\Delta) u_n)(a)| > 0.$$

Choosing $t_n > 0$ and $a_n \in \mathbb{R}^d$ such that

$$t_n^{(d-2s)/4} |(\chi(-t_n \Delta) u_n)(a_n)| \geq \tfrac{1}{2} \sup_{t > 0, \, a \in \mathbb{R}^d} t^{(d-2s)/4} |(\chi(-t\Delta) u_n)(a)|,$$

we see that

$$\limsup_{n \to \infty} t_n^{(d-2s)/4} |(\chi(-t_n \Delta) u_n)(a_n)| > 0.$$

Thus, the translated and dilated functions

$$\tilde{u}_n(x) := t_n^{(d-2s)/4} u_n(t_n^{1/2}(x - a_n))$$

satisfy

$$\|(-\Delta)^{s/2} \tilde{u}_n\|_2 = \|(-\Delta)^{s/2} u_n\|_2, \qquad \|\tilde{u}_n\|_q = \|u_n\|_q$$

and, with g defined in (1.6),

$$\int_{\mathbb{R}^d} g(x)\widetilde{u}_n(x)\,dx = t_n^{(d-2s)/4}(\chi(-t_n\Delta)u_n)(a_n).$$

By weak compactness, we obtain a subsequence and a $\widetilde{u} \in \dot{H}^s(\mathbb{R}^d)$ such that $\widetilde{u}_{n_k} \rightharpoonup \widetilde{u}$ in $\dot{H}^s(\mathbb{R}^d)$. We can choose the subsequence in such a way that in addition

$$\liminf_{k\to\infty}\left|\int_{\mathbb{R}^d} g(x)\widetilde{u}_{n_k}(x)\,dx\right| > 0.$$

By the Sobolev inequality, $\widetilde{u}_{n_k} \rightharpoonup u$ in $L^q(\mathbb{R}^d)$ and, since $g \in L^{q'}(\mathbb{R}^d)$, we find

$$\left|\int_{\mathbb{R}^d} g(x)\widetilde{u}(x)\,dx\right| = \liminf_{k\to\infty}\left|\int_{\mathbb{R}^d} g(x)\widetilde{u}_{n_k}(x)\,dx\right| > 0.$$

Thus, $\widetilde{u} \neq 0$, as we set out to prove.

1.2.2 Step 2: There Is Nothing Else Anywhere Else

Let (u_n) be a minimizing sequence for (1.4). We normalize the sequence as in (1.7). After translations, dilations and passing to a subsequence, we may assume that

$$u_n \rightharpoonup u \text{ in } \dot{H}^s(\mathbb{R}^d) \qquad \text{with } u \neq 0.$$

The fact that $u \neq 0$ follows from Step 1. We write

$$u_n = u + r_n \qquad \text{with } r_n \rightharpoonup 0 \text{ in } \dot{H}^s(\mathbb{R}^d).$$

From the Hilbert space structure of $\dot{H}^s(\mathbb{R}^d)$ and the normalization (1.7) we immediately deduce that

$$t := \lim_{n\to\infty} \|(-\Delta)^{s/2} r_n\|_2^2 \qquad \text{exists and satisfies} \qquad 1 = \|(-\Delta)^{s/2} u\|_2^2 + t. \tag{1.9}$$

We now argue that

$$m := \lim_{n\to\infty} \|r_n\|_q^q \qquad \text{exists and satisfies} \qquad S_{d,s}^{-q/2} = \|u\|_q^q + m. \tag{1.10}$$

1 The Sharp Sobolev Inequality and Its Stability: An Introduction

Indeed, from the weak convergence $u_n \rightharpoonup u$ in $\dot{H}^s(\mathbb{R}^d)$ one can deduce that $u_n \to u$ in $L^2_{\text{loc}}(\mathbb{R}^d)$ (arguing as in [74, Theorem 8.6]) and then, after passing to a subsequence, $u_n \to u$ almost everywhere. Thus, by the Brezis–Lieb lemma [74, Theorem 1.9],

$$\lim_{n\to\infty} \int_{\mathbb{R}^d} \big| |u_n|^q - |u|^q - |u_n - u|^q \big| \, dx = 0. \tag{1.11}$$

The optimizing property of (u_n) and the normalization (1.7) imply that $\|u_n\|_q^2 \to S_{d,s}^{-1}$. Inserting this information into (1.11), we obtain (1.10) along a subsequence. By a standard argument it holds in fact along the full sequence. (Otherwise, there existed a subsequence such that $\lim_{k\to\infty} \|r_{n_k}\|_q$ exists and is different from $S_{d,s}^{-q/2} - \|u\|_q^q$. Repeating the above argument for this subsequence, we arrive at a contradiction.) This proves (1.10).

From the Sobolev inequality, we know that $\|(-\Delta)^{s/2} r_n\|_2^2 \geq S_{d,s} \|r_n\|_q^2$ and therefore

$$t \geq S_{d,s} m^{2/q}. \tag{1.12}$$

Putting (1.9), (1.10) and (1.12) together, we find

$$1 = \|(-\Delta)^{s/2} u\|_2^2 + t$$
$$\geq \|(-\Delta)^{s/2} u\|_2^2 + S_{d,s} m^{2/q}$$
$$= \|(-\Delta)^{s/2} u\|_2^2 + \left(1 - S_{d,s}^{q/2} \|u\|_q^q\right)^{2/q}$$
$$\geq \|(-\Delta)^{s/2} u\|_2^2 + 1 - S_{d,s} \|u\|_q^2.$$

In the last inequality we used the elementary fact that

$$(a+b)^{2/q} \leq a^{2/q} + b^{2/q} \quad \text{for all } a, b \geq 0, \tag{1.13}$$

which relies on the fact that $q \geq 2$. Thus, we have shown that $\|(-\Delta)^{s/2} u\|_2^2 \leq S_{d,s} \|u\|_q^2$, which, taking into account that $u \neq 0$, implies that u is an optimizer for $S_{d,s}$. Thus, we have accomplished our first goal, namely showing the existence of an optimizer.

To reach our second goal, namely showing relative compactness of optimizing sequences, we observe that, since $q > 2$, equality in (1.13) occurs only when a or b is zero. Since in our application $b = S_{d,s}^{q/2} \|u\|_q^q$ is nonzero, we conclude that $a = 1 - S_{d,s}^{q/2} \|u\|_q^q$ is zero. According to (1.10) this means that $m = 0$. Since we also need to have equality in (1.12), we conclude that $t = 0$, that is, $r_n \to 0$ in $\dot{H}^s(\mathbb{R}^d)$. Note that this is *strong* convergence. Thus, we have shown $u_n \to u$ in $\dot{H}^s(\mathbb{R}^d)$, as claimed.

1.2.3 Appendix: The Hardy–Littlewood–Sobolev Inequality

Several results mentioned in this series of lectures were originally proved for a family of functional inequalities called *Hardy–Littlewood–Sobolev inequalities*, which is in a certain sense dual to the family of Sobolev inequalities considered here. While we have consistently used the latter formulation, it is worthwhile to explain this connection.

The family of Hardy–Littlewood–Sobolev (HLS) inequalities is a two-parameter family of inequalities, depending on parameters $0 < \lambda < d$ and $1 < p < \frac{d}{d-\lambda}$, and states that

$$\left\| |x|^{-\lambda} * f \right\|_q \lesssim \|f\|_p \qquad \text{with } \tfrac{1}{q} = \tfrac{1}{p} - \tfrac{d-\lambda}{d}.$$

This is a generalization of Young's convolution inequality, where the functions $|x|^{-\lambda}$ do not belong to the Lebesgue space $L^{d/\lambda}(\mathbb{R}^d)$, but only to its weak counterpart.

Relevant for us are two one-parameter families, namely those corresponding to $p = 2$ and to $q = 2$. The inequalities in these cases are dual to each other, which means, in particular, that their optimal constants coincide. In this appendix we will explain what this duality implies for the questions of existence and characterization of optimizers, as well as the relative compactness of optimizing sequences. (There is yet another family, corresponding to $q = p'$, that is equivalent, but we will not discuss it here.)

The relation between the Sobolev inequalities discussed in the main part of these lectures and the HLS inequalities discussed in this appendix comes from the well-known fact (see, e.g., [74, Theorem 5.9]) that the operator $(-\Delta)^{-s}$ has integral kernel

$$(-\Delta)^{-s}(x, x') = 2^{-2s} \pi^{-d/2} \frac{\Gamma(\frac{d}{2} - s)}{\Gamma(s)} |x - x'|^{-d+2s}. \tag{1.14}$$

Thus, writing the Sobolev inequality

$$\|(-\Delta)^{s/2} u\|_2^2 \geq S_{d,s} \|u\|_q^2 \qquad \text{for all } u \in \dot{H}^s(\mathbb{R}^d),$$

in the equivalent form

$$\left\|(-\Delta)^{-s/2} f \right\|_q \leq S_{d,s}^{-1/2} \|f\|_2 \qquad \text{for all } f \in L^2(\mathbb{R}^d), \tag{1.15}$$

we obtain the HLS inequality with $p = 2$. Moreover, $S_{d,s}$ being the sharp constant in the Sobolev inequality means that $S_{d,s}^{-1/2}$ is the norm of the operator $(-\Delta)^{-s/2}$ from $L^2(\mathbb{R}^d)$ to $L^q(\mathbb{R}^d)$, and therefore, up to the prefactor in the integral kernel, the optimal constant in the HLS inequality with the exponent 2 on the right side.

1 The Sharp Sobolev Inequality and Its Stability: An Introduction

We apply duality and pass to the HLS inequality with the exponent 2 on the left side of the inequality. Duality implies that $S_{d,s}^{-1/2}$ is equal to the norm of the operator $(-\Delta)^{-s/2}$ from $L^{q'}(\mathbb{R}^d)$ to $L^2(\mathbb{R}^d)$,

$$\left\|(-\Delta)^{-s/2} g\right\|_2 \leq S_{d,s}^{-1/2} \|g\|_{q'} \qquad \text{for all } g \in L^{q'}(\mathbb{R}^d). \tag{1.16}$$

This is the form of the HLS inequality in which it is most naturally studied in connection with sharp constants, compactness and conformal invariance. Note also that, by (1.14),

$$\left\|(-\Delta)^{-s/2} g\right\|_2^2 = 2^{-2s} \pi^{-d/2} \frac{\Gamma(\frac{d}{2} - s)}{\Gamma(s)} \iint_{\mathbb{R}^d \times \mathbb{R}^d} \frac{g(x) g(x')}{|x - x'|^{d-2s}} \, dx \, dx'.$$

We now show that optimizing sequences for (1.15) and (1.16) are in one-to-one correspondence with each other and that convergence of optimizing sequences is equivalent for both problems. We carry this out in a more general setting.

Lemma 1.3 *Let $\mathcal{H}$ be a Hilbert space, let X be a measure space and $1 < q < \infty$. Let $A : \mathcal{H} \to L^q(X)$ be a bounded linear operator, let $A^* : L^{q'}(X) \to \mathcal{H}$ be its adjoint and let $\alpha := \|A\| = \|A^*\|$.*

(a1) *If $f \in \mathcal{H}$ satisfies $\|f\|_{\mathcal{H}} = 1$ and $\|Af\|_q = \alpha$, then*

$$g := \|Af\|_q^{1-q} |Af|^{q-2} Af$$

satisfies $\|g\|_{q'} = 1$ and $\|A^* g\|_{\mathcal{H}} = \alpha$.

(a2) *If $(f_n) \subset \mathcal{H}$ satisfies $\|f_n\|_{\mathcal{H}} = 1$ and $\|Af_n\|_q \to \alpha$, then*

$$g_n := \|Af_n\|_q^{1-q} |Af_n|^{q-2} Af_n$$

satisfies $\|g_n\|_{q'} = 1$ and $\|A^* g_n\|_{\mathcal{H}} \to \alpha$. If, in addition, $g_n \to g$ in $L^{q'}(X)$, then $f_n \to \|A^* g\|_{\mathcal{H}}^{-1} A^* g$ in $\mathcal{H}$.

(b1) *If $g \in L^{q'}(X)$ satisfies $\|g\|_{q'} = 1$ and $\|A^* g\|_{\mathcal{H}} = \alpha$, then*

$$f := \|A^* g\|_{\mathcal{H}}^{-1} A^* g$$

satisfies $\|f\|_{\mathcal{H}} = 1$ and $\|Af\|_q = \alpha$.

(b2) *If $(g_n) \subset L^{q'}(X)$ satisfies $\|g_n\|_{q'} = 1$ and $\|A^* g_n\|_{\mathcal{H}} \to \alpha$, then*

$$f_n := \|A^* g_n\|_{\mathcal{H}}^{-1} A^* g_n$$

satisfies $\|f_n\|_{\mathcal{H}} = 1$ and $\|Af_n\|_q \to \alpha$. If, in addition, $f_n \to f$ in $\mathcal{H}$, then $g_n \to \|Af\|_q^{1-q} |Af|^{q-2} Af$ in $L^{q'}(X)$.

We apply this lemma with $\mathcal{H} = L^2(\mathbb{R}^d)$, $X = \mathbb{R}^d$ and $A = (-\Delta)^{-s/2}$. We infer that the optimal constant in (1.15) is attained if and only if that in (1.16) is attained, and that optimizers are in one-to-one correspondence. Moreover, convergence of an optimizing sequence for one inequality is equivalent to that for the other and, in particular, relative compactness (up to symmetries) for one inequality implies the same for the other. This explains what we mean by the 'equivalence' of the two optimization problems.

Proof The lemma is valid both when the underlying field is that of real and complex numbers. So, while in the rest of these lectures we deal exclusively with real-valued functions, here we will use complex notation.

The proof of (a1) and (b1) is a variation of the proof of the first part of (a2) and (b2), so we only prove the latter. For (a2) we have, clearly, $\|g_n\|_{q'} = 1$ and $\|A^*g_n\|_{\mathcal{H}} \leq \alpha \|g_n\|_{q'} = \alpha$. Meanwhile, since $\|f_n\|_{\mathcal{H}} = 1$,

$$\|A^*g_n\|_{\mathcal{H}} \geq \langle f_n, A^*g_n \rangle_{\mathcal{H}} = \int_X \overline{(Af_n)} \, g_n \, dx = \|Af_n\|_q = \alpha + o(1).$$

Thus, $\|A^*g_n\|_{\mathcal{H}} \to \alpha$, as claimed. Now assume that $g_n \rightharpoonup g$ in $L^{q'}(X)$. Then $A^*g_n \rightharpoonup A^*g$ in $\mathcal{H}$ and, passing to the limit in the above chain of inequalities $\alpha \geq \|A^*g_n\|_{\mathcal{H}} \geq \langle f_n, A^*g_n \rangle = \alpha + o(1)$, we see that any weak limit point f of (f_n) satisfies $\|A^*g\|_{\mathcal{H}} = \langle f, A^*g \rangle$. Since $\|f\|_{\mathcal{H}} \leq 1$, we see that we have equality in the Schwarz inequality and consequently $f = \|A^*g\|_{\mathcal{H}}^{-1} A^*g$. In particular, $\|f\|_{\mathcal{H}} = 1 = \|f_n\|_{\mathcal{H}}$, which implies that the convergence to f is strong. Uniqueness of the limit point proves that in fact the full sequence converges to f. This completes the proof of (a2).

The proof of (b2) is similar to that of (a2). The assumption $1 < q < \infty$ implies that we still have weak compactness; see [74, Theorem 2.18] or [91, Proposition 4.49]. We also make use of the characterization of equality in Hölder's inequality [74, Theorem 2.3]. We omit the details. □

1.2.4 Bibliographic Remarks

The existence of an optimizer for (1.4) for general s is due to Lieb [72] in the dual formulation of an HLS inequality. Lieb's proof uses the technique of symmetric decreasing rearrangement. Even if this argument does not yield the relative compactness of general optimizing sequences, several ingredients of it are still crucial for the latter problem.

The relative compactness of optimizing sequences is due to Lions; see [75] for the case $s = 1$ and [76] for the case of general s in the dual formulation.

The proof presented here is different from Lions's original one, although there are some similarities in the overall structure. In Lions's terminology, showing that there is something somewhere is excluding 'vanishing' and showing that there is nothing else anywhere else is excluding 'dichotomy'.

The first step in the proof of Theorem 1.1 that we presented is close to an argument that appears in [64] and has its roots in the work of Gérard [56]. Both [64] and [56] iterate the argument of extracting a weak limit to obtain a so-called profile decomposition, which is of importance in several areas of analysis. As shown in [53] and here, to prove the relative compactness up to symmetries of optimizing sequences, a full profile decomposition is not necessary and it suffices to extract one profile. The refined Sobolev inequality in Proposition 1.2 is due to Gérard, Meyer and Oru [57] and our presentation of the proof follows [5, Theorem 1.43]. For an alternative proof for $s = 1$, which extends to the p-norm of the gradient, we refer to [69].

Instead of the refined Sobolev inequality in terms of Besov spaces, one can also use an improvement of the Sobolev inequality in the scale of Lorentz spaces, namely,

$$\|(-\Delta)^{s/2} u\|_2 \gtrsim \|u\|_{L^{q,2}}.$$

Since $\|u\|_{L^q} = \|u\|_{L^{q,q}} \lesssim \|u\|_{L^{q,2}}^{2/q} \|u\|_{L^{q,\infty}}^{1-2/q}$, we obtain that, along a minimizing sequence, $\|u_n\|_{L^{q,\infty}} \gtrsim 1$. From this one can deduce the existence of a nontrivial weak limit point. Indeed, for $H^1(\mathbb{R}^d)$ this is a result of Lieb [73], but a similar proof works for $\dot{H}^1(\mathbb{R}^d)$, $d \geq 3$. For general s, see [10]. A Lorentz space improvement is also used in [72] for a similar, but slightly different purpose.

Yet another proof, based on a different kind of refined inequality, will be presented in the appendix to the next lecture.

The second step in the proof of Theorem 1.1 that we presented, including the Brezis–Lieb lemma and the use of the elementary inequality (1.13), is taken from Lieb's proof of the existence of an optimizer [72]. The final argument, upgrading weak convergence to strong convergence, is attributed to Browder in [18].

For a recent review of compactness methods similar to those employed in this lecture we refer to [86].

1.3 Lecture 2: Optimizers

Our main goal in this lecture is to solve the optimization problem

$$S_{d,s} := \inf_{0 \neq u \in \dot{H}^s(\mathbb{R}^d)} \frac{\|(-\Delta)^{s/2} u\|_2^2}{\|u\|_q^2} \qquad (1.17)$$

where, as always in this series of lectures,

$$0 < s < \frac{d}{2} \qquad \text{and} \qquad q = \frac{2d}{d-2s}.$$

By 'solving the optimization problem' we mean that we will compute the number $S_{d,s}$ explicitly and characterize all $u \in \dot{H}^s(\mathbb{R}^d)$ for which the infimum is achieved. It is quite remarkable that this is possible. The following theorem is due to Lieb.

Theorem 1.4 *Let $0 < s < \frac{d}{2}$. Then*

$$S_{d,s} = \frac{\Gamma(\frac{d}{2}+s)}{\Gamma(\frac{d}{2}-s)} |\mathbb{S}^d|^{2s/d} .$$

Moreover, the infimum in (1.17) is attained if and only if there are $a \in \mathbb{R}^d$, $\lambda > 0$ and $c \in \mathbb{R} \setminus \{0\}$ such that

$$u(x) = c\lambda^{-(d-2s)/2} Q(\lambda^{-1}(x-a)), \tag{1.18}$$

where

$$Q(x) = \left(\frac{2}{1+|x|^2}\right)^{(d-2s)/2} .$$

In fact, we will present the proof of a stronger result, which says that the optimization problem (1.17) does not have any local minimizers except for those stated in the theorem. Here, a function $0 \neq u_* \in \dot{H}^s(\mathbb{R}^d)$ is called a *local minimizer* of (1.17) if for all $\varphi \in \dot{H}^s(\mathbb{R}^d)$

$$\frac{d}{dt}\bigg|_{t=0} \frac{\|(-\Delta)^{s/2}(u_*+t\varphi)\|_2^2}{\|u_*+t\varphi\|_q^2} = 0 \quad \text{and} \quad \frac{d^2}{dt^2}\bigg|_{t=0} \frac{\|(-\Delta)^{s/2}(u_*+t\varphi)\|_2^2}{\|u_*+t\varphi\|_q^2} \geq 0 .$$

Theorem 1.5 *A function $0 \neq u_* \in \dot{H}^s(\mathbb{R}^d)$ is a local minimizer of (1.17) if and only if it is of the form (1.18) for some $a \in \mathbb{R}^d$, $\lambda > 0$ and $c \in \mathbb{R} \setminus \{0\}$.*

The proof of this theorem that we present in this lecture relies on a 'hidden' symmetry. This symmetry will be discussed next in detail.

1.3.1 Conformal Invariance

In the previous lecture we have already discussed the invariance of our optimization problem under translations and dilations. Another obvious invariance concerns that by orthogonal transformations of $\mathbb{R}^d$. There is a nonobvious invariance as well, namely under the *inversion on the unit sphere* $x \mapsto x/|x|^2$, which is implemented on functions u on $\mathbb{R}^d$ by

$$\tilde{u}(x) := |x|^{-d+2s} u(|x|^{-2}x) . \tag{1.19}$$

1 The Sharp Sobolev Inequality and Its Stability: An Introduction   15

Clearly, $\tilde{u}$ belongs to $L^q(\mathbb{R}^d)$ if and only if u does, and we have

$$\|\tilde{u}\|_q = \|u\|_q. \tag{1.20}$$

The important observation is that $\tilde{u}$ belongs to $\dot{H}^s(\mathbb{R}^d)$ if and only if u does, and that in this case

$$\|(-\Delta)^{s/2}\tilde{u}\|_2 = \|(-\Delta)^{s/2}u\|_2. \tag{1.21}$$

For $s = 1$, this can be proved directly by replacing $(-\Delta)^{1/2}$ under the norm by ∇. For general s, in particular noninteger ones, a direct proof is more tedious and it is preferable to deduce this result from the discussion below. In the following we will not consider (1.21) as proved, but rather use it as a motivation.

We recall a theorem of Liouville (see, e.g., [11, Theorem A.3.7] or [78, Appendix A]) that says that the Euclidean motions, together with dilations and the inversion on the unit sphere, generate the so-called *conformal group*, that is, the group of deformations of $\mathbb{R}^d \cup \{\infty\}$ that preserve angles. Thus, if $\Phi : \mathbb{R}^d \cup \{\infty\} \to \mathbb{R}^d \cup \{\infty\}$ is conformal with Jacobian $J_\Phi := |\det D\Phi|$ and if $u \in \dot{H}^s(\mathbb{R}^d)$, then

$$u_\Phi(x) := J_\Phi(x)^{1/q} u(\Phi(x))$$

belongs to $\dot{H}^s(\mathbb{R}^d)$ and

$$\|(-\Delta)^{s/2}u_\Phi\|_2 = \|(-\Delta)^{s/2}u\|_2, \qquad \|u_\Phi\|_q = \|u\|_q. \tag{1.22}$$

We emphasize that, by Liouville's theorem, (1.22) is a consequence of (1.20) and (1.21). Therefore, the first equality in (1.22) is not considered proved at this point of the lecture (unless for $s = 1$). A proof will be provided later on.

We should stress that our lectures do not really rely on Liouville's theorem. If we call a *Möbius transformation* any element of the subgroup of the conformal group generated by Euclidean motions, dilations and the inversion on the unit sphere, then everything we say remains valid when we substitute 'conformal' by 'Möbius' and, in the setting of the sphere that will appear momentarily, 'conformal' by 'conjugate of Möbius under stereographic projection'. We have opted for the use of 'conformal' for the sake of simplicity of the terminology and adherence to tradition in this field.

The conformal invariance allows us to reformulate the variational problem on the unit sphere $\mathbb{S}^d$ in $\mathbb{R}^{d+1}$. The inverse stereographic projection $\mathcal{S} : \mathbb{R}^d \to \mathbb{S}^d$ is given by

$$\mathcal{S}_j(x) := \frac{2x_j}{1+|x|^2}, \quad j = 1, \ldots, d, \qquad \mathcal{S}_{d+1}(x) := \frac{1-|x|^2}{1+|x|^2}.$$

This map is conformal and has Jacobian

$$J_{\mathcal{S}}(x) = \left(\frac{2}{1+|x|^2}\right)^d.$$

Sometimes we will extend $\mathcal{S}$ by $\mathcal{S}(\infty) := (0, \ldots, 0, -1)^T$ to a map $\mathbb{R}^d \cup \{\infty\} \to \mathbb{S}^d$.

Assume that a function u on $\mathbb{R}^d$ and a function U on $\mathbb{S}^d$ are related via

$$u(x) = J_{\mathcal{S}}(x)^{1/q} U(\mathcal{S}(x)). \tag{1.23}$$

Then clearly $U \in L^q(\mathbb{S}^d)$ if and only if $u \in L^q(\mathbb{R}^d)$, and

$$\|U\|_q = \|u\|_q. \tag{1.24}$$

Here the L^q-norm on $\mathbb{S}^d$ is defined with respect to the (unnormalized) surface measure; moreover, integration with respect to this measure is denoted by $d\omega$.

The crucial point is that $u \in \dot{H}^s(\mathbb{R}^d)$ if and only if $U \in H^s(\mathbb{S}^d)$, and that in this case

$$\mathcal{E}_s[U] = \|(-\Delta)^{s/2} u\|_2^2 \tag{1.25}$$

for a certain energy functional $\mathcal{E}_s$ that we are about to introduce.

We recall that the space $L^2(\mathbb{S}^d)$ is the orthogonal direct sum of subspaces of spherical harmonics; see, e.g., [88, Section IV.2] or [50, Subsection 3.8.2]. For $\ell \in \mathbb{N}_0$ we denote by P_ℓ the orthogonal projection in $L^2(\mathbb{R}^d)$ onto the subspace of spherical harmonics of degree ℓ. We define for $U \in H^s(\mathbb{S}^d)$,

$$\mathcal{E}_s[U] := \sum_{\ell=0}^{\infty} \frac{\Gamma(\ell + \frac{d}{2} + s)}{\Gamma(\ell + \frac{d}{2} - s)} \|P_\ell U\|_2^2.$$

Since the quotient of gamma functions in this definition is positive and grows like ℓ^{2s} (by Stirling's formula), we see that $\mathcal{E}_s[U]$ is equivalent to $\|U\|_{H^s(\mathbb{S}^d)}^2$. Moreover, for $s = 1$ we see that

$$\mathcal{E}_1[U] = \int_{\mathbb{S}^d} \left(|\nabla U|^2 + \frac{d(d-2)}{4} U^2\right) d\omega, \tag{1.26}$$

where ∇ denotes the gradient in the sense of Riemannian geometry. Identity (1.26) follows from the functional equation of the gamma function and the fact that $(-\Delta)P_\ell = \ell(\ell + d - 1)P_\ell$, where $-\Delta$ is the Laplace–Beltrami operator.

1 The Sharp Sobolev Inequality and Its Stability: An Introduction

Identity (1.25) for $s = 1$ (with the left side replaced by the right side of (1.26)) follows by a straightforward computation. As a preparation for the proof for general s we introduce the operator

$$A_s := \sum_{\ell=0}^{\infty} \frac{\Gamma(\ell + \frac{d}{2} + s)}{\Gamma(\ell + \frac{d}{2} - s)} P_\ell . \quad (1.27)$$

This is an unbounded, selfadjoint operator in $L^2(\mathbb{S}^d)$, which is positive definite and has form domain $H^s(\mathbb{S}^d)$ and operator domain $H^{2s}(\mathbb{S}^d)$. The operator A_s is connected with the quadratic form $\mathcal{E}_s$ by

$$\langle U, A_s U \rangle = \mathcal{E}_s[U],$$

valid for $U \in H^{2s}(\mathbb{S}^d)$ (or even for $U \in H^s(\mathbb{S}^d)$, provided one interprets the left side as the duality pairing between H^s and H^{-s}). Note also that

$$A_1 = -\Delta + \frac{d(d-2)}{4} .$$

This operator is called the *conformal Laplacian*. The operator A_2 is called the *Paneitz operator*. The family of operators A_s with integer s is referred to as the family of *GJMS operators* on the sphere.

Having introduced the necessary objects, we can now show the claimed identity (1.25).

Proof of (1.25) We denote by T the operator $U \mapsto u$ and by T^* its L^2-adjoint, that is,

$$TU = J_{\mathcal{S}}^{1/q} (U \circ \mathcal{S}), \qquad T^* u = J_{\mathcal{S}^{-1}}^{1/q'} (u \circ \mathcal{S}^{-1}) .$$

Then (1.25) can be written as $A_s = T^*(-\Delta)^s T$, which is equivalent to

$$A_s^{-1} = T^{-1}(-\Delta)^{-s} T^{-*} . \quad (1.28)$$

Combining the form (1.14) of the integral kernel of $(-\Delta)^{-s}$ with the fact that

$$|\mathcal{S}(x) - \mathcal{S}(x')|^2 = \frac{2}{1+|x|^2} |x - x'|^2 \frac{2}{1+|x'|^2} ,$$

we obtain

$$T^{-1}(-\Delta)^{-s} T^{-*}(\omega, \omega') = 2^{-2s} \pi^{-d/2} \frac{\Gamma(\frac{d}{2} - s)}{\Gamma(s)} |\omega - \omega'|^{-d+2s} .$$

Since this kernel only depends on $\omega \cdot \omega'$, the latter operator is diagonal with respect to the decomposition of $L^2(\mathbb{S}^d)$ into spherical harmonics and, according to Lemma 1.6

below, its eigenvalue on the space of spherical harmonics of degree ℓ is equal to

$$\frac{\Gamma(\ell + \frac{d}{2} - s)}{\Gamma(\ell + \frac{d}{2} + s)},$$

which is the same as the eigenvalue of A_s^{-1} on that space. This implies (1.28). □

Lemma 1.6 *Let* $0 < \alpha < \frac{d}{2}$ *and* $\ell \in \mathbb{N}_0$. *The eigenvalue of the operator in* $L^2(\mathbb{S}^d)$ *with kernel* $(1 - \omega \cdot \omega')^{-\alpha}$ *on the subspace* ran P_ℓ *is given by*

$$(4\pi)^{d/2} \, 2^{-\alpha} \, \frac{\Gamma(\frac{d}{2} - \alpha)}{\Gamma(\alpha)} \, \frac{\Gamma(\ell + \alpha)}{\Gamma(\ell + d - \alpha)}.$$

Proof This is a computation based on the Funk–Hecke formula and facts about Gegenbauer polynomials. Its details can be found in [52, Corollary 4.3]. Here we only explain why formula [52, (4.7)] is the same as that in the lemma. First, using the duplication formula for the gamma function, we see that $\kappa_N = (4\pi)^{N/2}$ for all $N \geq 1$. Second, using the reflection formula for the gamma function twice, we see that

$$\frac{(-1)^\ell \, \Gamma(1 - \alpha)}{\Gamma(-\ell + 1 - \alpha)} = \frac{(-1)^\ell \, \sin \pi(\ell + \alpha)}{\sin \pi \alpha} \, \frac{\Gamma(\ell + \alpha)}{\Gamma(\alpha)} = \frac{\Gamma(\ell + \alpha)}{\Gamma(\alpha)}.$$

This leads to the formula in the lemma. □

The identities (1.24) and (1.25) allow us to reformulate our optimization problem (1.17) on Euclidean space as an optimization problem on the sphere,

$$S_{d,s} = \inf_{0 \neq U \in H^s(\mathbb{S}^d)} \frac{\mathcal{E}_s[U]}{\|U\|_q^2}. \tag{1.29}$$

Moreover, optimizers for the problems on $\mathbb{R}^d$ and on $\mathbb{S}^d$ are in one-to-one correspondence via (1.23).

At this point we can give the long delayed proof of the invariance of (1.17) under inversions on the unit sphere.

Proof of (1.21) If $\tilde{U}$ is related to $\tilde{u}$ as in (1.23), then

$$\tilde{U}(\omega) = U(\omega_1, \ldots, \omega_d, -\omega_{d+1}).$$

That is, the inversion on the unit sphere for the $\mathbb{R}^d$-problem corresponds to the reflection on $\{\omega_{d+1} = 0\}$ for the $\mathbb{S}^d$-problem. Since $\|P_\ell \tilde{U}\|_2 = \|P_\ell U\|_2$ we deduce that $\mathcal{E}_s[\tilde{U}] = \mathcal{E}_s[U]$. The claimed equality (1.21) is therefore a consequence of (1.25). □

At this point the conformal invariance of the problem on $\mathbb{R}^d$, namely (1.22), is completely proved.

We also obtain the conformal invariance on $\mathbb{S}^d$. That is, if $\Psi : \mathbb{S}^d \to \mathbb{S}^d$ is a conformal transformation and if $U \in H^s(\mathbb{S}^d)$, then

$$U_\Psi(\omega) := J_\Psi(\omega)^{1/q} U(\Psi(\omega))$$

belongs to $H^s(\mathbb{S}^d)$ and

$$\mathcal{E}_s[U_\Psi] = \mathcal{E}_s[U], \qquad \|U_\Psi\|_q = \|U\|_q.$$

This follows from the corresponding result on $\mathbb{R}^d$ by noting that Ψ is a conformal transformation of $\mathbb{S}^d$ if and only if $\Phi := \mathcal{S}^{-1} \circ \Psi \circ \mathcal{S}$ is a conformal transformation of $\mathbb{R}^d$.

Remark We emphasize that for $s = 1$ the argument given above is unnecessarily complicated. In this case we first verify directly the invariance under inversions (1.21) and deduce the conformal invariance (1.22) on $\mathbb{R}^d$. Then we verify directly identity (1.25) with left side given by the right side of (1.26), and obtain as a consequence of the conformal invariance on $\mathbb{R}^d$ that on $\mathbb{S}^d$. In particular, Lemma 1.6 is not needed.

Example Let Q be as in Theorem 1.4 and let $u(x) = c\lambda^{-(d-2s)/2} Q(\lambda^{-1}(x-a))$ with $a \in \mathbb{R}^d$, $\lambda > 0$ and $c \in \mathbb{R}$. Then the corresponding U on $\mathbb{S}^d$ is given by

$$U(\omega) = c \left(\frac{\sqrt{1-|\zeta|^2}}{1 - \zeta \cdot \omega} \right)^{(d-2s)/2}$$

with $\zeta := (2\eta - \lambda^2(1+\eta_{d+1})e_{d+1})/(2 + \lambda^2(1+\eta_{d+1}))$ and $\eta := \mathcal{S}(a)$. This follows by a direct computation. Moreover, it is a simple exercise to show that the map $\mathbb{R}^d \times \mathbb{R}_+ \ni (a,\lambda) \mapsto \zeta \in \{z \in \mathbb{R}^{d+1} : |z| < 1\}$ is a bijection.

As a final preliminary we note that

$$\left\{ J_\Psi : \Psi \text{ conformal transformation of } \mathbb{S}^d \right\} = \left\{ \left(\frac{\sqrt{1-|\zeta|^2}}{1-\zeta \cdot \omega} \right)^d : |\zeta| < 1 \right\}. \tag{1.30}$$

To prove this, we note that $\Phi = \mathcal{S}^{-1} \circ \Psi \circ \mathcal{S}$ gives a bijection between conformal transformations Ψ of $\mathbb{S}^d$ and Φ of $\mathbb{R}^d$. Therefore, the claim is that

$$J_\mathcal{S}(\Phi(\mathcal{S}^{-1}(\omega))) J_\Phi(\mathcal{S}^{-1}(\omega)) J_{\mathcal{S}^{-1}}(\omega) = \left(\frac{\sqrt{1-|\zeta|^2}}{1-\zeta \cdot \omega} \right)^d,$$

in the sense that, as Φ runs through the conformal group, ζ runs through the unit ball. Equivalently,

$$J_\mathcal{S}(\Phi(x))\, J_\Phi(x) = \left(\frac{\sqrt{1-|\zeta|^2}}{1-\zeta\cdot\mathcal{S}(x)}\right)^d J_\mathcal{S}(x).$$

By Liouville's theorem it suffices to verify the latter identity separately for Euclidean motions, dilations and the inversion on the unit sphere. This is a tedious, but straightforward computation.

1.3.2 An Equivalent Formulation of the Theorem

After all these preparations, we now formulate the analogue of Theorem 1.5 on the sphere. A function $0 \neq U_* \in H^s(\mathbb{S}^d)$ is called a *local minimizer* of (1.29) if for all $\varphi \in H^s(\mathbb{S}^d)$,

$$\frac{d}{dt}\bigg|_{t=0} \frac{\mathcal{E}_s[U_*+t\varphi]}{\|U_*+t\varphi\|_q^2} = 0 \quad \text{and} \quad \frac{d^2}{dt^2}\bigg|_{t=0} \frac{\mathcal{E}_s[U_*+t\varphi]}{\|U_*+t\varphi\|_q^2} \geq 0. \quad (1.31)$$

The conformal invariance discussed above shows that if u and U are related by (1.23), then U is a local minimizer of (1.29) if and only if u is a local minimizer of (1.17).

Theorem 1.7 *A function $0 \neq U_* \in H^s(\mathbb{S}^d)$ is a local minimizer of (1.29) if and only if $U_* = c J_\Psi^{1/q}$ for a conformal transformation Ψ of $\mathbb{S}^d$ and a constant $c \in \mathbb{R}\setminus\{0\}$.*

Theorem 1.5 is an immediate consequence of Theorem 1.7. The equivalence of the characterization of optimizers follows from the above example and (1.30).

To prepare for the proof of Theorem 1.7, we compute the derivatives appearing in the definition of a local minimizer. We begin with the case $s=1$, where we use the form (1.26) of the functional. We have

$$\frac{d}{dt}\bigg|_{t=0} \frac{\mathcal{E}_1[U_*+t\varphi]}{\|U_*+t\varphi\|_q^2}$$
$$= \frac{2}{\|U_*\|_q^2} \int_{\mathbb{S}^d} \left(\nabla\varphi\cdot\nabla U_* + \tfrac{d(d-2)}{4}\varphi U_* - \frac{\mathcal{E}_1[U_*]}{\|U_*\|_q^q}\varphi|U_*|^{q-2}U_*\right)d\omega,$$

so the first condition in (1.31) is satisfied if and only if U_* solves the equation

$$-\Delta U_* + \tfrac{d(d-2)}{4} U_* - \frac{\mathcal{E}_1[U_*]}{\|U_*\|_q^q}|U_*|^{q-2}U_* = 0 \quad \text{on } \mathbb{S}^d. \quad (1.32)$$

1 The Sharp Sobolev Inequality and Its Stability: An Introduction

Moreover, for U_* for which the first derivative of $\mathcal{E}_1$ vanishes, we compute

$$\frac{d^2}{dt^2}\bigg|_{t=0} \frac{\mathcal{E}_1[U_*+t\varphi]}{\|U_*+t\varphi\|_q^2}$$

$$= \frac{2}{\|U_*\|_q^2}\left(\int_{\mathbb{S}^d}\left(|\nabla\varphi|^2 + \tfrac{d(d-2)}{4}\varphi^2 - (q-1)\tfrac{\mathcal{E}_1[U_*]}{\|U_*\|_q^q}|U_*|^{q-2}\varphi^2\right)d\omega\right.$$

$$\left.+ (q-2)\tfrac{\mathcal{E}_1[U_*]}{\|U_*\|_q^{2q}}\left(\int_{\mathbb{S}^d}|U_*|^{q-2}U_*\varphi\, d\omega\right)^2\right).$$

Thus, the second condition in (1.31) is satisfied if and only if the operator

$$-\Delta + \tfrac{d(d-2)}{4} - (q-1)\tfrac{\mathcal{E}_1[U_*]}{\|U_*\|_q^q}|U_*|^{q-2} + (q-2)\tfrac{\mathcal{E}_1[U_*]}{\|U_*\|_q^{2q}}\left||U_*|^{q-2}U_*\right\rangle\!\left\langle|U_*|^{q-2}U_*\right| \tag{1.33}$$

in $L^2(\mathbb{S}^d)$ is positive semidefinite. (Here $|f\rangle\langle f|$ denotes the rank one operator $\varphi \mapsto \langle f,\varphi\rangle f$.) This operator is considered as an unbounded, selfadjoint operator in $L^2(\mathbb{S}^d)$. It is bounded from below and has form domain $H^1(\mathbb{S}^d)$.

The computation for general s is similar. We recall that the operator A_s was introduced in (1.27). We see that in terms of this operator the first condition in (1.31) is equivalent to the equation

$$A_s U_* - \tfrac{\mathcal{E}_s[U_*]}{\|U_*\|_q^q}|U_*|^{q-2}U_* = 0 \qquad \text{on } \mathbb{S}^d \tag{1.34}$$

and, for U_* satisfying (1.34), the second condition in (1.31) is equivalent to the positive semidefiniteness of the operator

$$A_s - (q-1)\tfrac{\mathcal{E}_s[U_*]}{\|U_*\|_q^q}|U_*|^{q-2} + (q-2)\tfrac{\mathcal{E}_s[U_*]}{\|U_*\|_q^{2q}}\left||U_*|^{q-2}U_*\right\rangle\!\left\langle|U_*|^{q-2}U_*\right|. \tag{1.35}$$

1.3.3 Local Minimality of Constants

We turn to the proof of the first part of Theorem 1.7. Let Ψ be a conformal transformation of $\mathbb{S}^d$, $c \in \mathbb{R}\setminus\{0\}$ and $U_* = cJ_\Psi^{1/q}$. We wish to show that U_* is a local minimizer of (1.29). By homogeneity of the problem it suffices to consider $c=1$ and by conformal invariance it suffices to consider $\Psi = \mathrm{id}_{\mathbb{S}^d}$.

We begin with the case $s=1$, where we need to show that $U_*=1$ satisfies equation (1.32) and that the operator in (1.33) is positive semidefinite. Verification of (1.32) is straightforward. The operator in (1.33) becomes

$$\mathcal{L}_1 := -\Delta - d + d\,|\mathbb{S}^d|^{-1}\,|1\rangle\langle 1|.$$

We recall (see, e.g., [50, Theorem 3.49]) that the spectrum of the Laplace–Beltrami operator $-\Delta$ in $L^2(\mathbb{S}^d)$ consists of the discrete eigenvalues $\ell(\ell+d-1)$, $\ell \in \mathbb{N}_0$, (of certain known multiplicities which, however, are irrelevant for us at this point). The lowest eigenvalue is 0 and the corresponding eigenfunctions are precisely the constants. The operator $|\mathbb{S}^d|^{-1}|1\rangle\langle 1|$ is the projection P_0 onto constants in $L^2(\mathbb{S}^d)$. Since this operator commutes with $-\Delta$, we can use the above facts to describe the spectrum of $\mathcal{L}_1$. It consists precisely of the eigenvalues $\ell(\ell+d-1)-d$, $\ell \in \mathbb{N}$. In particular, its spectrum is contained in $\overline{\mathbb{R}_+}$, and therefore the operator $\mathcal{L}_1$ is positive semidefinite, as we wanted to show.

For later purposes we recall that the eigenvalue d of $-\Delta$ in $L^2(\mathbb{S}^d)$ has multiplicity $d+1$ and a basis of corresponding eigenfunctions is given by the coordinate functions ω_j, $j = 1, \ldots, d+1$. Therefore, the eigenvalue 0 of $\mathcal{L}_1$ has multiplicity $d+2$ and a basis of eigenfunctions is given by constants and the coordinate functions. These $d+2$ zero modes of $\mathcal{L}_1$ reflect the invariances of the variational problem: the coordinate functions come from the translation and dilation invariance, while the constant function comes from the homogeneity of the problem. In this sense the constant function 1 is a *nondegenerate* local minimizer: the only zero modes come from the invariances.

The argument for general s is similar. Equation (1.34) for $U_* = 1$ follows immediately from

$$\frac{\mathcal{E}_s[1]}{\|1\|_q^q} = \frac{\Gamma(\frac{d}{2}+s)}{\Gamma(\frac{d}{2}-s)},$$

which also shows that the operator in (1.35) becomes

$$\mathcal{L}_s := A_s - (q-1)\frac{\Gamma(\frac{d}{2}+s)}{\Gamma(\frac{d}{2}-s)} + (q-2)\frac{\Gamma(\frac{d}{2}+s)}{\Gamma(\frac{d}{2}-s)}|\mathbb{S}^d|^{-1}|1\rangle\langle 1|$$

$$= \sum_{\ell=2}^{\infty}\left(\frac{\Gamma(\ell+\frac{d}{2}+s)}{\Gamma(\ell+\frac{d}{2}-s)} - \frac{\Gamma(1+\frac{d}{2}+s)}{\Gamma(1+\frac{d}{2}-s)}\right)P_\ell. \tag{1.36}$$

The second equality here uses the functional equation of the gamma function. We emphasize that the terms with $\ell = 0$ and $\ell = 1$ vanish. Thus, spherical harmonics of degrees 0 and 1 are in the kernel of $\mathcal{L}_s$. Moreover, by the log-convexity of the gamma function, $\frac{d}{dt}\ln\frac{\Gamma(t+s)}{\Gamma(t-s)} = (\ln\Gamma)'(t+s) - (\ln\Gamma)'(t-s) > 0$ for all $t > s > 0$, so

$$\ell \mapsto \frac{\Gamma(\ell+\frac{d}{2}+s)}{\Gamma(\ell+\frac{d}{2}-s)} \quad \text{is increasing.} \tag{1.37}$$

It follows that $\mathcal{L}_s$ is positive definite on the orthogonal complement of the range of $P_0 + P_1$. Thus, we have shown that $\mathcal{L}_s$ is positive semidefinite, as we wanted to show.

This completes the proof of the first part of Theorem 1.7.

1.3.4 Classification of Local Minimizers

It remains to prove the second part of Theorem 1.7. As a preparation for the proof we first establish the following lemma, which will allow us to fix the center of mass by a conformal transformation.

Lemma 1.8 *Let $f \in L^1(\mathbb{S}^d)$ with $\int_{\mathbb{S}^d} f(\omega)\, d\omega \neq 0$. Then there is a conformal transformation Ψ of $\mathbb{S}^d$ such that*

$$\int_{\mathbb{S}^d} \Psi^{-1}(\omega) f(\omega)\, d\omega = 0.$$

Proof

Step 1. In this preliminary step we define a family of conformal transformations $\gamma_{\delta,\xi}$ of $\mathbb{S}^d$ depending on two parameters $\delta > 0$ and $\xi \in \mathbb{S}^d$. To do so, we denote dilations on $\mathbb{R}^d$ by $\mathcal{D}_\delta$, that is, $\mathcal{D}_\delta(x) = \delta x$. Moreover, for any $\xi \in \mathbb{S}^d$ we choose an orthogonal $(d+1) \times (d+1)$ matrix O such that $O\xi = (0,\ldots,0,1)^T$ and we put

$$\gamma_{\delta,\xi}(\omega) := \begin{cases} O^T \mathcal{S}\left(\mathcal{D}_\delta\left(\mathcal{S}^{-1}(O\omega)\right)\right) & \text{if } \omega \neq -\xi, \\ -\xi & \text{if } \omega = -\xi. \end{cases}$$

This transformation depends only on ξ and δ and not on the particular choice of O. Indeed, a straightforward computation shows that

$$\gamma_{\delta,\xi}(\omega) = \frac{2\delta}{(1+\omega\cdot\xi)+\delta^2(1-\omega\cdot\xi)} (\omega - (\omega\cdot\xi)\xi)$$
$$+ \frac{(1+\omega\cdot\xi)-\delta^2(1-\omega\cdot\xi)}{(1+\omega\cdot\xi)+\delta^2(1-\omega\cdot\xi)} \xi.$$

Since $\gamma_{\delta,\xi}$ is a composition of conformal transformations, it is conformal.

Step 2. We now turn to the main part of the proof, where we may assume that $f \in L^1(\mathbb{S}^d)$ is normalized by $\int_{\mathbb{S}^d} f(\omega)\, d\omega = 1$. We will show that the $\mathbb{R}^{d+1}$-valued map

$$F(r\xi) := \int_{\mathbb{S}^d} \gamma_{1-r,\xi}(\omega) f(\omega)\, d\omega, \qquad 0 < r < 1,\ \xi \in \mathbb{S}^d,$$

has a zero. Once we have shown this, we deduce the assertion of the lemma by taking $\Psi = \gamma_{1-r_0,\xi_0}^{-1}$, where $r_0\xi_0$ is the zero of F.

First, note that because of $\gamma_{1,\xi}(\omega) = \omega$ for all ξ and all ω, the limit of $F(r\xi)$ as $r \to 0$ is independent of ξ, so F extends to a continuous function on the open unit ball of $\mathbb{R}^{d+1}$. In order to understand its boundary behavior, one easily checks that for any $\omega \neq -\xi$ one has $\lim_{\delta \to 0} \gamma_{\delta,\xi}(\omega) = \xi$, and that this convergence is uniform on $\{(\omega, \xi) \in \mathbb{S}^d \times \mathbb{S}^d : 1 + \omega \cdot \xi \geq \varepsilon\}$ for any $\varepsilon > 0$. This implies that

$$\lim_{r \to 1} F(r\xi) = \xi \qquad \text{uniformly in } \xi.$$

Hence, F is a continuous function on the *closed* unit ball and is the identity on the boundary. The assertion is now a consequence of one of the equivalent forms of Brouwer's fixed point theorem; see, e.g., [83, Appendix]. □

We now prove the second part of Theorem 1.7. Again we begin with the case $s = 1$. Let $0 \neq U_* \in H^1(\mathbb{S}^d)$ be a local minimizer of (1.29). We wish to show that there is a conformal transformation Ψ of $\mathbb{S}^d$ and a constant $c \in \mathbb{R} \setminus \{0\}$ such that $U_* = c J_\Psi^{1/q}$.

According to Lemma 1.8 we can choose the conformal transformation Ψ in such a way that

$$0 = \int_{\mathbb{S}^d} \Psi^{-1}(\omega) |U_*(\omega)|^q \, d\omega = \int_{\mathbb{S}^d} \omega |(U_*)_\Psi(\omega)|^q \, d\omega.$$

Note that by conformal invariance $(U_*)_\Psi$ is also a local minimizer. Thus, by replacing U_* by $(U_*)_\Psi$, it suffices to show that if U_* is a local minimizer satisfying

$$\int_{\mathbb{S}^d} \omega |U_*|^q \, d\omega = 0, \tag{1.38}$$

then U_* is a constant. To prove this, we make use of the positive semidefiniteness of the linear operator (1.33). Evaluating the operator on the function $\omega_j U_*$, $j = 1, \ldots, d+1$, (that is, choosing $\varphi = \omega_j U_*$ in the second condition in (1.31)) we obtain

$$\int_{\mathbb{S}^d} \left(|\nabla(\omega_j U_*)|^2 + \tfrac{d(d-2)}{4} \omega_j^2 U_*^2 - (q-1) \tfrac{\mathcal{E}_1[U_*]}{\|U_*\|_q^q} \omega_j^2 |U_*|^q \right) d\omega \geq 0. \tag{1.39}$$

Here we used (1.38) to see that the rank-one term in the operator (1.33) vanishes on the chosen function. We have

$$|\nabla(\omega_j U_*)|^2 = \omega_j^2 |\nabla U_*|^2 + 2\omega_j U_* \nabla \omega_j \cdot \nabla U_* + U_*^2 |\nabla \omega_j|^2$$
$$= \omega_j^2 |\nabla U_*|^2 + U_* \nabla(\omega_j^2) \cdot \nabla U_* + U_*^2 \left(1 - \omega_j^2\right).$$

1 The Sharp Sobolev Inequality and Its Stability: An Introduction

(Recall that ∇ denotes the Riemannian gradient and that $\nabla \omega_j = e_j - \omega_j \omega$.) Inserting this into (1.39), we find

$$\int_{\mathbb{S}^d} \left(\omega_j^2 |\nabla U_*|^2 + U_* \nabla(\omega_j^2) \cdot \nabla U_* + U_*^2 \left(1 - \omega_j^2\right) \right.$$
$$\left. + \tfrac{d(d-2)}{4} \omega_j^2 U_*^2 - (q-1) \tfrac{\mathcal{E}_1[U_*]}{\|U_*\|_q^q} \omega_j^2 |U_*|^q \right) d\omega \geq 0 \,.$$

Summing these inequalities with respect to $j = 1, \ldots, d+1$ and using $\sum_j \omega_j^2 = 1$, we arrive at

$$0 \leq \int_{\mathbb{S}^d} \left(|\nabla U_*|^2 + d U_*^2 + \tfrac{d(d-2)}{4} U_*^2 - (q-1) \tfrac{\mathcal{E}_1[U_*]}{\|U_*\|_q^q} |U_*|^q \right) d\omega$$
$$= -(q-2) \int_{\mathbb{S}^d} |\nabla U_*|^2 \, d\omega \,.$$

Note that the coefficients of U_*^2 cancel. Since $q > 2$, we have shown $\int_{\mathbb{S}^d} |\nabla U_*|^2 \, d\omega \leq 0$, which implies that U_* is a constant, as claimed. This completes the proof of Theorem 1.7 for $s = 1$.

We now discuss the case of general s. As before we can use Lemma 1.8 to reduce the proof to showing that, if U_* is a local minimizer satisfying (1.38), then U_* is constant. Evaluating the operator (1.35) on the function $\omega_j U_*$, $j = 1, \ldots, d+1$, and recalling (1.38), we obtain

$$\mathcal{E}_s[\omega_j U_*] - (q-1) \tfrac{\mathcal{E}_s[U_*]}{\|U_*\|_q^q} \int_{\mathbb{S}^d} \omega_j^2 |U_*|^q \, d\omega \geq 0 \,.$$

Summing with respect to j gives

$$\sum_{j=1}^{d+1} \mathcal{E}_s[\omega_j U_*] - (q-1) \mathcal{E}_s[U_*] \geq 0 \,. \tag{1.40}$$

To simplify the first term on the left side, we need an auxiliary result about spherical harmonics.

Lemma 1.9 *For all $\ell \geq 0$,*

$$\sum_{j=1}^{d+1} \omega_j P_\ell \, \omega_j = \tfrac{\ell+1}{2\ell+d+1} P_{\ell+1} + \tfrac{\ell+d-2}{2\ell+d-3} P_{\ell-1} \,,$$

with the conventions that, if $\ell = 0$, $\tfrac{\ell+d-2}{2\ell+d-3} P_{\ell-1} = 0$ and that, if $d = 1$ and $\ell = 1$, $\tfrac{\ell+d-2}{2\ell+d-3} = \tfrac{1}{2}$.

Proof We prove the equality of integral kernels

$$(\omega \cdot \omega') P_\ell(\omega, \omega') = \tfrac{\ell+1}{2\ell+d+1} P_{\ell+1}(\omega, \omega') + \tfrac{\ell+d-2}{2\ell+d-3} P_{\ell-1}(\omega, \omega').$$

We have, for all ℓ,

$$P_\ell(\omega, \omega') = \frac{\nu_\ell}{|\mathbb{S}^d| C_\ell^{((d-1)/2)}(1)} C_\ell^{((d-1)/2)}(\omega \cdot \omega'),$$

where ν_ℓ is the dimension of the space of spherical harmonics of degree ℓ and where $C_\ell^{((d-1)/2)}$ is a Gegenbauer polynomial. For this formula, without the explicit value of the constant, see, e.g. [88, Theorem IV.2.14]. The value of the constant is determined by the relation $\operatorname{Tr} P_\ell = \nu_\ell$.

The claimed formula now follows from the recursion relation for Gegenbauer polynomials [1, (22.7.3)],

$$2(\ell + \alpha) t C_\ell^{(\alpha)}(t) = (\ell + 1) C_{\ell+1}^{(\alpha)}(t) + (\ell + 2\alpha - 1) C_{\ell-1}^{(\alpha)}(t)$$

together with the normalization [1, (22.2.3)]

$$C_\ell^{(\alpha)}(1) = \binom{\ell + 2\alpha - 1}{\ell} \text{ if } \alpha > 0, \qquad C_\ell^{(0)}(1) = \begin{cases} 1 & \text{if } \ell = 0, \\ \tfrac{2}{\ell} & \text{if } \ell \geq 1, \end{cases}$$

and the multiplicity formula [88, Section IV.2]

$$\nu_\ell = \frac{(d - 2 + \ell)! \, (d + 2\ell - 1)}{\ell! \, (d - 1)!}$$

(with the convention that $\nu_0 = 1$ if $d = 1$). □

It follows from Lemma 1.9 that

$$\sum_{j=1}^{d+1} \mathcal{E}_s[\omega_j U] = \sum_{\ell=0}^{\infty} \frac{\Gamma(\ell + \tfrac{d}{2} + s)}{\Gamma(\ell + \tfrac{d}{2} - s)} \sum_{j=1}^{d+1} \|P_\ell \omega_j U\|_2^2$$

$$= \sum_{\ell=0}^{\infty} \frac{\Gamma(\ell + \tfrac{d}{2} + s)}{\Gamma(\ell + \tfrac{d}{2} - s)} \left(\tfrac{\ell+1}{2\ell+d+1} \|P_{\ell+1} U\|_2^2 + \tfrac{\ell+d-2}{2\ell+d-3} \|P_{\ell-1} U\|_2^2 \right)$$

$$= \sum_{\ell=0}^{\infty} \left(\frac{\Gamma(\ell - 1 + \tfrac{d}{2} + s)}{\Gamma(\ell - 1 + \tfrac{d}{2} - s)} \frac{\ell}{2\ell+d-1} + \frac{\Gamma(\ell + 1 + \tfrac{d}{2} + s)}{\Gamma(\ell + 1 + \tfrac{d}{2} - s)} \frac{\ell+d-1}{2\ell+d-1} \right) \|P_\ell U\|_2^2$$

$$= \sum_{\ell=0}^{\infty} \frac{\Gamma(\ell + \tfrac{d}{2} + s)}{\Gamma(\ell + \tfrac{d}{2} - s)} \left(\frac{\ell - 1 + \tfrac{d}{2} - s}{\ell - 1 + \tfrac{d}{2} + s} \frac{\ell}{2\ell+d-1} + \frac{\ell + \tfrac{d}{2} + s}{\ell + \tfrac{d}{2} - s} \frac{\ell+d-1}{2\ell+d-1} \right) \|P_\ell U\|_2^2 .$$

Thus,

$$\sum_{j=1}^{d+1} \mathcal{E}_s[\omega_j U] - (q-1)\mathcal{E}_s[U] = -\sum_{\ell=0}^{\infty} \frac{\Gamma(\ell+\frac{d}{2}+s)}{\Gamma(\ell+\frac{d}{2}-s)} w_s(\ell) \|P_\ell U\|_2^2 \qquad (1.41)$$

with

$$w_s(\ell) := \frac{d+2s}{d-2s} - \frac{\ell-1+\frac{d}{2}-s}{\ell-1+\frac{d}{2}+s} \frac{\ell}{2\ell+d-1} - \frac{\ell+\frac{d}{2}+s}{\ell+\frac{d}{2}-s} \frac{\ell+d-1}{2\ell+d-1}.$$

A tedious but straightforward computation shows that

$$w_s(\ell) = \frac{4s}{d-2s} \frac{\ell(\ell+d-1)}{(\ell-1+\frac{d}{2}+s)(\ell+\frac{d}{2}-s)},$$

so

$$w_s(\ell) \geq 0 \quad \text{with equality if and only if } \ell = 0. \qquad (1.42)$$

Taking $U = U_*$ in (1.41), recalling the second variation inequality (1.40) and using (1.42), we deduce that $P_\ell U_* = 0$ for all $\ell \geq 1$, that is, U_* is a constant, as we wanted to prove. This completes the proof of Theorem 1.7.

1.3.5 Appendix: Subcritical Interpolation Inequalities

While the main focus of these lectures is on the Sobolev inequality with critical exponent, in this appendix we make a brief digression to the subcritical case and show the following result.

Theorem 1.10 *Let $2 \leq q < \infty$ if $d = 1, 2$ and $2 \leq q < \frac{2d}{d-2}$ if $d \geq 3$. Then for all $U \in H^1(\mathbb{S}^d)$,*

$$\int_{\mathbb{S}^d} \left(|\nabla U|^2 + \frac{d}{q-2} U^2 \right) d\omega \geq \frac{d}{q-2} |\mathbb{S}^d|^{1-2/q} \left(\int_{\mathbb{S}^d} |U|^q \, d\omega \right)^{2/q}. \qquad (1.43)$$

with equality if and only if U is constant.

Inequality (1.43) for $q = \frac{2d}{d-2}$ turns into the inequality $\mathcal{E}_1[U] \geq S_{d,1} \|U\|_q^2$, which we have shown in the main part of this lecture. We emphasize, however, that in this critical case the set of functions attaining equality is strictly larger than in the subcritical case.

Proof Given q we define $s := d(\frac{1}{2} - \frac{1}{q})$, so that $q = \frac{2d}{d-2s}$. Then, by the main theorem of this lecture, we have $S_{d,s} \|U\|_q^2 \leq \mathcal{E}_s[U]$, and our goal is to bound $\mathcal{E}_s[U]$

from above by a constant times the left side in (1.43). Expanding U into spherical harmonics, the task becomes to find the smallest constant C in the inequality

$$\frac{\Gamma(\ell+\frac{d}{2}+s)}{\Gamma(\ell+\frac{d}{2}-s)} = \frac{\Gamma(\ell+\frac{d}{2}+d(\frac{1}{2}-\frac{1}{q}))}{\Gamma(\ell+\frac{d}{2}-d(\frac{1}{2}-\frac{1}{q}))}$$

$$\leq C\left(\ell(\ell+d-1) + \frac{d}{q-2}\right) \quad \text{for all } \ell \in \mathbb{N}_0.$$

Using properties of gamma functions, one can show that the optimal constant is attained exactly at $\ell = 0$, which implies (1.43). □

1.3.6 Appendix: Optimizing Sequences

In this appendix we want to show that the technique of fixing the center of mass in Lemma 1.8 can also be useful to prove the relative compactness up to symmetries of optimizing sequences. We give the argument for $s = 1$. As before, $q = \frac{2d}{d-2}$.

The crucial ingredient is the following refined inequality of Aubin. For $d \geq 3$ and $\varepsilon > 0$ there is a $C_\varepsilon < \infty$ such that for all $U \in H^1(\mathbb{S}^d)$ with

$$\int_{\mathbb{S}^d} \omega |U|^q \, d\omega = 0$$

one has

$$\mathcal{E}_1[U] \geq (1-\varepsilon) 2^{2/d} S_{d,1} \|U\|_q^2 - C_\varepsilon \|U\|_2^2. \tag{1.44}$$

This inequality implies that for suitably normalized functions one obtains a Sobolev constant that is $> S_{d,1}$, at the expense of an L^2-term, which in many applications is harmless.

Let us use (1.44) to prove relative compactness of optimizing sequences. Let $(U_n) \subset H^1(\mathbb{S}^d)$ with $\mathcal{E}_1[U_n] = 1$ and $\|U_n\|_q^2 \to S_{d,1}^{-1}$. By Lemma 1.8 there is a conformal transformation Ψ_n of $\mathbb{S}^d$ such that $\widetilde{U}_n := (U_n)_{\Psi_n}$ satisfies $\int_{\mathbb{S}^d} \omega |\widetilde{U}_n|^q \, d\omega = 0$. Moreover, by conformal invariance, $\mathcal{E}_1[\widetilde{U}_n] = 1$ and $\|\widetilde{U}_n\|_q^2 \to S_{d,1}^{-1}$. After passing to a subsequence, we may assume that $\widetilde{U}_n \rightharpoonup \widetilde{U}$ in $H^1(\mathbb{S}^d)$. By Rellich's lemma, $\widetilde{U}_n \to \widetilde{U}$ in $L^2(\mathbb{S}^d)$. Therefore, applying (1.44) to $\widetilde{U}_n$ and passing to the limit, we obtain

$$1 \geq (1-\varepsilon) 2^{2/d} - C_\varepsilon \|\widetilde{U}\|_2^2.$$

This is valid for any fixed $\varepsilon > 0$. Choosing it so that $1 < (1-\varepsilon)2^{2/d}$, we deduce that $\tilde{U} \neq 0$. Thus, we have shown that there is something somewhere. The proof that there is nothing else anywhere else is as in the first lecture.

For the sake of completeness we present Aubin's proof of (1.44).

Proof of (1.44) Let $\delta > 0$ and set $h(t) := (t^2 + \delta^2)^{-(q-1)/(2q)} t$ for $t \in [-1, 1]$. For $\Omega, \omega \in \mathbb{S}^d$ we set $h_\Omega(\omega) := h(\Omega \cdot \omega)$ and note that

$$Z := \int_{\mathbb{S}^d} h_\Omega(\omega)^2 \, d\Omega = |\mathbb{S}^{d-1}| \int_0^\pi h(\cos\theta)^2 \sin^{d-1}\theta \, d\theta$$

is independent of ω. It follows that

$$\|U\|_q^2 = \|U^2\|_{q/2} = \left\| Z^{-1} \int_{\mathbb{S}^d} h_\Omega^2 U^2 \, d\Omega \right\|_{q/2} \leq Z^{-1} \int_{\mathbb{S}^d} \|h_\Omega^2 U^2\|_{q/2} \, d\Omega$$

$$= Z^{-1} \int_{\mathbb{S}^d} \|h_\Omega U\|_q^2 \, d\Omega.$$

We will show that there is a constant C, depending only on q, such that for each $\Omega \in \mathbb{S}^d$,

$$2^{-2/q} S_{d,1} \|h_\Omega U\|_q^2 \leq 2^{-1} \int_{\mathbb{S}^d} h_\Omega^2 \left(|\nabla U|^2 + \tfrac{d(d-2)}{4} |U|^2 \right) d\omega$$
$$+ 2^{-1} \int_{\mathbb{S}^d} |\nabla h_\Omega|^2 |U|^2 \, d\omega$$
$$+ 2^{-1} \int_{\mathbb{S}^d} (\nabla h_\Omega^2) \cdot U \nabla U \, d\omega + C^2 \delta^{2/q} \mathcal{E}_1[U]. \quad (1.45)$$

Integrating this bound with respect to Ω, we obtain

$$2^{-2/q} S_{d,1} Z^{-1} \int_{\mathbb{S}^d} \|h_\Omega U\|_q^2 \, d\Omega \leq \left(2^{-1} + C^2 Z^{-1} |\mathbb{S}^d| \delta^{2/q} \right) \mathcal{E}_1[U] + 2^{-1} \tilde{C} \|U\|_2^2$$

with $\tilde{C} := Z^{-1} \int_{\mathbb{S}^d} |\nabla h_\Omega(\omega)|^2 \, d\Omega$ (which is independent of ω). Noting that Z converges as $\delta \to 0$, we obtain the claimed bound in (1.44). (In contrast, we note that $\tilde{C}$ diverges as $\delta \to 0$, since $|t|^{-(q-1)/q} t$ is not H^1 near $t = 0$.)

It remains to prove (1.45). We may assume that $\mathcal{E}_1[(h_\Omega)_+ \nabla U] \geq \mathcal{E}_1[(h_\Omega)_- \nabla U]$, the opposite case being similar. We use $0 \leq t - h(t)^q \leq 2C^q \delta$ for all $t \in [0, 1]$ and some C, depending only on q, to bound

$$\|h_\Omega U\|_q^q \leq \int_{\mathbb{S}^d} (\Omega \cdot \omega)_+ |U|^q \, d\omega + \int_{\mathbb{S}^d} (h_\Omega)_-^q |U|^q \, d\omega$$
$$= \int_{\mathbb{S}^d} (\Omega \cdot \omega)_- |U|^q \, d\omega + \int_{\mathbb{S}^d} (h_\Omega)_-^q |U|^q \, d\omega$$

$$\leq 2 \int_{\mathbb{S}^d} \left((h_\Omega)_-^q + C^q \delta \right) |U|^q \, d\omega$$

$$\leq 2 \int_{\mathbb{S}^d} \left((h_\Omega)_-^2 + C^2 \delta^{2/q} \right)^{q/2} |U|^q \, d\omega.$$

The triangle inequality and Sobolev's inequality imply

$$2^{-2/q} S_{d,1} \|h_\Omega U\|_q^2 \leq S_{d,1} \left(\|(h_\Omega)_- U\|_q^2 + C^2 \delta^{2/q} \|U\|_q^2 \right)$$

$$\leq \mathcal{E}_1[(h_\Omega)_- U] + C^2 \delta^{2/q} \mathcal{E}_1[U]$$

$$\leq 2^{-1} \left(\mathcal{E}_1[(h_\Omega)_- U] + \mathcal{E}_1[(h_\Omega)_+ U] \right) + C^2 \delta^{2/q} \mathcal{E}_1[U].$$

Using $(h_\Omega)_-^2 + (h_\Omega)_+^2 = h_\Omega^2$ and

$$|\nabla (h_\Omega)_- U|^2 + |\nabla (h_\Omega)_+ U|^2 = h_\Omega^2 |\nabla U|^2 + |\nabla h_\Omega|^2 |U|^2 + (\nabla h_\Omega^2) \cdot U \nabla U,$$

we obtain the claimed bound (1.45). □

1.3.7 Bibliographic Remarks

There are a number of alternative proofs of Theorem 1.4, in particular for $s = 1$. Here we review some of them and give references.

The proof of Theorem 1.4 presented in this lecture in the case $s = 1$ is from [52]; see also [53]. The proof for general s is a new variant of the recent proof in [92]. It simplifies the corresponding argument in [52] (where duality was invoked and only positive functions were considered). The new ingredient in [92], compared to [52], is a commutator identity, which for integer s is due to [27]. Our Lemma 1.9 serves a similar purpose.

The presented proof may not be the most direct proof of Theorem 1.4, but it has the advantage of yielding Theorem 1.5, which may be new (at least for functions that are not necessarily nonnegative). Also, this proof is natural in this lecture series in view of the second variation analysis in the next lecture. Further, the approach presented in this lecture works in the setting of the Heisenberg group, where several other techniques mentioned below (for instance those based on symmetric decreasing rearrangement or the moving plane method) seem not to work. It has also been applied in the fully nonlinear setting [27, 28].

Lemma 1.8 is due to Hersch [63], where it was used in the problem of maximizing the first nontrivial eigenvalue of the Laplace–Beltrami operator on $\mathbb{S}^2$ over all metrics with fixed area and conformal to the standard metric.

As far as we know, the optimal value of the constant $S_{d,1}$ with $s = 1$ as well as the optimizing functions appeared for the first time in the unpublished preprint of Rodemich [84] and in the papers of Aubin [2] and Talenti [89]. These works deal with the more general situation of an L^p-norm of the gradient with $1 \leq p < d$ (and the correspondingly modified q). They use rearrangement techniques to reduce the problem to a one-dimensional problem that had been solved by Bliss [15]. Since the relevant rearrangement inequality for the gradient can be an equality without the functions being radial, it does not seem possible to derive the characterization of optimizers using these techniques.

We also mention [85] where local minimality of Q is shown for $s = 1$ and $d = 3$.

Theorem 1.4 for general s is due to Lieb [72], who found the optimal value of $S_{d,s}$ and characterized all optimizers. He carried this out in the dual formulation of the Hardy–Littlewood–Sobolev inequality. In this connection he observed and utilized the conformal invariance for general s. He also used a strict rearrangement inequality from [71]. The interplay between rearrangement and conformal invariance ('competing symmetries') is also crucial for the alternative proof of Carlen and Loss [24]. The role of conformal invariance was emphasized in [9]. There Lemma 1.6 appeared, albeit without proof. For GJMS operators on more general manifolds than $\mathbb{S}^d$ we refer to [60, 61].

The minimization problem $S_{d,1/2}$ with $s = 1/2$ is equivalent to finding the best constant in a Sobolev trace inequality on $\mathbb{R}^d \times \mathbb{R}_+$; see, e.g., [9]. The latter problem was solved by Escobar [45] by an adaptation of Obata's method mentioned below. For an alternative proof see [25].

Even before [2, 89], Obata [82] has characterized all 'sufficiently nice' solutions of the Euler–Lagrange equation corresponding to the minimization problem for $S_{d,1}$ on $\mathbb{S}^d$. Up to proving the existence of an optimizer and showing that it is 'sufficiently nice', this leads to the sharp value of the constant $S_{d,1}$ and the characterization of its optimizers. The fact that optimizers, which are weak solutions of the equation, are 'sufficiently nice' is due to Trudinger [90]. The method of Obata was extended in [14, 58]. A related method appears in [6, Theorem 6.10] in the setting of diffusion semigroups satisfying a curvature-dimension condition.

Another result concerning the Euler–Lagrange equation corresponding to the minimization problem for $S_{d,1}$ on $\mathbb{R}^d$ was obtained in [59]. There, using the method of moving planes, it was shown that any positive, classical and sufficiently fast decaying solution is necessarily radial about some point and decreasing with respect to the distance from that point. This reduces the classification of all solutions to a simple ODE analysis. We remark that the relevant ODE becomes autonomous in logarithmic coordinates. It was observed in [20] that the decay assumption in [59] can be removed by employing the invariance under inversions on the unit sphere.

The method of moving planes (and its relative, the method of moving spheres) has been adapted in [31, 70] to the Euler–Lagrange equation for the optimization problem $S_{d,s}$ in the dual formulation for general s. Combined with the conformal invariance this leads to a classification of all positive finite-energy solutions and, consequently, of all minimizers.

A related reflection/inversion technique was used in [50, 51] to give another proof of the characterization of optimizers under the additional assumption $s \leq 1$.

A proof of Theorem 1.4 via optimal transport theory appears in [36] (for $s = 1$) and in [81] (for $s = 1/2$).

Theorem 1.4 with $s = 1$ can also be proved using nonlinear (porous medium or fast diffusion) flows; see [37] and, in the dual setting of an HLS inequality, [22].

The subcritical Sobolev inequality in Theorem 1.10 is classical for $d = 1$. For general d it appeared around the same time in works of Bidaut-Véron–Véron [14, Appendix B], Bakry [6, Theorem 6.10] and Beckner [9, Theorem 4]. Our presentation follows the latter paper. The method of [6, 14] is related to that of Obata [82] (see also [58, Appendix B]) and also classifies solutions of the corresponding Euler–Lagrange equation. An alternative proof of the subcritical Sobolev inequality on the sphere is to first use symmetric decreasing rearrangement on the sphere and then to use the one-dimensional result in [7, pp. 204–205]. Yet another proof is by nonlinear flows [37]; for more on linear and nonlinear flows see [38–40]. These works, in particular, bring into evidence a relation between the 'elliptic' proofs of [6, 14, 58, 82] and the 'parabolic' proofs in [7, 37]. For a remarkable recent results obtained by elliptic methods, see [41].

Aubin's inequality (1.44) is from [4]. Inequalities of this type have recently attracted some attention; see, e.g., [29, 33, 62]. The idea of using Aubin's inequality to deduce that a weak limit is nonzero is implicit in [4, Lemme 3] and [30, Lemma 5.7] in the context of prescribing scalar curvature.

1.4 Lecture 3: Stability

Our goal in this lecture is to prove a stability result for the sharp Sobolev inequality

$$\|(-\Delta)^{s/2}u\|_2^2 \geq S_{d,s}\|u\|_q^2 \qquad \text{for all } u \in \dot{H}^s(\mathbb{R}^d).$$

That is, we want to prove that if $\|(-\Delta)^{s/2}u\|_2^2/\|u\|_q^2$ is close to $S_{d,s}$, then u is close in $\dot{H}^s(\mathbb{R}^d)$ to an optimizer. We denote by

$$\mathcal{G} := \left\{ g \in \dot{H}^s(\mathbb{R}^d) : \|(-\Delta)^{s/2}g\|_2^2 = S_{d,s}\|g\|_q^2 \right\}$$

the set of all optimizers (and zero).

The compactness theorem from Lecture 1 already gives a qualitative version of this stability. Specifically, it implies that for any $\varepsilon > 0$ there is a $\delta > 0$ such that, if $\|(-\Delta)^{s/2}u\|_2^2 \leq (1+\delta)S_{d,s}\|u\|_q^2$, then $\inf_{g \in \mathcal{G}} \|(-\Delta)^{s/2}(u-g)\|_2 \leq \varepsilon \|(-\Delta)^{s/2}u\|_2$.

In this lecture we are interested in a *quantitative* stability result, which shows an explicit dependence of δ on ε. That is, we want to bound the normalized Sobolev

1 The Sharp Sobolev Inequality and Its Stability: An Introduction

deficit $\|(-\Delta)^{s/2}u\|_2^2/\|u\|_q^2 - S_{d,s}$ from below by a power of the normalized distance $\inf_{g\in\mathcal{G}} \|(-\Delta)^{s/2}(u-g)\|_2/\|(-\Delta)^{s/2}u\|_2$,

$$\frac{\|(-\Delta)^{s/2}u\|_2^2}{\|u\|_q^2} - S_{d,s} \gtrsim \left(\inf_{g\in\mathcal{G}} \frac{\|(-\Delta)^{s/2}(u-g)\|_2}{\|(-\Delta)^{s/2}u\|_2}\right)^\alpha.$$

Since $\inf_{g\in\mathcal{G}} \|(-\Delta)^{s/2}(u-g)\|_2/\|(-\Delta)^{s/2}u\|_2 \leq 1$ (see (a) in Lemma 1.12 below), this inequality is stronger the smaller the power α is. Meanwhile, since we expect the left side to be sufficiently smooth and since minima of smooth functions are of quadratic or higher order, we do not expect a better power than $\alpha = 2$.

The following theorem provides such a bound with the desired power $\alpha = 2$. For $s = 1$ it is due to Bianchi and Egnell. (Strictly speaking, the following theorem proves the above bound with the right side multiplied by a factor $\|u\|_q^2/\|(-\Delta)^{s/2}u\|_2^2 \leq S_{d,s}^{-1}$. This difference, however, is immaterial as long as we do not care about constants: If $\|u\|_q^2/\|(-\Delta)^{s/2}u\|_2^2 \geq (2S_{d,s})^{-1}$, say, then the prefactor is harmless, while if $\|u\|_q^2/\|(-\Delta)^{s/2}u\|_2^2 < (2S_{d,s})^{-1}$, then the inequality is anyway trivially true in view of the bound $\inf_{g\in\mathcal{G}} \|(-\Delta)^{s/2}(u-g)\|_2/\|(-\Delta)^{s/2}u\|_2 \leq 1$.)

Theorem 1.11 *Let $0 < s < \frac{d}{2}$ and $q := \frac{2d}{d-2s}$. Then, for all $u \in \dot{H}^s(\mathbb{R}^d)$,*

$$\|(-\Delta)^{s/2}u\|_2^2 - S_{d,s}\|u\|_q^2 \gtrsim \inf_{g\in\mathcal{G}} \|(-\Delta)^{s/2}(u-g)\|_2^2.$$

The implicit constant in the inequality in the theorem depends on d and s. The argument that we present is via compactness and does not yield an explicit constant. For recent progress on the problem of giving a constructive proof, see the remarks at the end of this lecture.

We will also prove the reverse inequality

$$\inf_{u\notin\mathcal{G}} \frac{\|(-\Delta)^{s/2}u\|_2^2 - S_{d,s}\|u\|_q^2}{\inf_{g\in\mathcal{G}} \|(-\Delta)^{s/2}(u-g)\|_2^2} \leq \frac{4s}{d+2s+2}, \tag{1.46}$$

which shows, in particular, that the power two of the distance to $\mathcal{G}$ in the theorem cannot be replaced by a smaller power.

1.4.1 The Upper Bound

To get some intuition into the mechanism behind the proof of the theorem, we begin by proving (1.46). It is natural to approach this problem by taking $u = u_* + \varepsilon r$ with an optimizer u_* and a function r, to be determined, and by expanding the relevant quotient in ε. There will be a coefficient in front of the leading order in ε and this coefficient is a functional of r. The idea is to determine r so as to minimize this functional. It will turn out that the problem for r is a spectral problem that can be solved explicitly. This gives the constant $\frac{4s}{d+2s+2}$ in (1.46) as a certain spectral gap.

It is more convenient to carry out this idea in the equivalent setting of the inequality on the sphere, that is, to prove

$$\inf_{U \notin \mathcal{H}} \frac{\mathcal{E}_s[U] - S_{d,s}\|U\|_q^2}{\inf_{h \in \mathcal{H}} \mathcal{E}_s[U-h]} \leq \frac{4s}{d+2s+2}, \quad (1.47)$$

where

$$\mathcal{H} := \left\{ h \in H^s(\mathbb{S}^d) : \mathcal{E}_s[h] = S_{d,s}\|h\|_q^2 \right\}.$$

The ansatz is then $U = U_* + \varepsilon R$ and, by conformal invariance, we may assume $U_* = 1$. Similarly to the derivative computations in the previous lecture, we find

$$\lim_{\varepsilon \to 0} \varepsilon^{-2} \left(\mathcal{E}_s[1 + \varepsilon R] - S_{d,s}\|1 + \varepsilon R\|_q^2 \right) = \langle R, \mathcal{L}_s R \rangle \quad (1.48)$$

with the operator $\mathcal{L}_s$ introduced in (1.36). This gives the behavior of the numerator on the left side of (1.47). We would like to show that the denominator also behaves quadratically in ε and, more precisely, that under suitable assumptions on R we have

$$\inf_{h \in \mathcal{H}} \mathcal{E}_s[(1+\varepsilon R) - h] = \varepsilon^2 \mathcal{E}_s[R].$$

Note that here we always have $\leq$ since we can always choose $h = 1 \in \mathcal{H}$.

In the following lemma, we summarize some properties of the distance function

$$\delta[U] := \inf_{h \in \mathcal{H}} \sqrt{\mathcal{E}_s[U-h]}.$$

Lemma 1.12 *Let $0 < s < \frac{d}{2}$.*

(a) $\delta[U]^2 \leq \mathcal{E}_s[U]$ *with strict inequality if and only if $U \neq 0$.*
(b) *For any $U \in \dot{H}^s(\mathbb{S}^d)$, there is an $h \in \mathcal{H}$ such that $\mathcal{E}_s[U-h] = \delta[U]^2$. If $U \neq 0$ and $\tau_U := \delta[U]/\sqrt{\mathcal{E}_s[U] - \delta[U]^2}$, then*

$$\|U - h\|_q \leq \tau_U \|h\|_q.$$

(c) *If $\mathcal{E}_s[U - 1] = \delta[U]^2$, then $R := U - 1$ satisfies*

$$\int_{\mathbb{S}^d} R(\omega)\,d\omega = 0 \quad \text{and} \quad \int_{\mathbb{S}^d} \omega\, R(\omega)\,d\omega = 0. \quad (1.49)$$

(d) *There is an $\varepsilon_0 > 0$ such that, if $\mathcal{E}_s[R] < \varepsilon_0$ and (1.49) holds, then $\delta[1 + R] = \sqrt{\mathcal{E}_s[R]}$.*

Before proving this lemma, we use it to complete the proof of (1.47).

It follows from Lemma 1.12 that, if R satisfies (1.49), then for all sufficiently small ε we have $\inf_{h\in\mathcal{H}} \mathcal{E}_s[(1+\varepsilon R) - h] = \varepsilon^2 \mathcal{E}_s[R]$. Combining this with (1.48) we find

$$\lim_{\varepsilon \to 0} \frac{\mathcal{E}_s[1+\varepsilon R] - S_{d,s}\|1+\varepsilon R\|_q^2}{\inf_{h\in\mathcal{H}} \mathcal{E}_s[1+\varepsilon R - h]} = \frac{\langle R, \mathcal{L}_s R\rangle}{\mathcal{E}_s[R]}.$$

At this point we can choose R so as to minimize the right side. We will show below that

$$\inf\left\{\frac{\langle R, \mathcal{L}_s R\rangle}{\mathcal{E}_s[R]} \,:\, R \text{ satisfies } (1.49)\right\} = \frac{4s}{d+2s+2}.$$

The argument given there also shows that the infimum is attained if and only if R is a spherical harmonic of degree 2, so this is the optimal choice of R. This completes the proof of (1.47) and, therefore, of (1.46).

Proof of Lemma 1.12 We recall that the elements in $\mathcal{H}$ are of the form $c\, Q_\zeta$ with $c \in \mathbb{R}$ and $\zeta \in \mathbb{R}^{d+1}$ with $|\zeta| < 1$. Here we set $Q_\zeta(\omega) = (\sqrt{1-|\zeta|^2}/(1-\zeta \cdot \omega))^{(d-2s)/2}$. For later purposes we record the normalizations

$$\|Q_\zeta\|_q^q = |\mathbb{S}^d|, \qquad \mathcal{E}_s[Q_\zeta] = \frac{\Gamma(\frac{d}{2}+s)}{\Gamma(\frac{d}{2}-s)} |\mathbb{S}^d| \qquad (1.50)$$

and the Euler–Lagrange equation

$$\mathcal{E}_s[V, Q_\zeta] = \frac{\Gamma(\frac{d}{2}+s)}{\Gamma(\frac{d}{2}-s)} \int_{\mathbb{S}^d} V Q_\zeta^{q-1}\, d\omega \qquad \text{for all } V \in H^s(\mathbb{S}^d). \qquad (1.51)$$

Indeed, the first equality in (1.50) follows from the characterization of Q_ζ as a Jacobian in the previous lecture and the second one from the minimality, which also gives the Euler–Lagrange equation.

(a) We write $\mathcal{E}_s[\cdot, \cdot]$ for the bilinear form associated to the quadratic form $\mathcal{E}_s[\cdot]$. Since

$$\mathcal{E}_s[U - c\, Q_\zeta] = \mathcal{E}_s[U] - \frac{\mathcal{E}_s[Q_\zeta, U]^2}{\mathcal{E}_s[Q_\zeta]} + \mathcal{E}[Q_\zeta]\left(c - \frac{\mathcal{E}_s[Q_\zeta, U]}{\mathcal{E}_s[Q_\zeta]}\right)^2,$$

it follows that

$$\delta[U]^2 = \inf_{c,\zeta} \mathcal{E}_s[U - c\, Q_\zeta] = \mathcal{E}_s[U] - \sup_\zeta \frac{\mathcal{E}_s[Q_\zeta, U]^2}{\mathcal{E}_s[Q_\zeta]} \qquad (1.52)$$

and that for each ζ the optimal c is given by $c = \mathcal{E}_s[Q_\zeta, U]/\mathcal{E}_s[Q_\zeta]$. Item (a) follows immediately from (1.52) and the nondegeneracy of $\mathcal{E}_s$.

(b) In view of (1.50) and (1.51) we can write (1.52) as

$$\delta[U]^2 = \mathcal{E}_s[U] - \frac{\Gamma(\frac{d}{2}+s)}{\Gamma(\frac{d}{2}-s)} |\mathbb{S}^d|^{-1} \sup_\zeta \left(\int_{\mathbb{S}^d} Q_\zeta^{q-1} U \, d\omega \right)^2. \tag{1.53}$$

It is easy to see that $\zeta \mapsto \int_{\mathbb{S}^d} Q_\zeta^{q-1} U \, d\omega$ is continuous and tends to zero as $|\zeta| \to 1$, and therefore the supremum over ζ is attained, as claimed.

Moreover, if $U \neq 0$ and if the infimum is attained at $h = c_0 Q_{\zeta_0}$, then, recalling the expression for the optimal $c = c_0$,

$$\sup_\zeta \frac{\mathcal{E}_s[Q_\zeta, U]^2}{\mathcal{E}_s[Q_\zeta]} = \frac{\mathcal{E}_s[Q_{\zeta_0}, U]^2}{\mathcal{E}_s[Q_{\zeta_0}]} = c_0^2 \mathcal{E}_s[Q_{\zeta_0}] = \mathcal{E}_s[h].$$

Thus, by (1.52), $\delta[U]^2 = \mathcal{E}_s[U] - \mathcal{E}_s[h]$ and so

$$S_{d,s} \|U - h\|_q^2 \leq \mathcal{E}_s[U - h] = \delta[U]^2 = \tau_U^2 \left(\mathcal{E}_s[U] - \delta[U]^2 \right)$$
$$= \tau_U^2 \mathcal{E}_s[h] = \tau_U^2 S_{d,s} \|h\|_q^2,$$

as claimed.

(c) We assume now that the infimum is attained at $h = 1$, that is, at $(c, \zeta) = (1, 0)$. Then $1 = c = \mathcal{E}_s[1, U]/\mathcal{E}_s[1] = |\mathbb{S}^d|^{-1} \int_{\mathbb{S}^d} U \, d\omega$, where we used (1.50) and (1.51). This proves the first equality in (1.49). Moreover, by (1.53),

$$\nabla_\zeta \Big|_{\zeta=0} \int_{\mathbb{S}^d} Q_\zeta^{q-1} U \, d\omega = 0,$$

which gives the second equality in (1.49).

(d) We prove now conversely that for sufficiently small R, the validity of the orthogonality conditions implies that the distance is attained at the function 1.

To prove this, we apply the implicit function theorem and find $\varepsilon_1, \varepsilon_2 > 0$ such that for $R \in H^s(\mathbb{S}^d)$ with $\mathcal{E}_s[R] < \varepsilon_1$ there is a unique $\zeta \in B_{\varepsilon_2}(0)$ such that $\nabla_\zeta \int_{\mathbb{S}^d} Q_\zeta^{q-1}(1 + R) \, d\omega = 0$. (The invertibility of the relevant matrix in the application of the implicit function theorem follows by a lengthy, but straightforward computation. Indeed, $D_\zeta^2 \big|_{\zeta=0} \int_{\mathbb{S}^d} Q_\zeta^{q-1} \, d\omega$ is a nonzero multiple of the identity matrix.) Since, by assumption, the condition $\nabla_\zeta \int_{\mathbb{S}^d} Q_\zeta^{q-1}(1+R) \, d\omega = 0$ is satisfied at $\zeta = 0$, we infer that, when restricted to B_{ε_2}, the supremum in (1.53) is attained at $\zeta = 0$.

Let us show that, by decreasing ε_1 if necessary, we can ensure that it is not attained outside of B_{ε_2}. We note that

$$\eta := 1 - |\mathbb{S}^d|^{-1} \sup_{|\zeta| \geq \varepsilon_2} \left| \int_{\mathbb{S}^d} Q_\zeta^{q-1} d\omega \right| > 0.$$

Indeed, by Hölder we have $|\cdot| \leq \|Q_\zeta\|_q^{q-1} |\mathbb{S}^d|^{1/q} = |\mathbb{S}^d|$ with equality if and only if Q_ζ is a constant. Since $|\zeta| \geq \varepsilon_2$, Q_ζ cannot be a constant, and by continuity we deduce $\eta > 0$. (We note that this argument can be made quantitative via the quantitative version of Hölder's inequality in [23].) Thus, if $|\zeta| \geq \varepsilon_2$,

$$\left| \int_{\mathbb{S}^d} Q_\zeta^{q-1}(1+R) d\omega \right| \leq (1-\eta)|\mathbb{S}^d| + \|Q_\zeta\|_q^{q-1} \|R\|_q$$

$$\leq (1-\eta)|\mathbb{S}^d| + |\mathbb{S}^d|^{(q-1)/q} S_{d,s}^{-1/2} \mathcal{E}_s[R]^{1/2}$$

Thus, if $|\zeta| \geq \varepsilon_2$ and $\mathcal{E}_s[R] \leq S_{d,s} |\mathbb{S}^d|^{2/q} \eta^2$, then

$$\left| \int_{\mathbb{S}^d} Q_\zeta^{q-1}(1+R) d\omega \right| \leq |\mathbb{S}^d| = \left| \int_{\mathbb{S}^d} Q_0^{q-1}(1+R) d\omega \right|.$$

This means that in (1.53) the supremum can be restricted to $|\zeta| < \varepsilon_2$, where it is attained at the origin, as we have seen. This proves (d) with $\varepsilon_0 := \min\{\varepsilon_1, S_{d,s} |\mathbb{S}^d|^{2/q} \eta^2\}$. □

1.4.2 The Lower Bound

We now turn to the proof of Theorem 1.11. The main step of the proof is contained in the following proposition, where we abbreviate (suppressing the s-dependence)

$$\delta[u] := \inf_{g \in \mathcal{G}} \|(-\Delta)^{s/2}(u-g)\|_2.$$

Also, we write

$$\tau_u := \delta[u] / \sqrt{\|(-\Delta)^{s/2} u\|_2^2 - \delta[u]^2} \qquad \text{if } 0 \neq u \in \dot{H}^s(\mathbb{R}^d).$$

By (a) in Lemma 1.12 and conformal invariance, one sees that τ_u is well defined.

Proposition 1.13 *Let* $0 < s < \frac{d}{2}$ *and* $q := \frac{2d}{d-2s}$. *Then, for all* $0 \neq u \in \dot{H}^s(\mathbb{R}^d)$,

$$\|(-\Delta)^{s/2} u\|_2^2 - S_{d,s} \|u\|_q^2 - \frac{4s}{d+2s+2} \delta[u]^2 \gtrsim -\tau_u^{\min\{q-2,1\}} \delta[u]^2.$$

Proof Using the stereographic projection, we cast the inequality in the proposition into an equivalent inequality on the sphere. Namely, for $0 \neq U \in H^s(\mathbb{S}^d)$,

$$\mathcal{E}_s[U] - S_{d,s} \|U\|_q^2 - \frac{4s}{d+2s+2} \delta[U]^2 \gtrsim -\tau_U^{\min\{q-2,1\}} \delta[U]^2. \tag{1.54}$$

By Lemma 1.12 the infimum $\delta[U]$ is attained, and by conformal invariance we may assume that it is attained at a constant function c. We write

$$U = c + R$$

and recall that R satisfies the orthogonality conditions (1.49).

Using the elementary inequality

$$\left||a|^q - |b|^q - q|b|^{q-2}b(a-b) - \tfrac{1}{2}q(q-1)|b|^{q-2}(a-b)^2\right|$$
$$\lesssim |b|^{(q-3)_+}|a-b|^{\min\{q,3\}} + |a-b|^q,$$

valid for all $a, b \in \mathbb{R}$, together with the first condition in (1.49), we find

$$\left|\int_{\mathbb{S}^d} |U|^q \, d\omega - |c|^q |\mathbb{S}^d| - \tfrac{1}{2}q(q-1)|c|^{q-2} \int_{\mathbb{S}^d} R^2 \, d\omega\right|$$
$$\lesssim |c|^{(q-3)_+} \|R\|_q^{\min\{q,3\}} + \|R\|_q^q.$$

Using the elementary inequality

$$(1+t)^{2/q} \leq 1 + \tfrac{2}{q}t,$$

valid for all $t \geq 0$, we deduce

$$\|U\|_q^2 \leq |\mathbb{S}^d|^{2/q} c^2 + (q-1)|\mathbb{S}^d|^{-1+2/q} \int_{\mathbb{S}^d} R^2 \, d\omega$$
$$+ \text{const } |c|^{2-q} \left(|c|^{(q-3)_+} \|R\|_q^{\min\{q,3\}} + \|R\|_q^q\right).$$

Meanwhile, by (1.51) and (1.49),

$$\mathcal{E}_s[U] = c^2 \mathcal{E}_s[1] + \mathcal{E}_s[R].$$

Combining the previous two relations and recalling $\mathcal{E}_s[1] = S_{d,s} |\mathbb{S}^d|^{2/q}$ yields

$$\mathcal{E}_s[U] - S_{d,s} \|U\|_q^2 \geq \mathcal{E}_s[R] - (q-1)\mathcal{E}_s[1]|\mathbb{S}^d|^{-1} \int_{\mathbb{S}^d} R^2 \, d\omega$$
$$- \text{const } |c|^{2-q} \left(|c|^{(q-3)_+} \|R\|_q^{\min\{q,3\}} + \|R\|_q^q\right). \tag{1.55}$$

1 The Sharp Sobolev Inequality and Its Stability: An Introduction

In terms of the decomposition of R into spherical harmonics, the quadratic terms on the right side are equal to

$$\mathcal{E}_s[R] - (q-1)\mathcal{E}_s[1]|\mathbb{S}^d|^{-1}\int_{\mathbb{S}^d} R^2\, d\omega$$

$$= \sum_{\ell=0}^{\infty}\left(\frac{\Gamma(\ell+\frac{d}{2}+s)}{\Gamma(\ell+\frac{d}{2}-s)} - (q-1)\mathcal{E}_s[1]|\mathbb{S}^d|^{-1}\right)\|P_\ell R\|_2^2$$

$$= \sum_{\ell=0}^{\infty}\left(\frac{\Gamma(\ell+\frac{d}{2}+s)}{\Gamma(\ell+\frac{d}{2}-s)} - \frac{\Gamma(1+\frac{d}{2}+s)}{\Gamma(1+\frac{d}{2}-s)}\right)\|P_\ell R\|_2^2\,.$$

We now recall that, according to (1.49),

$$P_0 R = P_1 R = 0\,,$$

so the above sum can be restricted to $\ell \geq 2$. It follows from (1.37) that

$$\frac{\Gamma(\ell+\frac{d}{2}+s)}{\Gamma(\ell+\frac{d}{2}-s)} - \frac{\Gamma(1+\frac{d}{2}+s)}{\Gamma(1+\frac{d}{2}-s)}$$

$$\geq \left(1 - \frac{\Gamma(2+\frac{d}{2}-s)\,\Gamma(1+\frac{d}{2}+s)}{\Gamma(2+\frac{d}{2}+s)\,\Gamma(1+\frac{d}{2}-s)}\right)\frac{\Gamma(\ell+\frac{d}{2}+s)}{\Gamma(\ell+\frac{d}{2}-s)}$$

$$= \frac{2s}{1+\frac{d}{2}+s}\frac{\Gamma(\ell+\frac{d}{2}+s)}{\Gamma(\ell+\frac{d}{2}-s)}\,.$$

To summarize, we have shown that

$$\mathcal{E}_s[R] - (q-1)\mathcal{E}_s[1]|\mathbb{S}^d|^{-1}\int_{\mathbb{S}^d} R^2\, d\omega \geq \frac{2s}{1+\frac{d}{2}+s}\mathcal{E}_s[R] = \frac{2s}{1+\frac{d}{2}+s}\delta[U]^2\,.$$

It remains to deal with the remainder terms in (1.55). Using Sobolev's inequality and the inequality $\|R\|_q \leq \tau_U |\mathbb{S}^d|^{1/q}|c|$ from (b) in Lemma 1.12, we find

$$|c|^{2-q}\left(|c|^{(q-3)_+}\|R\|_q^{\min\{q,3\}} + \|R\|_q^q\right)$$

$$\lesssim |c|^{2-q}\left(|c|^{(q-3)_+}\|R\|_q^{\min\{q-2,1\}} + \|R\|_q^{q-2}\right)\delta[U]^2$$

$$\lesssim \left(\tau_U^{\min\{q-2,1\}} + \tau_U^{q-2}\right)\delta[U]^2\,.$$

This leads to the bound

$$\mathcal{E}_s[U] - S_{d,s}\|U\|_q^2 - \frac{4s}{d+2s+2}\delta[U]^2 \geq -C_s\left(\tau_U^{\min\{q-2,1\}} + \tau_U^{q-2}\right)\delta[U]^2$$

for some constant C_s depending on s. When $\tau_U \leq M$ for some constant $M > 0$ to be determined, then the right side can be bounded from below by $-C_s(1 + M^{(q-3)_+})\tau_U^{\min\{q-2,1\}}\delta[U]^2$, which is (1.54). Meanwhile, when $C_s(1 + M^{(q-3)_+})\tau_U^{\min\{q-2,1\}} \geq \frac{4s}{d+2s+2}$, we have

$$\mathcal{E}_s[U] - S_{d,s}\|U\|_q^2 \geq 0 \geq \left(\frac{4s}{d+2s+2} - C_s(1 + M^{(q-3)_+})\tau_U^{\min\{q-2,1\}}\right)\delta[U]^2,$$

so again (1.54) holds. This covers all possibly values of τ_U, provided M is chosen such that

$$C_s M^{\min\{q-2,1\}}(1 + M^{(q-3)_+}) \leq \frac{4s}{d+2s+2}$$

which is clearly possible. This completes the proof of the proposition. □

Proof of Theorem 1.11 We argue by contradiction, assuming the claimed inequality would not hold. Then there is a sequence $(u_n) \subset \dot{H}^s(\mathbb{R}^d)$ such that

$$\frac{\|(-\Delta)^{s/2}u_n\|_2^2 - S_{d,s}\|u_n\|_q^2}{\delta[u_n]^2} \to 0. \tag{1.56}$$

By homogeneity we may assume that $\|(-\Delta)^{s/2}u_n\|_2 = 1$. Then the Sobolev inequality and the inequality $\delta[u_n] \leq \|(-\Delta)^{s/2}u_n\|_2 = 1$ imply

$$0 \leq 1 - S_{d,s}\|u_n\|_q^2 \leq \frac{\|(-\Delta)^{s/2}u_n\|_2^2 - S_{d,s}\|u_n\|_q^2}{\delta[u_n]^2}.$$

By (1.56) we deduce that $\|u_n\|_q^2 \to S_{d,s}^{-1}$. It follows from Lions's theorem proved in the first lecture that $\delta[u_n] \to 0$. Using this information in the inequality in Proposition 1.13, we obtain

$$\liminf_{n\to\infty} \frac{\|(-\Delta)^{s/2}u_n\|_2^2 - S_{d,s}\|u_n\|_q^2}{\delta[u_n]^2} \geq \frac{4s}{d+2s+2}.$$

This contradicts (1.56) and completes the proof of Theorem 1.11. □

1.4.3 Bibliographical Remarks

The main theorem in this lecture for $s = 1$, as well as the strategy employed for general s, are due to Bianchi and Egnell [12]. They answered a question posed by Brezis and Lieb [17]. The result for general s, as well as the observation to use conformal invariance, appeared in [32]; earlier results in the local case (that is, for integer s) are in [8, 77].

The basic ingredients of the Bianchi–Egnell method are a compactness theorem for optimizing sequences and the fact that all zero modes of the Hessian come from symmetries. For functional inequalities for which these two ingredients are available there is a good chance that the Bianchi–Egnell method can be applied. This has been carried out in a large number of cases; see, for instance, the introduction of [49] for references.

We do not attempt to give an overview over the works on the stability problem of functional inequalities in the last two decades. Let us just mention the works [35, 48, 55] on the isoperimetric inequality, which had a huge impact on the field, as well as the surveys [42, 46, 47]. Related to the topic of these lectures, we mention the stability result [21] for the HLS inequality. Indeed, this is deduced from the main result of the present lecture together with a quantitative version of the duality argument used in the proof of Lemma 1.3 in the appendix of the first lecture.

A stability result for the Sobolev inequality in the case $s = 1$ appears in [16]. It is of a somewhat different flavor than Theorem 1.11 and has an explicit constant in the bound, at the expense of being only valid for functions u with sufficiently fast decay.

After the lectures at the summer school on which these notes are based, there have been some developments concerning the stability theorem, which we briefly describe. Let us denote the optimal constant in the stability theorem by

$$c_{d,s}^{\mathrm{BE}} := \inf_{u \notin \mathcal{G}} \frac{\|(-\Delta)^{s/2} u\|_2^2 - S_{d,s} \|u\|_q^2}{\inf_{g \in \mathcal{G}} \|(-\Delta)^{s/2}(u - g)\|_2^2}.$$

Note that the proof presented in this lecture used compactness and only showed that $c_{d,s}^{\mathrm{BE}}$ is positive, without giving any lower bound. In the case $s = 1$, the paper [43] provided for the first time an explicit lower bound on $c_{d,1}^{\mathrm{BE}}$. This was achieved by replacing the use of Lions's compactness theorem by a more precise argument based on rearrangement methods, using both a discrete and a continuous symmetrization flow. In fact, in [43] it was shown that

$$c_{d,1}^{\mathrm{BE}} \gtrsim d^{-1},$$

which is optimal with respect to its large d-behavior in view of the upper bound $c_{d,1}^{\mathrm{BE}} \leq 4/(d+4)$ from (1.46). This was achieved by cutting the remainder R in a suitable way and using the above Taylor expansion of the nonlinearity only where R is sufficiently small compared to c. The first part of the argument, namely that giving an explicit lower bound, extends to general s, as shown in [34]. The optimal behavior of the constant with respect to d for $s = 1$ leads to a quantitative version of the logarithmic Sobolev inequality [43].

Another development occurred in [65–67], where it was shown that for general s and $d \geq 2$ the constant $c_{d,s}^{\text{BE}}$ is attained. This involves, among other things, showing that the inequality in (1.46) is strict and uses a compactness argument, reminiscent of but much more intricate than those presented in Lecture 1. The case $d = 1$ appears to be different.

1.5 Lecture 4: Nondegenerate and Degenerate Stability

In this final lecture we discuss the following one-parameter family of Sobolev-type inequalities

$$\int_0^T \int_{\mathbb{S}^{d-1}} \left(|\partial_t u|^2 + |\nabla_\omega u|^2 + \tfrac{(d-2)^2}{4} u^2 \right) d\omega\, dt$$

$$\geq S_d(T) \left(\int_0^T \int_{\mathbb{S}^{d-1}} |u|^q\, d\omega\, dt \right)^{2/q},$$

valid for functions $u \in H^1((\mathbb{R}/T\mathbb{Z}) \times \mathbb{S}^{d-1})$. Here, as always,

$$d \geq 3 \quad \text{and} \quad q = \tfrac{2d}{d-2}.$$

Moreover, $T > 0$ is a parameter. We abbreviate $\Sigma_T := (\mathbb{R}/T\mathbb{Z}) \times \mathbb{S}^{d-1}$, $dv_g = d\omega\, dt$, $\|u\|_q := \|u\|_{L^q(\Sigma_T, v_g)}$ and, for $u \in H^1(\Sigma_T)$,

$$\mathcal{E}_T[u] := \int_{\Sigma_T} \left(|\partial_t u|^2 + |\nabla_\omega u|^2 + \tfrac{(d-2)^2}{4} u^2 \right) dv_g.$$

(There should be no risk of confusing this with $\mathcal{E}_s[U]$ from the previous two lectures.) By $S_d(T)$ we denote the optimal constant in the above inequality, that is,

$$S_d(T) := \inf_{0 \neq u \in H^1(\Sigma_T)} \frac{\mathcal{E}_T[u]}{\|u\|_q^2}. \tag{1.57}$$

Attention to these inequalities was drawn by Schoen in connection with the Yamabe problem. A remarkable feature is the existence of a critical parameter

$$T_* := \tfrac{2\pi}{\sqrt{d-2}}$$

1 The Sharp Sobolev Inequality and Its Stability: An Introduction

such that for $T \leq T_*$ the infimum in (1.57) is attained precisely at constants, while for $T > T_*$ it is attained at a nonconstant function that is independent of the variable ω. Our goal is to show that for $T \neq T_*$ one has a quadratic stability similar as in the previous lecture, while for $T = T_*$ one only has a quartic stability.

We let

$$\mathcal{G}_T := \left\{ g \in H^1(\Sigma_T) : \mathcal{E}_T[g] = S_d(T) \|g\|_q^2 \right\}$$

denote the set of all optimizers (and zero).

The main result of this lecture is the following stability theorem for (1.57).

Theorem 1.14 *Let $d \geq 3$, $q = \frac{2d}{d-2}$ and $T > 0$. Then, for all $u \in H^1(\Sigma_T)$,*

$$\mathcal{E}_T[u] - S_d(T) \|u\|_q^2 \gtrsim \begin{cases} \inf_{g \in \mathcal{G}_T} \mathcal{E}_T[u-g] & \text{if } T \neq T_*, \\ \inf_{g \in \mathcal{G}_T} \frac{\mathcal{E}_T[u-g]^2}{\mathcal{E}_T[u]} & \text{if } T = T_*. \end{cases}$$

Of course, the constant implicit in the $\gtrsim$ depends on T.

We will also show that the order of vanishing given by the theorem is optimal. That is, we will show that

$$\inf_{u \notin \mathcal{G}_T} \frac{\mathcal{E}_T[u] - S_d(T) \|u\|_q^2}{\inf_{g \in \mathcal{G}_T} \mathcal{E}_T[u-g]} \leq c_T \tag{1.58}$$

with a certain constant $c_T < \infty$ defined in (1.69) below. Moreover, at $T = T_*$ we have $c_{T_*} = 0$ and we show

$$\inf_{u \notin \mathcal{G}_{T_*}} \frac{\mathcal{E}_{T_*}[u] \left(\mathcal{E}_{T_*}[u] - S_d(T_*) \|u\|_q^2 \right)}{\inf_{g \in \mathcal{G}_{T_*}} \mathcal{E}_{T_*}[u-g]^2} \leq \frac{(q+2)(q-2)}{12(q-1)}. \tag{1.59}$$

The bounds (1.58) and (1.59) imply that one cannot have a better stability result than a quadratic one if $T \neq T_*$ and a quartic one if $T = T_*$.

We emphasize that we refer to the first (resp. second) bound in Theorem 1.14 as quadratic (resp. quartic) stability, since the term $\inf_{g \in \mathcal{G}_T} \mathcal{E}_T[u-g]$ vanishes quadratically as u approaches $\mathcal{G}_T$.

The basic strategy to prove Theorem 1.14 is the same as that in the previous lectures: We prove a compactness theorem, classify the optimizers and the zero modes of their Hessian and then we put these ingredients together. (We will not give a full proof of the classification of optimizers, but refer to the literature at some points.) For $T \neq T_*$ this works in a straightforward way. The case $T = T_*$, however, is different since there is a zero mode of the Hessian that does *not* come from symmetries. This is responsible for the quartic behavior and on a technical level necessitates a certain iteration of the basic strategy, which we will explain.

1.5.1 Optimizing Sequences

We begin by proving relative compactness of optimizing sequences for the optimization problem (1.57).

Proposition 1.15 *Let $T > 0$. Let $(u_n) \subset H^1(\Sigma_T)$ with $\mathcal{E}_T[u_n] = 1$ and $\|u_n\|_q^2 \to S_d(T)^{-1}$. Then there is a subsequence that converges in $H^1(\Sigma_T)$ to an optimizer of (1.57).*

Note that, in contrast to the corresponding theorem in Lecture 1, here there are no noncompact symmetries that could lead to a loss of compactness.

The proof of the proposition relies on the following strict upper bound on $S_d(T)$. We denote by $S_d := S_{d,1}$ the constant from the first three lectures.

Lemma 1.16 *For all $T > 0$, $S_d(T) < S_d$.*

Proof of Lemma 1.16 Let $Q(t) := \cosh^{-(d-2)/2} t$ and note that

$$-Q'' + \left(\tfrac{d-2}{2}\right)^2 Q = \tfrac{d(d-2)}{4} Q^{(d+2)/(d-2)} \qquad \text{in } \mathbb{R}.$$

Taking $(0, T) \times \mathbb{S}^{d-1} \ni (t, \omega) \mapsto Q(t - T/2)$, considered as a function on Σ_T, as a trial function, we obtain

$$S_d(T) \leq \frac{2|\mathbb{S}^{d-1}| \int_0^{T/2} ((Q')^2 + (\tfrac{d-2}{2})^2 Q^2)\, dt}{\left(2|\mathbb{S}^{d-1}| \int_0^{T/2} Q^q\, dt\right)^{2/q}}$$

$$= \frac{d(d-2)}{4} \left(2|\mathbb{S}^{d-1}| \int_0^{T/2} Q^q\, dt\right)^{1-2/q} + \frac{2|\mathbb{S}^{d-1}| Q'(T/2) Q(T/2)}{\left(2|\mathbb{S}^{d-1}| \int_0^{T/2} Q^q\, dt\right)^{2/q}}.$$

The boundary term is < 0 since Q is positive and decreasing on $(0, \infty)$, and the bulk term is

$$< \frac{d(d-2)}{4} \left(2|\mathbb{S}^{d-1}| \int_0^{\infty} Q^q\, dt\right)^{1-2/q} = S_d\,.$$

The last equality can be seen either by explicit computation and comparison with the value of S_d, or by noting that the Q in this proof coincides with the Q in the characterization of optimizers of S_d in logarithmic coordinates (see, for instance, (1.65) below). □

Proof of Proposition 1.15 We proceed similarly as in Step 2 of the proof of the main theorem in Lecture 1. After passing to a subsequence, we may assume that $u_n \rightharpoonup u$ in $H^1(\Sigma_T)$. We write

$$u_n = u + r_n \qquad \text{with } r_n \rightharpoonup 0 \text{ in } H^1(\Sigma_T).$$

1 The Sharp Sobolev Inequality and Its Stability: An Introduction

By the same arguments as in Lecture 1, we deduce that

$$t := \lim_{n\to\infty} \mathcal{E}_T[r_n] \quad \text{exists and satisfies} \quad 1 = \mathcal{E}_T[u] + t \tag{1.60}$$

and

$$m := \lim_{n\to\infty} \|r_n\|_q^q \quad \text{exists and satisfies} \quad S_d(T)^{-q/2} = \|u\|_q^q + m. \tag{1.61}$$

In contrast to Lecture 1, we will argue more carefully when estimating t from below by m. We will show that

$$t \geq S_d m^{2/q}, \tag{1.62}$$

where $S_d := S_{d,1}$ is the constant from the first three lectures. We emphasize that (1.62) with $S_d(T)$ instead of S_d would be immediate. Since $S_d > S_d(T)$ by Lemma 1.16, (1.62) as it stands is an improvement of this immediate bound, and this improvement will be crucial in our proof.

Inequality (1.62) follows from an inequality of Aubin, which says that for any $\varepsilon > 0$ there is a $C_{\varepsilon,T} < \infty$ such that for all $v \in H^1(\Sigma_T)$,

$$\mathcal{E}_T[v] \geq (1-\varepsilon) S_d \|v\|_q^2 - C_{\varepsilon,T} \|v\|_2^2. \tag{1.63}$$

In fact, Aubin's inequality is valid on any closed Riemannian manifold $(\mathcal{M}, g)$ of dimension $d \geq 3$ with the left side replaced by $\int_{\mathcal{M}} \left(|\nabla v|_g^2 + \frac{d-2}{4(d-1)} R_g v^2 \right) dv_g$, where R_g denotes the scalar curvature. For the proof of (1.63) one covers the manifold by finitely many balls, whose radii are so small that in each ball the metric is Euclidean 'up to an ε'. Then one localizes to these balls using a partition of unity and applies in each ball the Euclidean Sobolev inequality. The error term involving $C_{\varepsilon,T}$ comes from the localization error.

For the proof of (1.62) we apply (1.63) to $v = r_n$. Since $r_n \rightharpoonup 0$ in $H^1(\Sigma_T)$ implies $r_n \to 0$ in $L^2(\Sigma_T)$, we deduce that $t \geq (1-\varepsilon) S_d m^{2/q}$. Since $\varepsilon > 0$ is arbitrary, we obtain (1.62).

We now deduce the proposition from (1.60), (1.61) and (1.62). We find

$$1 = \mathcal{E}_T[u] + t \geq \mathcal{E}_T[u] + S_d(T) m^{2/q} + (1 - S_d(T)/S_d) t$$
$$= \mathcal{E}_T[u] + (1 - S_d(T)^{q/2} \|u\|_q^q)^{2/q} + (1 - S_d(T)/S_d) t$$
$$\geq \mathcal{E}_T[u] + 1 - S_d(T) \|u\|_q^2 + (1 - S_d(T)/S_d) t,$$

where we used the same elementary inequality (1.13) as in Lecture 1. Using the strict inequality $S_d(T) < S_d$ we deduce that

$$\mathcal{E}_T[u] = S_d(T) \|u\|_q^2 \quad \text{and} \quad t = 0. \tag{1.64}$$

Because of (1.60) we deduce from the second condition in (1.64) that $r_n \to 0$ in $H^1(\Sigma_T)$, that is, $u_n \to u$ in $H^1(\Sigma_T)$. In particular, $u \neq 0$ and, by the first condition in (1.64), u is an optimizer. This completes the proof. $\square$

Note that there is a certain analogy between the above proof of Proposition 1.15 and the proof of Theorem 1.1 that was presented in the appendix to Lecture 2. In both cases the compactness comes from an improved constant in front of the L^q term at the expense of adding an L^2 term.

1.5.2 Optimizers

According to the previous proposition, for any T there is an optimizer u_* for the optimization problem (1.57).

We claim that either $u_* \geq 0$ or $u_* \leq 0$. To see this, we recall that by Sobolev space theory, the positive and negative parts $(u_*)_\pm$ of u_* belong to $H^1(\Sigma_T)$ and $\|\nabla u_*\|_2^2 = \|\nabla (u_*)_+\|_2^2 + \|\nabla (u_*)_-\|_2^2$. Thus, if neither $(u_*)_+$ nor $(u_*)_-$ vanish almost everywhere, then

$$S_d(T) = \frac{\mathcal{E}_T[u_*]}{\|u_*\|_q^2} = \theta^{2/q} \frac{\mathcal{E}_T[(u_*)_+]}{\|(u_*)_+\|_q^2} + (1-\theta)^{2/q} \frac{\mathcal{E}_T[(u_*)_-]}{\|(u_*)_-\|_q^2}$$
$$\geq \left(\theta^{2/q} + (1-\theta)^{2/q}\right) S_d(T)$$

with

$$\theta := \frac{\|(u_*)_+\|_q^q}{\|(u_*)_+\|_q^q + \|(u_*)_-\|_q^q}.$$

Since $0 < \theta < 1$ and $q > 2$ we have $\theta^{2/q} + (1-\theta)^{2/q} > 1$, a contradiction. Thus, after changing the sign of u_* if necessary, we may assume that $u_* \geq 0$.

The Euler–Lagrange equation of the optimization problem is

$$-\partial_t^2 u_* - \Delta_\omega u_* + \frac{(d-2)^2}{4} u_* - \frac{\mathcal{E}_T[u_*]}{\|u_*\|_q^q} u_*^{q-1} = 0 \qquad \text{on } \Sigma_T.$$

If we define a function U_* on $\mathbb{R}^d \setminus \{0\}$ by

$$U_*(x) := |x|^{-(d-2)/2} u_*(\ln|x|, x/|x|), \qquad (1.65)$$

then, by a straightforward computation,

$$-\Delta U_* = \frac{\mathcal{E}_T[u_*]}{\|u_*\|_q^q} U_*^{q-1} \qquad \text{in } \mathbb{R}^d \setminus \{0\}.$$

1 The Sharp Sobolev Inequality and Its Stability: An Introduction

Moreover, the singularity at the origin is nonremovable in the sense that

$$\int_{B_\varepsilon(0)} U_*^q \, dx = \infty \qquad \text{for all } \varepsilon > 0.$$

(Indeed, the integral

$$\int_{r<|x|<e^T r} U_*^q \, dx = \|u_*\|_q^q$$

is independent of $r > 0$.) Therefore, one is in position to apply a theorem of Caffarelli, Gidas and Spruck [20], which implies that U_* is radially symmetric about the origin. Equivalently, the function u_* is independent of ω.

Normalizing u_* such that $\frac{\mathcal{E}_T[u_*]}{\|u_*\|_q^q} = \frac{d(d-2)}{4}$, one is therefore led to the ODE

$$-\partial_t^2 u + \frac{(d-2)^2}{4} u - \frac{d(d-2)}{4} u^{q-1} = 0 \qquad \text{in } \mathbb{R},$$

which can be studied by phase-plane analysis. The equation has the constant solution

$$u_0 := ((d-2)/d)^{(d-2)/4}$$

as well as the homoclinic solution $\cosh^{-(d-2)/2}$. For any $\alpha \in (u_0, 1)$ there is a unique solution $u = u_\alpha$ with $u(0) = \alpha$ and $u'(0) = 0$. This solution is positive and periodic with a certain minimal period $\tau(\alpha)$ and it is symmetric about $t = 0$ and decreasing on $[0, \tau(\alpha)/2]$. It is known that $\alpha \mapsto \tau(\alpha)$ is continuous and monotone increasing (see, e.g., [26]) with

$$\lim_{\alpha \to u_0} \tau(\alpha) = \frac{2\pi}{\sqrt{d-2}} = T_* \qquad \text{and} \qquad \lim_{\alpha \to 1} \tau(\alpha) = \infty.$$

Returning to our solution u_*, which is T-periodic, we conclude immediately that $u_* = u_0$ if $T \leq T_*$. Assume now $T > T_*$. We deduce from the above analysis that there is an $s \in \mathbb{R}/T\mathbb{Z}$ and a $k \in \mathbb{N}$ with $k < T/T_*$ such that $u_*(t) = u_\alpha(t-s)$, where $\alpha \in (u_0, 1)$ is uniquely determined by k via $T = k\tau(\alpha)$. If $T \leq 2T_*$, we have necessarily $k = 1$. If $T > 2T_*$, a priori more than one value of k is possible, but, using a variational argument based on the stability of the solutions [87, Section 2, p. 134], one can show that the minimizer necessarily has $k = 1$. To summarize, we have shown that

$$\mathcal{G}_T = \begin{cases} \{c : c \in \mathbb{R}\} & \text{if } T \leq T_*, \\ \{c \, u_{\tau^{-1}(T)}(\cdot - s) : c \in \mathbb{R}, \, s \in \mathbb{R}/T\mathbb{Z}\} & \text{if } T > T_*. \end{cases}$$

Here $\tau^{-1} : (T_*, \infty) \to (u_0, 1)$ denotes the inverse of τ, which exists by the strict monotonicity of τ.

1.5.3 Zero Modes of the Hessian

Having characterized the optimizers, we next turn our attention to the zero modes of the Hessian. As before we work in the normalization $u_* \geq 0$ and $\frac{\mathcal{E}_T[u_*]}{\|u_*\|_q^q} = \frac{d(d-2)}{4}$, so that u_* solves

$$-\partial_t^2 u_* + \frac{(d-2)^2}{4} u_* - \frac{d(d-2)}{4} u_*^{q-1} = 0 \quad \text{in } \mathbb{R}. \tag{1.66}$$

The Hessian of our optimization problem is the operator

$$\mathcal{L}_T := -\partial_t^2 - \Delta_\omega + \frac{(d-2)^2}{4} - (q-1)\frac{d(d-2)}{4} u_*^{q-2}$$
$$+ (q-2)\frac{d(d-2)}{4} \|u\|_q^{-q} \left| u_*^{q-1} \right\rangle\!\left\langle u_*^{q-1} \right|$$
$$= -\partial_t^2 - \Delta_\omega + \frac{(d-2)^2}{4} - \frac{d(d+2)}{4} u_*^{q-2} + d \|u\|_q^{-q} \left| u_*^{q-1} \right\rangle\!\left\langle u_*^{q-1} \right|,$$

considered as a selfadjoint, lower bounded operator in $L^2(\Sigma_T)$ with form domain $H^1(\Sigma_T)$. Our goal is to show that

$$\ker \mathcal{L}_T = \begin{cases} \{u_*\} & \text{if } T < T_*, \\ \operatorname{span}\{u_*, \sin(\frac{2\pi}{T_*} \cdot), \cos(\frac{2\pi}{T_*} \cdot)\} & \text{if } T = T_*, \\ \operatorname{span}\{u_*, \partial_t u_*\} & \text{if } T > T_*. \end{cases} \tag{1.67}$$

The fact that u_* is in the kernel comes from the homogeneity of the optimization problem. The fact that $\partial_t u_*$ is in the kernel for $T > T_*$ comes from the fact that translates of an optimizer are again optimizers and that optimizers are not constant. From that perspective the claimed elements in the kernel for $T > T_*$ are natural and the thrust of the assertion in this case lies in the fact that there are no other, linearly independent elements in the kernel. Similarly, in the case $T < T_*$ it is shown that there is no element linearly independent from the natural one. What might be surprising is that at the critical value $T = T_*$ the sine and cosine in the kernel are not related to any symmetry of the problem. This is ultimately the reason why in the main theorem in this lecture the stability exponent is 4 for $T = T_*$, while it is 2 for $T \neq T_*$.

Proof of (1.67) For $T \leq T_*$ we have $u_* = u_0 = ((d-2)/d)^{(d-2)/4}$ and therefore

$$\mathcal{L}_T = -\partial_t^2 - \Delta_\omega - (d-2) + (d-2)|\Sigma_T|^{-1}|1\rangle\langle 1|. \tag{1.68}$$

The assertion follows easily from the spectral properties of $-\partial_t^2$ and $-\Delta_\omega$.

Now assume $T > T_*$. We have to classify all $\varphi \in H^2(\Sigma_T)$ such that $\mathcal{L}_T\varphi = 0$. We first note that, as a consequence of (1.66), $\mathcal{L}_T u_* = 0$ and $\mathcal{L}_T \partial_t u_* = 0$. Now given φ as above, we consider

$$\widetilde{\varphi} := \varphi + c u_* \quad \text{with} \quad c := \|u_*\|_q^{-q} \int_{\Sigma_T} u_*^{q-1} \varphi \, dv_g.$$

Then a simple computation shows that

$$\widetilde{\mathcal{L}}_T \widetilde{\varphi} = 0 \quad \text{with} \quad \widetilde{\mathcal{L}}_T := -\partial_t^2 - \Delta_\omega + \tfrac{(d-2)^2}{4} - \tfrac{d(d+2)}{4} u_*^{q-2}.$$

To solve this equation, we can expand $\widetilde{\varphi}$ with respect to spherical harmonics in the ω-variable and solve the equation for each fixed degree. This leads to the equations

$$\widetilde{\mathcal{L}}_{T,\ell}\widetilde{\varphi}_\ell = 0 \quad \text{with} \quad \widetilde{\mathcal{L}}_{T,\ell} := -\partial_t^2 + \ell(\ell + d - 2) + \tfrac{(d-2)^2}{4} - \tfrac{d(d+2)}{4} u_*^{q-2},$$

parametrized by $\ell \in \mathbb{N}_0$, where now $\widetilde{\mathcal{L}}_{T,\ell}$ is an operator in $L^2(\mathbb{R}/T\mathbb{Z})$ with form domain $H^1(\mathbb{R}/T\mathbb{Z})$.

We begin with $\ell = 0$. Differentiating (1.66) with respect to either t or α (recall that $u_* = u_\alpha$ where $\tau(\alpha) = T$), we find two solutions $\partial_t u_*$ and $\partial_\alpha u_*$ of the equation

$$-v'' + \tfrac{(d-2)^2}{4} v - \tfrac{d(d+2)}{4} u_*^{q-2} v = 0.$$

Since $\partial_t u_*(0) = 0$ and $\partial_\alpha u_*(0) = 1$ (since $u_*(0) = \alpha$), these two solutions are linearly independent. Thus $\widetilde{\varphi}_0$ is a linear combination of these two functions. Differentiating the equation $u_*(t+T) = u_*(t)$ with respect to α, we obtain $\partial_\alpha u_*(t+T) = \partial_\alpha u_*(t) - u_*'(t)\tau'(\alpha)$. Since $\tau'(\alpha) \neq 0$ (see, e.g., [26]), we see that $\partial_\alpha u_*$ is not periodic, so in fact $\widetilde{\varphi}_0$ is a multiple of $\partial_t u_*$.

Now we consider $\ell = 1$. A computation shows that the two functions

$$e^{\pm t}(u_*' \pm \tfrac{d-2}{2} u_*)$$

satisfy the equation

$$-v'' + (d-1)v + \tfrac{(d-2)^2}{4} v - \tfrac{d(d+2)}{4} u_*^{q-2} v = 0.$$

At the two infinities, one of them is exponentially growing and one is exponentially decaying. They are clearly linearly independent. Since no nontrivial linear combination of them is periodic, we conclude that $\widetilde{\varphi}_1 = 0$.

Finally, we consider $\ell \geq 2$. Since u_* is a minimizer, we know that $\mathcal{L}_T$ is positive semidefinite. Since the rank-one contribution to $\mathcal{L}_T$ only affects $\ell = 0$, we deduce that the operators $\widetilde{\mathcal{L}}_{T,\ell}$ are positive semidefinite for $\ell \geq 1$. Since $\widetilde{\mathcal{L}}_{T,1}$ has compact resolvent, the fact that its kernel is trivial implies that it is positive definite. Since $\widetilde{\mathcal{L}}_{T,\ell}$ with $\ell \geq 2$ differs from $\widetilde{\mathcal{L}}_{T,1}$ by a positive constant, we deduce that $\widetilde{\mathcal{L}}_{T,\ell}$ is positive definite as well and, in particular, has trivial kernel.

To summarize, we have shown that $\widetilde{\varphi} = C\partial_t u_*$ for some $C \in \mathbb{R}$, that is, $\varphi = -cu_* + C\partial_t u_*$, as claimed. □

For $T > 0$, let

$$c_T := \inf \left\{ \frac{\langle v, \mathcal{L}_T v \rangle}{\mathcal{E}_T[v]} : v \in H^1(\Sigma_T), \ \mathcal{E}_T[u_*, v] = \mathcal{E}_T[\partial_t u_*, v] = 0 \right\}. \quad (1.69)$$

Here $\mathcal{E}_T[\cdot, \cdot]$ denotes the bilinear form associated to the quadratic form $\mathcal{E}_T[\cdot]$ and $\langle \cdot, \cdot \rangle$ denotes both the L^2-inner product and the $H^1 \times H^{-1}$ duality pairing. We note that for $T \leq T_*$ the function u_* is a constant, so the second orthogonality condition in (1.69) is trivially satisfied in this case.

Lemma 1.17 *If $T \leq T_*$, then*

$$c_T = \frac{\min\{(\frac{2\pi}{T})^2, d-1\} - (d-2)}{\min\{(\frac{2\pi}{T})^2, d-1\} + (\frac{d-2}{2})^2}.$$

In particular, $c_T > 0$ if $T < T_*$, and $c_{T_*} = 0$. If $T > T_*$, then $c_T > 0$.

Proof For $T \leq T_*$, u_* is a constant, so the orthogonality conditions in (1.69) reduce to $\int_{\Sigma_T} v \, dv_g = 0$ and the operator $\mathcal{L}_T$ takes the form (1.68). Diagonalizing $-\partial_t^2$ and $-\Delta_\omega$, we see that for $T \leq T_*$

$$c_T = \inf_{k \in \mathbb{Z}, \ \ell \in \mathbb{N}_0, \ (k,\ell) \neq (0,0)} \frac{(\frac{2\pi}{T})^2 k^2 + \ell(\ell + d - 2) - (d-2)}{(\frac{2\pi}{T})^2 k^2 + \ell(\ell + d - 2) + (\frac{d-2}{2})^2}.$$

The claimed result follows by a simple computation.

For $T \geq T_*$ we argue more qualitatively. It is easy to see that the infimum defining c_T is attained by some $v_* \neq 0$. If we had $c_T = 0$, then $\langle v_*, \mathcal{L}_T v_* \rangle = 0$ and therefore, since $\mathcal{L}_T \geq 0$, $\mathcal{L}_T v_* = 0$. By (1.67), v_* is a linear combination of u_* and $\partial_t u_*$. Using the equation for u_*, we find $\mathcal{E}_T[u_*, v] = \frac{d(d-2)}{4} \langle u_*^{q-1}, v \rangle$ and $\mathcal{E}_T[\partial_t u_*, v] = \frac{d(d+2)}{4} \langle u_*^{q-2} \partial_t u_*, v \rangle$ for any $v \in H^1(\Sigma_T)$. In particular, v_* is L^2-orthogonal to u_*^{q-1} and $u_*^{q-2} \partial_t u_*$, which implies that $v_* = 0$, a contradiction. □

1.5.4 Nondegenerate Stability: The Upper Bound

We turn to the question of stability and prove Theorem 1.14. We begin with the simpler case $T \neq T_*$, where we will show nondegenerate stability in the form of a quadratic bound.

As in the previous lecture, it is instructive to first prove the upper bound, namely (1.58). As there, we make the ansatz $u = u_* + \varepsilon r$ with r to be determined, and we find

$$\lim_{\varepsilon \to 0} \varepsilon^{-2} \left(\mathcal{E}_T[u_* + \varepsilon r] - S_d(T) \|u_* + \varepsilon r\|_q^2 \right) = \langle r, \mathcal{L}_T r \rangle.$$

Moreover, arguing as in the proof of Lemma 1.12, we find that, if r satisfies

$$\mathcal{E}_T[u_*, r] = \mathcal{E}_T[\partial_t u_*, r] = 0$$

and if ε is sufficiently small, depending on r, then

$$\inf_{g \in \mathcal{G}_T} \mathcal{E}_T[u_* + \varepsilon r - g] = \varepsilon^2 \mathcal{E}_T[r].$$

Thus,

$$\lim_{\varepsilon \to 0} \frac{\mathcal{E}_T[u_* + \varepsilon r] - S_d(T)\|u_* + \varepsilon r\|_q^2}{\inf_{g \in \mathcal{G}_T} \mathcal{E}_T[u_* + \varepsilon r - g]} = \frac{\langle r, \mathcal{L}_T r \rangle}{\mathcal{E}_T[r]}.$$

By definition, the infimum over the right side with respect to r gives the constant c_T defined in (1.69).

This proves the expected result that stability cannot hold with a smaller exponent than 2. (Concerning our counting of the vanishing exponent, we note that $\inf_{g \in \mathcal{G}_T} \mathcal{E}_T[u - g]$ vanishes quadratically—that is, with exponent 2—as u approaches $\mathcal{G}_T$.) Moreover, since $c_{T_*} = 0$, the above argument shows that at $T = T_*$ no quadratic stability can hold. We will discuss an upper bound for $T = T_*$ later, but first we prove that for $T \neq T_*$ one does indeed have quadratic stability.

1.5.5 Nondegenerate Stability: The Lower Bound

We are now ready to give the proof of Theorem 1.14 for $T \neq T_*$. We abbreviate, suppressing the T-dependence,

$$\delta[u] := \inf_{g \in \mathcal{G}_T} \sqrt{\mathcal{E}_T[u - g]}.$$

Proposition 1.18 *Let $T \neq T_*$. Then, for all $0 \neq u \in H^1(\Sigma_T)$,*

$$\mathcal{E}_T[u] - S_d(T)\|u\|_q^2 - c_T \delta[u]^2 \gtrsim -\tau_u^{\min\{q-2,1\}} \delta[u]^2,$$

where $\tau_u := \delta[u]/\sqrt{\mathcal{E}_T[u] - \delta[u]^2}$ and where c_T is defined in (1.69).

Once we have proved this proposition, we obtain Theorem 1.14 by the same argument as in the previous lecture, based on the relative compactness of optimizing sequences.

Proof It is easy to see that the infimum $\delta[u]$ is attained. After a possible sign change and, in case $T > T_*$, a translation, we may assume that it is attained at cu_*, with the normalization $u_* \geq 0$ and $\frac{\mathcal{E}_T[u_*]}{\|u_*\|_q^q} = \frac{d(d-2)}{4}$. We write

$$U = cu_* + R$$

and observe the orthogonality conditions

$$\mathcal{E}_T[u_*, R] = \mathcal{E}_T[\partial_t u_*, R] = 0. \tag{1.70}$$

Of course, the second condition here is trivial if $T < T_*$ (in which case u_* is constant).

Expanding the q-norm as in the previous lecture, we arrive at

$$\mathcal{E}_T[u] - S_d(T)\|u\|_q^2 \geq \mathcal{E}_T[R] - (q-1)\int_{\Sigma_T} u_*^{q-2} R^2 \, dv_g$$
$$- \text{const}\, |c|^{2-q} \left(|c|^{(q-3)_+} \|R\|_q^{\min\{q-3\}} + \|R\|_q^q\right).$$

For the quadratic term we have by definition (1.69) and the orthogonality conditions (1.70)

$$\mathcal{E}_T[R] - (q-1)\int_{\Sigma_T} u_*^{q-2} R^2 \, dv_g = \langle R, \mathcal{L}_T R \rangle \geq c_T \mathcal{E}_T[R] = c_T \delta[u]^2.$$

For the remainder term we bound, just like in the previous lecture,

$$\|R\|_q \leq \tau_u \|u_*\|_q |c|.$$

Thus, again as in the previous lecture,

$$|c|^{2-q}\left(|c|^{(q-3)_+} \|R\|_q^{\min\{q-3\}} + \|R\|_q^q\right) \lesssim \left(\tau_u^{\min\{q-2,1\}} + \tau_u^{q-2}\right)\delta[u]^2.$$

1 The Sharp Sobolev Inequality and Its Stability: An Introduction

This proves that

$$\mathcal{E}_T[u] - S_d(T)\|u\|_q^2 - c_T \,\delta[u]^2 \gtrsim -\left(\tau_u^{\min\{q-2,1\}} + \tau_u^{q-2}\right)\delta[u]^2.$$

By the same argument as at the end of the proof of Proposition 1.13 this implies the claimed inequality. □

1.5.6 Degenerate Stability: The Upper Bound

In the remainder of this lecture, we discuss the degenerate stability in the case $T = T_*$.

We begin with the proof of the upper bound (1.59), which shows that stability cannot hold with an exponent smaller than 4. If we were only interested in showing this, we could simply take the same trial function $u_* + \varepsilon r$ as in the case $T \neq T_*$ and expand the q-norm to higher order than before. It is however instructive and helpful for the understanding of the following proof of Theorem 1.14 to consider a more general family of trial states

$$u = u_* + \varepsilon r_* + \varepsilon^2 s.$$

We recall that $u_* = u_0 = ((d-2)/d)^{(d-2)/4}$. The function r_* is chosen so as to minimize $\langle r, \mathcal{L}_{T_*} r\rangle / \mathcal{E}_{T_*}[r]$ under the orthogonality condition $\mathcal{E}_{T_*}[u_*, r] = 0$. As shown in (1.67), this leads to the choice of r_* as a linear combination of $\sin(\frac{2\pi}{T_*} \cdot)$ and $\cos(\frac{2\pi}{T_*} \cdot)$. By translation invariance the choice of the linear combination parameters is immaterial, and we choose to take

$$r_* = \cos(\frac{2\pi}{T_*} \cdot).$$

The function s is our variational parameter that we will optimize over at the end. We assume that

$$\mathcal{E}_{T_*}[u_*, s] = 0 \tag{1.71}$$

(which is the same as $\int_{\Sigma_T} s\, dv_g = 0$) and that

$$\mathcal{E}_{T_*}[r_*, s] = \mathcal{E}_{T_*}[\partial_t r_*, s] = 0 \tag{1.72}$$

(which, using (1.71), is the same as $\int_{\Sigma_T} r_* s\, dv_g = \int_{\Sigma_T} (\partial_t r_*) s\, dv_g = 0$).

The motivation for requiring the first orthogonality condition in (1.72) is that, if s contained 'a part of r_*', we could simply absorb this part into εr_* by redefining ε.

The motivation for the second condition is that, if s contained 'a part of $\partial_t r_*$', we could essentially absorb this term by translating r_*.

The orthogonality condition (1.71) guarantees that $\mathcal{E}_{T_*}[u_*, r_* + \varepsilon s] = 0$, so as in the proof of Lemma 1.12 we find that, if ε is sufficiently small depending on (r_* and) s,

$$\inf_{g \in \mathcal{G}_{T_*}} \mathcal{E}_{T_*}[u_* + \varepsilon r_* + \varepsilon^2 s - g] = \mathcal{E}_{T_*}[\varepsilon r_* + \varepsilon^2 s] = \varepsilon^2 \mathcal{E}_{T_*}[r_*] + \varepsilon^4 \mathcal{E}_{T_*}[s].$$

The second equality uses the first orthogonality condition in (1.72).

A lengthy, but straightforward computation shows that

$$\lim_{\varepsilon \to 0} \varepsilon^{-4} \left(\mathcal{E}_{T_*}[u_* + \varepsilon r_* + \varepsilon^2 s] - S_d(T_*) \|u_* + \varepsilon r_* + \varepsilon^2 s\|_q^2 \right)$$
$$= \langle s, \mathcal{L}_{T_*} s \rangle - \frac{(d-2)^2}{4}(q-1)(q-2)u_0^{-1} \langle r_*^2, s \rangle + C_*$$

with

$$C_* := \frac{(d-2)^2}{4} \frac{(q-1)(q-2)}{4} u_0^{-2} |\Sigma_T| \left(-\frac{1}{3}(q-3) \int_{\Sigma_T} r_*^4 \frac{dv_g}{|\Sigma_T|} \right.$$
$$\left. + (q-1) \left(\int_{\Sigma_T} r_*^2 \frac{dv_g}{|\Sigma_T|} \right)^2 \right).$$

At this point, in order to obtain an upper bound that is as small as possible, we need to solve the optimization problem

$$\inf \left\{ \langle s, \mathcal{L}_{T_*} s \rangle - 2\langle f_*, s \rangle : s \text{ satisfies (1.71) and (1.72)} \right\}$$

with $f_* := \frac{(d-2)^2}{8}(q-1)(q-2)u_0^{-1} r_*^2$. According to (1.67), the orthogonality conditions on s mean that s is L^2-orthogonal to the kernel of $\mathcal{L}_{T_*}$. Therefore, denoting by $P^\perp$ the orthogonal projection onto the orthogonal complement of this kernel and by $\mathcal{L}_{T_*}^\perp$ the restriction of $\mathcal{L}_{T_*}$ to the range of $P^\perp$, where it is invertible, we can write

$$\langle s, \mathcal{L}_{T_*} s \rangle - 2\langle f_*, s \rangle = \left\| \left(\mathcal{L}_{T_*}^\perp \right)^{1/2} s - \left(\mathcal{L}_{T_*}^\perp \right)^{-1/2} P^\perp f_* \right\|_2^2 - \left\langle P^\perp f_*, (\mathcal{L}_{T_*}^\perp)^{-1} P^\perp f_* \right\rangle.$$

This is minimized by taking $s = (\mathcal{L}_{T_*}^\perp)^{-1} P^\perp f_*$. With this choice one obtains

$$\lim_{\varepsilon \to 0} \varepsilon^{-4} \left(\mathcal{E}_{T_*}[u_* + \varepsilon r_* + \varepsilon^2 s] - S_d(T_*) \|u_* + \varepsilon r_* + \varepsilon^2 s\|_q^2 \right)$$
$$= C_* - \left\langle P^\perp f_*, (\mathcal{L}_{T_*}^\perp)^{-1} P^\perp f_* \right\rangle.$$

1 The Sharp Sobolev Inequality and Its Stability: An Introduction

Consequently,

$$\lim_{\varepsilon \to 0} \frac{\mathcal{E}_{T_*}[u_* + \varepsilon r_* + \varepsilon^2 s]\left(\mathcal{E}_{T_*}[u_* + \varepsilon r_* + \varepsilon^2 s] - S_d(T_*)\|u_* + \varepsilon r_* + \varepsilon^2 s\|_q^2\right)}{\inf_{g \in \mathcal{G}_{T_*}} \mathcal{E}_{T_*}[u_* + \varepsilon r_* + \varepsilon^2 s - g]^2}$$

$$= \frac{\mathcal{E}_{T_*}[u_*]\left(C_* - \left\langle P^\perp f_*, (\mathcal{L}_{T_*}^\perp)^{-1} P^\perp f_* \right\rangle\right)}{\mathcal{E}_{T_*}[r_*]^2}.$$

The fact that u_* is a minimizer implies that $C_* - \left\langle P^\perp f_*, (\mathcal{L}_{T_*}^\perp)^{-1} P^\perp f_* \right\rangle \geq 0$. If this difference vanished, then quartic stability would be violated. A direct computation, however, shows that this difference is strictly positive,

$$C_* - \left\langle P^\perp f_*, (\mathcal{L}_{T_*}^\perp)^{-1} P^\perp f_* \right\rangle > 0. \tag{1.73}$$

We do not give the details of this computation, but as an intermediate step we mention that

$$C_* = \tfrac{(d-2)^2}{4} \tfrac{1}{32}(q-1)(q-2)(q+1) |\Sigma_{T_*}| u_0^{-2},$$

as well as

$$(\mathcal{L}_{T_*}^\perp)^{-1} P^\perp f_*(t,\omega) = \tfrac{d-2}{48}(q-1)(q-2) u_0^{-1} \cos \tfrac{4\pi}{T_*} t,$$

so

$$\left\langle P^\perp f_*, (\mathcal{L}_{T_*}^\perp)^{-1} P^\perp f_* \right\rangle = \tfrac{(d-2)^2}{4} \tfrac{1}{96}(q-1)^2(q-2) |\Sigma_{T_*}| u_0^{-2}$$

In this way we obtain the claimed value in the upper bound (1.59).

As we will see in the proof of the lower bound given momentarily, the positivity in (1.73) is the reason why we have quartic stability, rather than only stability of a higher order. We think of (1.73) as a *secondary nondegeneracy condition*. For $T = T_*$ the primary nondegeneracy condition, which says that elements in the kernel of $\mathcal{L}_{T_*}$ come from symmetries, is violated, and therefore no quadratic stability can hold. The secondary nondegeneracy condition (1.73), however, is satisfied and therefore one does have quartic stability. It is conceivable that there is a Sobolev-type functional inequality where both the primary and secondary nondegeneracy conditions fail and where the validity of a sextic stability result depends on a tertiary nondegeneracy condition, although no such example is known to the author.

1.5.7 Degenerate Stability: The Lower Bound

Finally, we sketch the proof of Theorem 1.14 in the case $T = T_*$. As in the previous lecture, given the precompactness of optimizing sequences, it suffices to prove the following asymptotic lower bound, where we set

$$\tau_u := \frac{\delta[u]}{\sqrt{\mathcal{E}_{T_*}[u] - \delta[u]^2}} = \frac{\delta[u]}{\frac{d-2}{2}|\Sigma_{T_*}|^{-1/2}|\int_{\Sigma_{T_*}} u\, dv_g|}.$$

Proposition 1.19 *Let $d \geq 3$ and $T = T_*$. Then, for all $u \in H^1(\Sigma_{T_*})$ with $\int_{\Sigma_{T_*}} u\, dv_g \neq 0$,*

$$\mathcal{E}_{T_*}[u]\left(\mathcal{E}_{T_*}[u] - S_d(T_*)\|u\|_q^2\right) - \frac{(q+2)(q-2)}{12(q-1)}\delta[u]^4 \gtrsim -\tau_u\, \delta[u]^4.$$

Proof We denote by u_* the constant function $u_0 = ((d-2)/d)^{(d-2)/4}$ and set $c := u_0^{-1}|\Sigma_{T_*}|^{-1}\int_{\Sigma_{T_*}} u\, dv_g$ and $r := u - cu_*$, so that

$$u = cu_* + r, \qquad \int_{\Sigma_{T_*}} r\, dv_g = 0 \quad \text{and} \quad \delta[u]^2 = \mathcal{E}_{T_*}[r].$$

We may assume that $\tau_u \leq \tau_*$ for some $\tau_* > 0$ depending only on d, for the claimed inequality holds trivially for $\tau_u > \tau_*$ after adjusting the implicit constant.

Step 1. We show that, by choosing $\tau_* > 0$ sufficiently small, depending only on d, we may assume that r depends only on t and not on ω. To do so, we decompose

$$r = r_0 + r' \qquad \text{with } r_0(t) := |\mathbb{S}^{d-1}|^{-1}\int_{\mathbb{S}^{d-1}} r(t, \omega)\, d\omega.$$

Similarly as in the proof of Lemma 1.12 we have

$$S_d(T_*)\left(\|r_0\|_q^2 + \|r'\|_q^2\right) \leq \mathcal{E}_{T_*}[r_0] + \mathcal{E}_{T_*}[r'] = \mathcal{E}_{T_*}[r] = \tau_u^2\left(\mathcal{E}_{T_*}[u] - \mathcal{E}_{T_*}[r]\right)$$

$$= \tau_u^2 \mathcal{E}_{T_*}[cu_*] = \tau_u^2 S_d(T_*)c^2\|u_*\|_q^2$$

Thus, by choosing $\tau_* > 0$ small, we can ensure that both $\|r_0\|_q$ and $\|r'\|_q$ are as small as we wish with respect to $|c|$.

We have

$$\mathcal{E}_{T_*}[u] = \mathcal{E}_{T_*}[cu_* + r_0] + \mathcal{E}_{T_*}[r']$$

1 The Sharp Sobolev Inequality and Its Stability: An Introduction 57

and, by a second-order Taylor expansion,

$$\left| \|u\|_q^2 - \|cu_* + r_0\|_q^2 - (q-1)|\Sigma_{T_*}|^{-1+2/q} \int_{\Sigma_{T_*}} r'^2 \, dv_g \right|$$
$$\lesssim |c|^{\max\{2-q,-1\}} \left(\|r'\|_q^{\min\{q,3\}} + \|r_0\|_q^{\min\{q-2,1\}} \|r'\|_q^2 \right).$$

Here we used the smallness of τ_* in order to bound the difference between $\|cu_* + r_0\|_q^{2-q}$ and $\|cu_*\|_q^{2-q}$. We use this smallness again to bound

$$|c|^{\max\{2-q,-1\}} \left(\|r'\|_q^{\min\{q,3\}} + \|r_0\|_q^{\min\{q-2,1\}} \|r'\|_q^2 \right) \lesssim \tau_*^{\min\{q-2,1\}} \|r'\|_q^2.$$

Thus, we obtain

$$\mathcal{E}_{T_*}[u] - S_d(T_*)\|u\|_q^2 \geq \mathcal{E}_{T_*}[cu_* + r_0] - S_d(T_*)\|cu_* + r_0\|_q^2$$
$$+ \langle r', \mathcal{L}_{T_*} r' \rangle - \text{const } \tau_*^{\min\{q-2,1\}} \|r'\|_q^2.$$

Since r' is L^2-orthogonal to the kernel of $\mathcal{L}_{T_*}$, we have $\langle r', \mathcal{L}_{T_*} r' \rangle \gtrsim \mathcal{E}_{T_*}[r']$, and therefore, if $\tau_* > 0$ is sufficiently small,

$$\mathcal{E}_{T_*}[u] - S_d(T_*)\|u\|_q^2 \geq \mathcal{E}_{T_*}[cu_* + r_0] - S_d(T_*)\|cu_* + r_0\|_q^2 + \gamma \, \mathcal{E}_{T_*}[r']$$

with some constant $\gamma > 0$.

Assume now that we can prove the desired bound for the function $cu_* + r_0$, which only depends on the t-variable. Then the right side above is

$$\geq \frac{(q+2)(q-2)}{12(q-1)} (1 - \text{const } \tau_u) \frac{\mathcal{E}_{T_*}[r_0]^2}{\mathcal{E}_{T_*}[cu_* + r_0]} + \gamma \, \mathcal{E}_{T_*}[r'].$$

This gives the desired bound for the function u, since for all sufficiently small $\tau_* > 0$,

$$\frac{\mathcal{E}_{T_*}[r_0]^2}{\mathcal{E}_{T_*}[cu_* + r_0]} + \gamma' \mathcal{E}_{T_*}[r'] \geq \frac{\mathcal{E}_{T_*}[r]^2}{\mathcal{E}_{T_*}[cu_* + r_0]} \geq \frac{\mathcal{E}_{T_*}[r]^2}{\mathcal{E}_{T_*}[u]}.$$

Indeed, the first inequality here is equivalent to $\gamma' \mathcal{E}_{T_*}[cu_* + r_0] \geq 2\mathcal{E}_{T_*}[r_0] + \mathcal{E}_{T_*}[r']$, and this holds since $\mathcal{E}_{T_*}[r_0]$ and $\mathcal{E}_{T_*}[r']$ can be chosen as small as we wish with respect to $|c|$.

Step 2. According to Step 1, we may assume that r is independent of t. We decompose further

$$r = r_\| + r_\perp \quad \text{with } r_\| \in \ker \mathcal{L}_{T_*}, \ r_\perp \in (\ker \mathcal{L}_{T_*})^\perp.$$

In this step we will show that, by choosing $\tau_* > 0$ sufficiently small, depending only on d, we may assume that $\mathcal{E}_{T_*}[r_\parallel] \geq \mathcal{E}_{T_*}[r_\perp]$.

Indeed, by a second-order Taylor expansion as in Step 1, one finds

$$\mathcal{E}_{T_*}[u] - S_d(T_*)\|u\|_q^2 \geq \langle r, \mathcal{L}_{T_*} r\rangle - \text{const } \tau_*^{\min\{q-2,1\}} \|r\|_q^2.$$

Note that

$$\langle r, \mathcal{L}_{T_*} r\rangle = \langle r_\perp, \mathcal{L}_{T_*} r_\perp\rangle \gtrsim \mathcal{E}_{T_*}[r_\perp].$$

Moreover, assuming $\mathcal{E}_{T_*}[r_\parallel] \leq \mathcal{E}_{T_*}[r_\perp]$, we have

$$\|r\|_q^2 \lesssim \mathcal{E}_{T_*}[r] = \mathcal{E}_{T_*}[r_\parallel] + \mathcal{E}_{T_*}[r_\perp] \leq 2\mathcal{E}_{T_*}[r_\perp].$$

Thus, if τ_* is sufficiently small, we have

$$\mathcal{E}_{T_*}[u] - S_d(T_*)\|u\|_q^2 \gtrsim \mathcal{E}_{T_*}[r_\perp] \geq \tfrac{1}{2} \mathcal{E}_{T_*}[r]$$

Since $\mathcal{E}_{T_*}[u]/\delta[u]^2 = \tau_u^{-2} + 1$, this is a stronger bound than the claimed one.

Step 3. We now come to the main part of the proof. We recall that r depends only on t and decompose it as in Step 2 with $\mathcal{E}_{T_*}[r_\perp] \leq \mathcal{E}_{T_*}[r_\parallel]$. Note that

$$r_\parallel = a\cos(\tfrac{2\pi}{T_*}\cdot) + b\sin\cos(\tfrac{2\pi}{T_*}\cdot) = \sqrt{a^2+b^2}\cos(\tfrac{2\pi}{T_*}(\cdot - \tilde{t}))$$

and after a translation of u we may assume that $\tilde{t} = 0$. Thus, $r_\parallel = \varepsilon r_*$ with $r_* = \cos(\tfrac{2\pi}{T_*}\cdot)$ and $\varepsilon := \sqrt{a^2+b^2}$.

As we are dealing with functions of a single variable, we have a Sobolev embedding into L^∞, which gives

$$\varepsilon^2 + \|r_\perp\|_\infty^2 \lesssim \mathcal{E}_{T_*}[r_\parallel] + \mathcal{E}_{T_*}[r_\perp] = \mathcal{E}_{T_*}[r] = \tau_u^2 \mathcal{E}_{T_*}[cu_*] = \tau_u^2 S_d(T_*) c^2 \|u_*\|_q^2.$$

Thus, by choosing $\tau_* > 0$ small, we can ensure that both ε and $\|r_\perp\|_\infty$ are as small as we wish with respect to $|c|$.

Thanks to this L^∞-bound, we can Taylor expand the norm to fourth order (note that $q < 4$ if $d > 4$) and obtain eventually

$$\|u\|_q^2 = c^2\|u_*\|_q^2 + (q-1)\|u_*\|_q^{2-q} u_0^{q-2} \int_{\Sigma_{T_*}} (\varepsilon^2 r_*^2 + r_\perp^2)\,dv_g$$

$$+ (q-1)(q-2)\|u_*\|_q^{2-q} u_0^{q-3} |c|^{-1} \varepsilon^2 \int_{\Sigma_{T_*}} r_*^2 r_\perp \,dv_g$$

$$+ \tfrac{1}{12}(q-1)(q-2)(q-3)\|u_*\|_q^{2-q} u_0^{q-4} c^{-2} \varepsilon^4 \int_{\Sigma_{T_*}} r_*^4 \,dv_g$$

$$- \tfrac{1}{4}(q-1)^2(q-2)\|u_*\|_q^{2-2q} u_0^{2q-4} c^{-2}\varepsilon^4 \left(\int_{\Sigma_{T_*}} r_*^2 \, dv_g\right)^2$$

$$+ \mathcal{O}(|c|^{-3}\varepsilon^5 + |c|^{-1}\varepsilon\|r_\perp\|_\infty^2).$$

(In this bound we controlled a term $|c|^{-1}\|r_\perp\|_\infty^3$ by $|c|^{-1}\varepsilon\|r_\perp\|_\infty^2$ using our assumption $\|r_\perp\|_\infty^2 \lesssim \mathcal{E}_{T_*}[r_\perp] \leq \mathcal{E}_{T_*}[r_\parallel] = \text{const } \varepsilon^2$. Moreover, we controlled a term $c^{-2}\varepsilon^3\|r_\perp\|_\infty$ by $|c|^{-3}\varepsilon^5 + |c|^{-1}\varepsilon\|r_\perp\|_\infty^2$ using Schwarz.) Thus,

$$\mathcal{E}_{T_*}[u] - S_d(T_*)\|u\|_q^2 = \langle r_\perp, \mathcal{L}_{T_*} r_\perp \rangle - 2|c|^{-1}\varepsilon^2 \langle f_*, r_\perp \rangle + c^{-2}\varepsilon^4 C_*$$
$$+ \mathcal{O}(|c|^{-3}\varepsilon^5 + |c|^{-1}\varepsilon\|r_\perp\|_\infty^2),$$

where C_* and f_* are as in the proof of the upper bound. By completing a square, much like in the proof of the upper bound,

$$\langle r_\perp, \mathcal{L}_{T_*} r_\perp \rangle - 2|c|^{-1}\varepsilon^2 \langle f_*, r_\perp \rangle - \text{const } |c|^{-1}\varepsilon\|r_\perp\|_\infty^2$$
$$\geq -c^{-2}\varepsilon^4 \langle P^\perp f_*, (\mathcal{L}_{T_*}^\perp)^{-1} P^\perp f_* \rangle - \text{const } |c|^{-3}\varepsilon^5.$$

This shows that we are almost in the situation of the upper bound, and we find

$$\frac{\mathcal{E}_{T_*}[u]\left(\mathcal{E}_{T_*}[u] - S_d(T_*)\|u\|_q^2\right)}{\mathcal{E}_{T_*}[\varepsilon r_*]^2} \geq \frac{(q+2)(q-2)}{12(q-1)} - \text{const } \tau_u.$$

This is almost the claimed bound, except that we have $\mathcal{E}_{T_*}[\varepsilon r_*]^2$ in the denominator instead of $\mathcal{E}_{T_*}[\varepsilon r_* + r_\perp]^2$. If $\mathcal{E}_{T_*}[r_\perp] \lesssim c^{-2}\varepsilon^4$ (with any fixed implied constant), we have $\mathcal{E}_{T_*}[\varepsilon r_* + r_\perp]^2 \leq (1 + \text{const } \tau_u^2)\mathcal{E}_{T_*}[\varepsilon r_*]^2$, so this difference is harmless.

Meanwhile, if $\mathcal{E}_{T_*}[r_\perp] \geq Mc^{-2}\varepsilon^4$ with a sufficiently large constant $M > 0$, we argue slightly differently and use

$$\langle r_\perp, \mathcal{L}_{T_*} r_\perp \rangle - 2|c|^{-1}\varepsilon^2 \langle f_*, r_\perp \rangle \geq \gamma \, \mathcal{E}_{T_*}[r_\perp]$$

with some constant $\gamma > 0$. Thus,

$$\mathcal{E}_{T_*}[u] - S_d(T_*)\|u\|_q^2 \geq \left(\gamma \, \mathcal{E}_{T_*}[r_\perp] + c^{-2}\varepsilon^4 C_*\right)(1 - \text{const } \tau_u).$$

Here, as before, we want to replace $\varepsilon^4 = \mathcal{E}_{T_*}[\varepsilon r_*]^2/\mathcal{E}_{T_*}[r_*]^2$ by $\mathcal{E}_{T_*}[\varepsilon r_* + r_\perp]^2/\mathcal{E}_{T_*}[r_*]^2$. (We could also use the fact that C_* is strictly larger than the constant we want to obtain, but we do not need this.) Thus the claimed bound will follow if we can show that

$$\gamma \, \mathcal{E}_{T_*}[r_\perp] + c^{-2}\varepsilon^4 C_* \geq c^{-2} \frac{\mathcal{E}_{T_*}[\varepsilon r_* + r_\perp]^2}{\mathcal{E}_{T_*}[r_*]^2} C_*.$$

This is equivalent to $c^2(\gamma/C_*)\mathcal{E}_{T_*}[r_*]^2 \geq 2\mathcal{E}_{T_*}[\varepsilon r_*] + \mathcal{E}_{T_*}[r_\perp]$ and, since $\mathcal{E}_{T_*}[r_\perp] \leq \mathcal{E}_{T_*}[\varepsilon r_*]$, this follows from $\varepsilon \lesssim \tau_u |c|$ by choosing $\tau_* > 0$ small enough. This concludes the proof of Theorem 1.14 in case $T = T_*$. □

Remark 1.20 The above proof actually yields the stronger stability result

$$\mathcal{E}_{T_*}[u] - S_d(T_*)\|u\|_q^2 \gtrsim \frac{\mathcal{E}_{T_*}[\pi u]^2}{\mathcal{E}_{T_*}[u]} + \mathcal{E}_{T_*}[\pi^\perp u],$$

where π denotes the orthogonal projection in $L^2(\Sigma_{T_*})$ onto the span of $\sin(\frac{2\pi}{T_*} \cdot)$ and $\cos(\frac{2\pi}{T_*} \cdot)$ and where $\pi^\perp = 1 - \pi$. Thus the quartic behavior appears only on a low-dimensional subspace, while on its orthogonal complement we have quadratic stability.

1.5.8 Bibliographic Remarks

Quantitative stability for minimizing Yamabe metrics on closed manifolds was studied by Engelstein, Neumayer and Spolaor [44]. Their result specialized to Σ_T, plus the known fact that the optimizers are, in their terminology, nondegenerate (if $T < T_*$) or integrable (if $T > T_*$), implies the main theorem of this lecture in the case $T \neq T_*$. In the critical case $T = T_*$ their paper yields an inequality with an unspecified power ≥ 4. The fact that this power can be chosen to be equal to 4 is from [49].

The explicit study of the Yamabe problem on Σ_T is due to Schoen [87], where also the phase-plane analysis appears; for more details see also [26], including a proof of monotonicity of the period map following [13]. The classification of zero modes of the Hessian uses ideas from [79, 80]. The analysis of the solutions on Σ_T and of their linearization plays a role in the description of the asymptotic behavior of positive solutions of $-\Delta u = u^{(d+2)/(d-2)}$ near isolated singularities; see, e.g., [20, 68].

The fact that there is an optimizer for $S_d(T)$ under the assumption $S_d(T) < S_d$ is a special case of a result of Aubin [3]. Our proof is different and yields relative compactness of optimizing sequences. It is related to Lieb's proof in [18, Lemma 1.2].

An optimal, quartic stability result for the subcritical Sobolev inequality (1.43) on $\mathbb{S}^d$, which we discussed in the appendix of Lecture 2, can be obtained by the same method as in this lecture [49]. The fact that for this inequality one has quadratic stability away from a low-dimensional subspace was observed in [19], where in addition explicit constants were obtained by avoiding the use of compactness. An optimal, quartic stability result for a certain subfamily of Caffarelli–Kohn–Nirenberg inequalities was obtained in [54]. The overall strategy employed there is that of [49], but the fact that optimizers are nonconstant leads to additional difficulties that need to be overcome.

References

1. Abramowitz, M., Stegun, I.A.: Handbook of Mathematical Functions with Formulas, Graphs, and Mathematical Tables. Reprint of the 1972 edition. Dover Publications, New York (1992)
2. Aubin, T.: Problèmes isopérimétriques et espaces de Sobolev (in French). J. Differ. Geom. **11**(4), 573–598 (1976)
3. Aubin, T.: Équations différentielles non linéaires et problème de Yamabe concernant la courbure scalaire (in French). J. Math. Pures Appl. **55**(3), 269–296 (1976)
4. Aubin, T.: Meilleures constantes dans le théorème d'inclusion de Sobolev et un théorème de Fredholm non linéaire pour la transformation conforme de la courbure scalaire (in French). J. Funct. Anal. **32**(2), 148–174 (1979)
5. Bahouri, H., Chemin, J.Y., Danchin, R.: Fourier Analysis and Nonlinear Partial Differential Equations. Grundlehren der mathematischen Wissenschaften, vol. 343. Springer, Heidelberg (2011)
6. Bakry, D.: L'hypercontractivité et son utilisation en théorie des semigroupes (in French). In: Lectures on Probability Theory (Saint-Flour, 1992). Lecture Notes in Mathematics, vol. 1581, pp. 1–114. Springer, Berlin (1994)
7. Bakry, D., Émery, M.: Diffusions hypercontractives. In: Séminaire de probabilités, XIX, 1983/84. Lecture Notes in Mathematics, vol. 1123, pp. 177–206. Springer, Berlin (1985)
8. Bartsch, T., Weth, T., Willem, M.: A Sobolev inequality with remainder term and critical equations on domains with topology for the polyharmonic operator. Calc. Var. Partial Differ. Equ. **18**(3), 253–268 (2003)
9. Beckner, W.: Sharp Sobolev inequalities on the sphere and the Moser–Trudinger inequality. Ann. Math. **138**(1), 213–242 (1993)
10. Bellazzini, J., Frank, R.L., Visciglia, N.: Maximizers for Gagliardo-Nirenberg inequalities and related non-local problems. Math. Ann. **360**(3–4), 653–673 (2014)
11. Benedetti, R., Petronio, C.: Lectures on Hyperbolic Geometry. Universitext. Springer, Berlin (1992)
12. Bianchi, G., Egnell, H.: A note on the Sobolev inequality. J. Funct. Anal. **100**(1), 18–24 (1991)
13. Bidaut-Véron, M.F., Bouhar, M.: On characterization of solutions of some nonlinear differential equations and applications. SIAM J. Math. Anal. **25**(3), 859–875 (1994)
14. Bidaut-Véron, M.F., Véron, L.: Nonlinear elliptic equations on compact Riemannian manifolds and asymptotics of Emden equations. Invent. Math. **106**(3), 489–539 (1991). Erratum: ibid. **112** (1993), no. 2, 445
15. Bliss, G.A.: An integral inequality. J. Lond. Math. Soc. **5**(1), 40–46 (1930)
16. Bonforte, M., Dolbeault, J., Nazaret, B., Simonov, N.: Stability in Gagliardo–Nirenberg–Sobolev inequalities: Flows, regularity and the entropy method. Memoirs Am. Math. Soc. (2020, to appear). arXiv:2007.03674
17. Brézis, H., Lieb, E.H.: Sobolev inequalities with remainder terms. J. Funct. Anal. **62**(1), 73–86 (1985)
18. Brézis, H., Nirenberg, L.: Positive solutions of nonlinear elliptic equations involving critical Sobolev exponents. Commun. Pure Appl. Math. **36**(4), 437–477 (1983)
19. Brigati, G., Dolbeault, J., Simonov, N.: Logarithmic Sobolev and interpolation inequalities on the sphere: constructive stability results. Ann. Inst. H. Poincaré C Anal. Non Linéaire **41**(5), 1289–1321 (2024)
20. Caffarelli, L.A., Gidas, B., Spruck, J.: Asymptotic symmetry and local behavior of semilinear elliptic equations with critical Sobolev growth. Commun. Pure Appl. Math. **42**(3), 271–297 (1989)
21. Carlen, E.A.: Duality and stability for functional inequalities. Ann. Fac. Sci. Toulouse Math. **26**(2), 319–350 (2017)
22. Carlen, E.A., Carrillo, J.A., Loss, M.: Hardy–Littlewood–Sobolev inequalities via fast diffusion flows. Proc. Natl. Acad. Sci. USA **107**(46), 19696–19701 (2010)

23. Carlen, E.A., Frank, R.L., Lieb, E.H.: Stability estimates for the lowest eigenvalue of a Schrödinger operator. Geom. Funct. Anal. **24**(1), 63–84 (2014)
24. Carlen, E.A., Loss, M.: Extremals of functionals with competing symmetries. J. Funct. Anal. **88**(2), 437–456 (1990)
25. Carlen, E.A., Loss, M.: On the minimization of symmetric functionals. Rev. Math. Phys. **6**(5A), 1011–1032 (1994)
26. Carlotto, A., Chodosh, O., Rubinstein, Y.A.: Slowly converging Yamabe flows. Geom. Topol. **19**(3), 1523–1568 (2015)
27. Case, J.S.: The Frank–Lieb approach to sharp Sobolev inequalities. Commun. Contemp. Math. **23**(3), 2050015, 16pp. (2021)
28. Case, J.S., Wang, Y.: Towards a fully nonlinear sharp Sobolev trace inequality. J. Math. Study **53**(4), 402–435 (2020)
29. Chang, S.Y.A., Hang, F.: Improved Moser-Trudinger-Onofri inequality under constraints. Commun. Pure Appl. Math. **75**(1), 197–220 (2022)
30. Chang, S.Y.A., Yang, P.C.: A perturbation result in prescribing scalar curvature on $\mathbb{S}^n$. Duke Math. J. **64**(1), 27–69 (1991)
31. Chen, W., Li, C., Ou, B.: Classification of solutions for an integral equation. Commun. Pure Appl. Math. **59**(3), 330–343 (2006). Corrigendum: ibid., no. 7, 1064
32. Chen, S., Frank, R.L., Weth, T.: Remainder terms in the fractional Sobolev inequality. Ind. Univ. Math. J. **62**(4), 1381–1397 (2013)
33. Chen, X., Wei, W., Wu, N.: Almost sharp Sobolev trace inequalities in the unit ball under constraints. (2021, Preprint). arXiv:2107.08647
34. Chen, L., Lu, G., Tang, H.: Stability of Hardy-Littlewood-Sobolev inequalities with explicit lower bounds (2023, preprint). arXiv:2301.04097
35. Cicalese, M., Leonardi, G.P.: A selection principle for the sharp quantitative isoperimetric inequality. Arch. Ration. Mech. Anal. **206**(2), 617–643 (2012)
36. Cordero-Erausquin, D., Nazaret, B., Villani, C.: A mass-transportation approach to sharp Sobolev and Gagliardo–Nirenberg inequalities. Adv. Math. **182**(2), 307–332 (2004)
37. Demange, J.: Improved Gagliardo–Nirenberg–Sobolev inequalities on manifolds with positive curvature. J. Funct. Anal. **254**(3), 593–611 (2008)
38. Dolbeault, J., Esteban, M.J., Kowalczyk, M., Loss, M.: Sharp interpolation inequalities on the sphere: new methods and consequences. Chin. Ann. Math. Ser. B **34**(1), 99–112 (2013)
39. Dolbeault, J., Esteban, M.J., Loss, M.: Nonlinear flows and rigidity results on compact manifolds. J. Funct. Anal. **267**(5), 1338–1363 (2014)
40. Dolbeault, J., Esteban, M.J., Loss, M.: Rigidity versus symmetry breaking via nonlinear flows on cylinders and Euclidean spaces. Invent. Math. **206**(2), 397–440 (2016)
41. Dolbeault, J., Esteban, M.J., Loss, M.: Interpolation inequalities on the sphere: linear vs. nonlinear flows. Ann. Fac. Sci. Toulouse Math. **26**(2), 351–379 (2017)
42. Dolbeault, J., Esteban, M.J.: Hardy–Littlewood–Sobolev and related inequalities: stability. In: The Physics and Mathematics of Elliott Lieb – the 90th Anniversary, vol. I, pp. 247–268. EMS Press, Berlin (2022)
43. Dolbeault, J., Esteban, M.J., Figalli, A., Frank, R.L., Loss, M.: Sharp stability for Sobolev and log-Sobolev inequalities, with optimal dimensional dependence (2022, preprint). arXiv:2209.08651
44. Engelstein, M., Neumayer, R., Spolaor, L.: Quantitative stability for minimizing Yamabe metrics. Trans. Am. Math. Soc. Ser. B **9**, 395–414 (2022)
45. Escobar, J.F.: Sharp constant in a Sobolev trace inequality. Ind. Univ. Math. J. **37**, 687–698 (1988)
46. Figalli, A.: Stability in Geometric and Functional Inequalities. European Congress of Mathematics, pp. 585–599. European Mathematical Society, Zürich (2013)
47. Figalli, A.: Quantitative stability results for the Brunn-Minkowski inequality. In: Proceedings of the International Congress of Mathematicians–Seoul 2014, vol. III, pp. 237–256. Kyung Moon Sa, Seoul (2014)

48. Figalli, A., Maggi, F., Pratelli, A.: A mass transportation approach to quantitative isoperimetric inequalities. Invent. Math. **182**(1), 167–211 (2010)
49. Frank, R.L.: Degenerate stability of some Sobolev inequalities. Ann. Inst. H. Poincaré C Anal. Non Linéaire **39**(6), 1459–1484 (2022)
50. Frank, R.L., Lieb, E.H.: Inversion positivity and the sharp Hardy–Littlewood–Sobolev inequality. Calc. Var. Partial Differ. Equ. **39**(1–2), 85–99 (2010)
51. Frank, R.L., Lieb, E.H.: Spherical reflection positivity and the Hardy–Littlewood–Sobolev inequality. In: Concentration, Functional Inequalities and Isoperimetry. Contemporary Mathematics, vol. 545, pp. 89–102. American Mathematical Society, Providence (2011)
52. Frank, R.L., Lieb, E.H.: A new, rearrangement-free proof of the sharp Hardy–Littlewood–Sobolev inequality. In: Spectral Theory, Function Spaces and Inequalities. Operator Theory: Advances and Applications, vol. 219, pp. 55–67 (Birkhäuser, Basel, 2012)
53. Frank, R.L., Lieb, E.H.: Sharp constants in several inequalities on the Heisenberg group. Ann. Math. **176**(1), 349–381 (2012)
54. Frank, R.L., Peteranderl, J.W.: Degenerate stability of the Caffarelli–Kohn–Nirenberg inequality along the Felli–Schneider curve. Calc. Var. Partial Differ. Equ. **63**(44), (2024)
55. Fusco, N., Maggi, F., Pratelli, A.: The sharp quantitative isoperimetric inequality. Ann. Math. **168**(3), 941–980 (2008)
56. Gérard, P.: Description du défaut de compacité de l'injection de Sobolev (in French). ESAIM Control Optim. Calc. Var. **3**, 213–233 (1998)
57. Gérard, P., Meyer, Y., Oru, F.: Inégalités de Sobolev précisées (in French). Séminaire sur les Équations aux Dérivés Partielles, 1996–1997, Exp. No. IV, 11 pp., École Polytech., Palaiseau (1997)
58. Gidas, B., Spruck, J.: Global and local behavior of positive solutions of nonlinear elliptic equations. Commun. Pure Appl. Math. **34**(4), 525–598 (1981)
59. Gidas, B., Ni, W.M., Nirenberg, L.: Symmetry of Positive Solutions of Nonlinear Elliptic Equations in $\mathbb{R}^n$. Mathematical Analysis and Applications, Part A. Advances in Mathematical Supplies Studies, vol. 7a, pp. 369–402. Academic Press, New York (1981)
60. Graham, C.R., Zworski, M.: Scattering matrix in conformal geometry. Invent. Math. **152**(1), 89–118 (2003)
61. Graham, C.R., Jenne, R., Mason, L.J., Sparling, G.A.J.: Conformally invariant powers of the Laplacian. I. Existence. J. Lond. Math. Soc. **46**(3), 557–565 (1992)
62. Hang, F., Wang, X.: Improved Sobolev inequality under constraints. Int. Math. Res. Not. IMRN **2022**(14), 10822–10857 (2022)
63. Hersch, J.: Quatre propriétés isopérimétriques de membranes sphériques homogènes. C. R. Acad. Sci. Paris Sér. A-B **270**, A1645–A1648 (1970)
64. Killip, R., Vişan, M.: Nonlinear Schrödinger equations at critical regularity. In: Evolution Equations. Clay Mathematics Proceedings, vol. 17, pp. 325–437. American Mathematical Society, Providence (2013)
65. König, T.: Stability for the Sobolev inequality: existence of a minimizer. J. Eur. Math. Soc. (2022, preprint). arXiv:2211.14185
66. König, T.: On the sharp constant in the Bianchi–Egnell stability inequality. Bull. Lond. Math. Soc. **55**(4), 2070–2075 (2023)
67. König, T.: An exceptional property of the one-dimensional Bianchi-Egnell inequality. Calc. Var. Partial Differ. Equ. **63**(123), (2024)
68. Korevaar, N., Mazzeo, R., Pacard, F., Schoen, R.: Refined asymptotics for constant scalar curvature metrics with isolated singularities. Invent. Math. **135**(2), 233–272 (1999)
69. Ledoux, M.: On improved Sobolev embedding theorems. Math. Res. Lett. **10**(5–6), 659–669 (2003)
70. Li, Y.Y.: Remark on some conformally invariant integral equations: the method of moving spheres. J. Eur. Math. Soc. **6**(2), 153–180 (2004)
71. Lieb, E.H.: Existence and uniqueness of the minimizing solution of Choquard's nonlinear equation. Stud. Appl. Math. **57**(2), 93–105 (1976/1977)

72. Lieb, E.H.: Sharp constants in the Hardy–Littlewood–Sobolev and related inequalities. Ann. Math. **118**(2), 349–374 (1983)
73. Lieb, E.H.: On the lowest eigenvalue of the Laplacian for the intersection of two domains. Invent. Math. **74**, 441–448 (1983)
74. Lieb, E.H., Loss, M.: Analysis. Graduate Studies in Mathematics, vol. 14, 2nd edn. American Mathematical Society, Providence (2001)
75. Lions, P.L.: The concentration-compactness principle in the calculus of variations. The limit case. I. Rev. Mat. Iberoamericana **1**(1), 145–201 (1985)
76. Lions, P.L.: The concentration-compactness principle in the calculus of variations. The limit case. II. Rev. Mat. Iberoamericana **1**(2), 45–121 (1985)
77. Lu, G., Wei, J.: On a Sobolev inequality with remainder terms. Proc. Am. Math. Soc. **128**(1), 75–84 (2000)
78. Luckhaus, S., Zemas, K.: Rigidity estimates for isometric and conformal maps from $\mathbb{S}^{n-1}$ to $\mathbb{R}^n$. Invent. Math. **230**(1), 375–461 (2022)
79. Mazzeo, R., Pacard, F.: Constant scalar curvature metrics with isolated singularities. Duke Math. J. **99**(3), 353–418 (1999)
80. Mazzeo, R., Pollack, D., Uhlenbeck, K.: Moduli spaces of singular Yamabe metrics. J. Am. Math. Soc. **9**(2), 303–344 (1996)
81. Nazaret, B.: Best constant in Sobolev trace inequalities on the half-space. Nonlinear Anal. **65**(10), 1977–1985 (2006)
82. Obata, M.: The conjectures on conformal transformations of Riemannian manifolds. J. Differ. Geom. **6**, 247–258 (1971/1972)
83. Penot, J.P.: Analysis. From Concepts to Applications. Universitext. Springer, Cham (2016)
84. Rodemich, E.: The Sobolev inequality with best possible constant. In: Analysis Seminar Caltech (1966)
85. Rosen, G.: Minimum value for c in the Sobolev inequality $\|\varphi^3\| \leq c \|\nabla\varphi\|^3$. SIAM J. Appl. Math. **21**, 30–32 (1971)
86. Sabin, J.: Compactness methods in Lieb's work. In: The Physics and Mathematics of Elliott Lieb – The 90th Anniversary, vol. II, pp. 219–251. EMS Press, Berlin (2022)
87. Schoen, R.M.: Variational theory for the total scalar curvature functional for Riemannian metrics and related topics. In: Topics in Calculus of Variations (Montecatini Terme, 1987). Lecture Notes in Mathematics, vol. 1365, pp. 120–154. Springer, Berlin (1989)
88. Stein, E.M., Weiss, G.: Introduction to Fourier Analysis on Euclidean Spaces. Princeton Mathematical Series, vol. 32. Princeton University Press, Princeton (1971)
89. Talenti, G.: Best constant in Sobolev inequality. Ann. Mat. Pura Appl. **110**, 353–372 (1976)
90. Trudinger, N.S.: Remarks concerning the conformal deformation of Riemannian structures on compact manifolds. Ann. Scuola Norm. Sup. Pisa Cl. Sci. **22**, 265–274 (1968)
91. van Neerven, J.: Functional Analysis. Cambridge Studies in Advanced Mathematics, vol. 201. Cambridge University Press, Cambridge (2022)
92. Yan, Z.: Improved Sobolev inequalities on CR sphere (2023, preprint). arXiv:2301.07170

Chapter 2
Nonlinear Potential Theoretic Methods in Nonuniformly Ellliptic Problems

Giuseppe Mingione

Abstract Nonuniform ellipticity is a classical topic in the theory of partial differential equations. While several results in regularity theory have been adding up over decades, many basic issues, as for instance the validity of Schauder theory and sharp dependence of regularity upon data, remained opened for a while. In these notes we give an overview of recent results and techniques about the topic, that, via a novel use of nonlinear potential theoretic methods, allow to answer several of the above questions.

2.1 About the Content

The aim of these notes is twofold, that is

- To briefly survey a few recent results in the regularity theory of minimizers of variational integrals of the type

$$W^{1,1}_{\text{loc}}(\Omega, \mathbb{R}^N) \ni w \mapsto \mathcal{F}(w, \Omega) := \int_\Omega [F(x, Dw) - \mu \cdot w] \, dx \qquad (2.1.1)$$

and of solutions to elliptic equations of the type

$$- \operatorname{div} a(x, Du) = \mu \qquad \text{in } \Omega \subset \mathbb{R}^n. \qquad (2.1.2)$$

In both cases, and for the rest of the paper, $\Omega \subset \mathbb{R}^n$ denotes a bounded open subset and $n \geq 2$ unless otherwise specified. We shall refer to the scalar case when $N = 1$ and to the vectorial one otherwise. In fact, unless otherwise specified, we shall always deal with the scalar case $N = 1$. In particular, this

G. Mingione (✉)
Dipartimento SMFI, Università di Parma, Parma, Italy
e-mail: giuseppe.mingione@unipr.it

will always happen when dealing with (2.1.2), i.e., we shall always deal with equations but in the case of Theorem 2.8.4.

As from the title, the emphasis will be put on *nonuniformly elliptic problems*.
- In connection to the previous point, we shall try to give a streamlined presentation of a few facts and techniques based on the use of Nonlinear Potential Theory in the setting of nonuniformly elliptic problems; see Sect. 2.10. In this respect, we shall sacrifice generality to readability. We shall indeed present a series of cases and results treated via a unified approach using a certain family of nonlinear potentials. Several of such cases can be then treated in a more efficient via specific variations of the general method presented, but still based on the use of nonlinear potentials. These variations will be then described in the commentaries of Sect. 2.10 and references will be given.

In (2.1.1) the integrand $F \colon \Omega \times \mathbb{R}^{N \times n} \to \mathbb{R}$ will be assumed to be non-negative and Carathéodory regular. Moreover, again when dealing with functionals (2.1.1), we shall assume (at least) that

$$\mu \in L^n(\Omega, \mathbb{R}^N) \tag{2.1.3}$$

so that $\mu \cdot w$ is locally integrable whenever $w \in W^{1,1}_{\text{loc}}(\Omega, \mathbb{R}^N)$ as an effect of Sobolev embedding theorem. As for (2.1.2), the vector field $a \colon \Omega \times \mathbb{R}^n \to \mathbb{R}^n$ will always be at least Carathéodory regular and μ will in the most general case be a Borel measure. Minimizers of the functional in (2.1.1) and solutions to (2.1.2) are defined as follows, respectively:

Definition 2.1.1 A map $u \in W^{1,1}_{\text{loc}}(\Omega, \mathbb{R}^N)$ is a (local) minimizer of the functional $\mathcal{F}$ in (2.1.1) with $\mu \in L^n_{\text{loc}}(\Omega, \mathbb{R}^N)$ if, for every open subset $\tilde{\Omega} \Subset \Omega$, we have $F(\cdot, Du) \in L^1(\tilde{\Omega})$ and if $\mathcal{F}(u, \tilde{\Omega}) \leq \mathcal{F}(w, \tilde{\Omega})$ holds for every competitor $w \in u + W^{1,1}_0(\tilde{\Omega}, \mathbb{R}^N)$.

Definition 2.1.2 A function $u \in W^{1,1}_{\text{loc}}(\Omega)$ is a distributional (local) solution to (2.1.2) iff $a(\cdot, Du) \in L^1_{\text{loc}}(\Omega, \mathbb{R}^n)$, μ is a Borel measure with locally finite total mass in Ω and

$$\int_\Omega a(x, Du) \cdot D\varphi \, dx = \int_\Omega \varphi \, d\mu \tag{2.1.4}$$

holds for every $\varphi \in C_0^\infty(\Omega)$.

These are not yet so-called energy solutions, for which we refer to Definition 2.2.1, and are actually called very weak solutions - see discussion after Theorem 2.6.2 in Sect. 2.6. In these notes we shall use no more than integrability properties of μ. As all the results presented here will be interior regularity results, without loss of generality we shall always assume that the integrability properties of μ in question will be global, as for instance in (2.1.3). Eventually letting $\mu \equiv 0$ outside Ω, we shall assume that μ is defined, with the same integrability properties, on the whole $\mathbb{R}^n$.

This paper finds its origins in a series of lectures given at the 2022 CIME school and, as mentioned before, they are mostly concerned with nonuniformly elliptic problems. For nonlinear potential theoretic results and related Nonlinear Calderón-Zygmund theory in the uniformly elliptic setting the reader might like to try the notes from the previous 2016 CIME lectures [182].

2.1.1 Notation

By c we shall denote generic constants larger than 1; such constants will change their specific value in different occurrences, although used in the same setting. What will really matter will be their dependence on various relevant parameters. This will be usually specified in parentheses. A similar role will be played by constants denoted by $\vartheta, \tilde{c}, \kappa$ and the like; this will be clear by the context and their exact values will not be important, but their dependence on parameters will. We shall use the symbol $\lesssim$ to mean an inequality that occurs up to a universal constant, i.e., a constant usually depending on n, or on a fixed number of parameters whose precise value is not relevant in the context. For instance, $a \lesssim b$ means that there exists a constant $c \geq 1$ such that $a \leq cb$. In the case c depends itself on parameters, for instance κ_1, κ_2, and it is useful to specify such a dependence, we shall denote $a \lesssim_{\kappa_1,\kappa_2} b$. Finally, $a \approx b$ means that both $a \lesssim b$ and $b \lesssim a$ occur. In a similar way, writing $a \equiv_{\kappa_1,\kappa_2} b$ means that $a = cb$ for a constant c depending on κ_1, κ_2 and so on. In the following

$$B_r(x_0) := \{x \in \mathbb{R}^n : |x - x_0| < r\}$$

denotes the open ball with center x_0 and radius $r > 0$. When not important, or when it will be clear from the context, we shall omit denoting the center as follows: $B_r \equiv B_r(x_0)$. As usual, the Sobolev embedding exponent 2^* is defined as

$$2^* := \begin{cases} \frac{2n}{n-2} & \text{if } n > 2 \\ \text{any number } > 2 & \text{if } n = 2 \,. \end{cases} \quad (2.1.5)$$

With $\mathcal{B} \subset \mathbb{R}^n$ being a measurable subset with positive measure, and with $f : \mathcal{B} \to \mathbb{R}^k$, $k \geq 1$, being an integrable map, we shall denote by

$$(f)_\mathcal{B} \equiv \fint_\mathcal{B} f \, dx := \frac{1}{|\mathcal{B}|} \int_\mathcal{B} f(x) \, dx$$

its integral average; here $|\mathcal{B}|$ denotes the Lebesgue measure of $\mathcal{B}$. Needless to say, we shall automatically identify L^1-functions μ with Borel measures, thereby denoting

$$|\mu|(\mathcal{B}) = \int_{\mathcal{B}} |\mu|\,dx \quad \text{for every measurable subset } \mathcal{B} \Subset \Omega\,.$$

In denoting several function spaces like $L^t(\Omega)$, $W^{1,t}(\Omega)$, we will denote the vector valued version by $L^t(\Omega, \mathbb{R}^N)$, $W^{1,p}(\Omega, \mathbb{R}^N)$ in the case the maps considered take values in $\mathbb{R}^N$, $N \in \mathbb{N}$. When clear from the contest, we will also abbreviate $L^t(\Omega, \mathbb{R}^N)$, $W^{1,t}(\Omega, \mathbb{R}^N) \equiv L^t(\Omega), W^{1,t}(\Omega)$ and so on. In the rest of the paper p, q, ν, L and s will denote numbers such that

$$1 < p \le q\,, \quad 0 < \nu \le L\,, \quad 0 \le s \le 1\,. \tag{2.1.6}$$

In relation to (2.1.6), we set

$$\mathtt{data} = (n, p, q, \nu, L)\,,$$

where n is the ambient dimension of the equations or the integrals we are considering, as indicated after (2.1.1)–(2.1.2). This notation will still be used when $p = q$ and in this case $\mathtt{data} = (n, p, \nu, L)$. Finally, we shall also denote

$$H_s(z) := |z|^2 + s^2\,, \quad z \in \mathbb{R}^n\,. \tag{2.1.7}$$

2.2 Notions of Nonuniform Ellipticity

In this section we discuss a few definitions of uniform and nonuniform ellipticity for functionals and equations of the type in (2.1.1) and (2.1.2), respectively. Major emphasis will be put on nonautonomous problems, that is, problems with coefficients.

2.2.1 Classical (Pointwise) Definitions

Let us start with an elliptic equation in divergence form of the type in (2.1.2), where the vector field $a \colon \Omega \times \mathbb{R}^n \to \mathbb{R}^n$ is initially assumed to be Carathéodory regular together with its derivatives $\partial_z a(\cdot)$ being Carathéodory regular in $\Omega \times \mathbb{R}^n \setminus \{0_{\mathbb{R}^n}\}$. Moreover, in the following we shall always assume that $z \mapsto a(\cdot, z) \in C^0(\mathbb{R}^n) \cap C^1(\mathbb{R}^n \setminus \{0_{\mathbb{R}^n}\})$. In order to give a quantitative description of its ellipticity properties, we shall consider two non-negative, Carathéodory functions $g_1, g_2 \colon \Omega \times (0, \infty) \to (0, \infty)$ that we are going to use as lower and upper bounds on the eigenvalues of

$z \mapsto \partial_z a(x, z)$, respectively. Specifically, we shall assume that

$$\begin{cases} g_1(x, |z|)\mathbb{I}_d \leq \partial_z a(x, z) \\ |\partial_z a(x, z)| \leq g_2(x, |z|) \end{cases} \quad (2.2.1)$$

holds whenever $z \in \mathbb{R}^n \setminus \{0_{\mathbb{R}^n}\}$ and $x \in \Omega$, with $\mathbb{I}_d$ denoting the identity. Uniform ellipticity of the vector field $a(\cdot)$ (or, equivalently, of the Eq. (2.1.2)) then requires that

---

The Ellipticity Ratio

$$\mathcal{R}_a(x, z) := \frac{g_2(x, |z|)}{g_1(x, |z|)} \quad (2.2.2)$$

---

remains uniformly bounded, i.e.,

$$\sup_{x \in \Omega, |z| \neq 0} \mathcal{R}_a(x, z) < \infty. \quad (2.2.3)$$

When $\partial_z a(\cdot)$ is symmetric this amounts to require that the ratio

$$\frac{\text{highest eigenvalue of } \partial_z a(x, z)}{\text{lowest eigenvalue of } \partial_z a(x, z)}$$

is still uniformly bounded in the above range of x and z; in this case the quantity in the last display provides an alternative, more tailored definition of ellipticity ratio. In the same line of thought, when considering integral functionals of the type in (2.1.1), we shall say that the integrand $F(\cdot)$ (or, equivalently, the functional $\mathcal{F}$) is uniformly elliptic if so is $\partial_z F(\cdot)$. That is, if and only if its Euler-Lagrange equation

$$- \operatorname{div} \partial_z F(x, Du) = \mu$$

is uniformly elliptic. In this case we shall often adopt the shortened notation $\mathcal{R}_F \equiv \mathcal{R}_{\partial_z F}$. Sometimes, we shall denote $\mathcal{R} \equiv \mathcal{R}_F$ or $\mathcal{R} \equiv \mathcal{R}_a$ when no ambiguity shall arise on the identity of $F(\cdot)$ or $a(\cdot)$.

On several occasions it is also sufficient to adopt a slightly different notion of uniform ellipticity, namely, instead of (2.2.3) one might require that

$$\sup_{x \in \Omega, |z| \geq 1} \mathcal{R}_a(x, z) < \infty \quad (2.2.4)$$

or even

$$\limsup_{|z|\to\infty} \mathcal{R}_a(x, z) < \infty, \quad \text{uniformly with respect to } x \in \Omega. \tag{2.2.5}$$

This happens especially when one is interested in proving local Lipschitz regularity of solutions so that only the behaviour of the operator for large values of $|z|$ (that is, of the gradient) matters. In this case (2.2.4)–(2.2.5) become more suitable definitions.

On the contrary, we shall say that the equation in (2.1.2) is nonuniformly elliptic if, for at least one point $x \in \Omega$, it happens that

$$\sup_{|z|>0} \mathcal{R}_a(x, z) = \infty. \tag{2.2.6}$$

The importance of conditions like (2.2.3) or (2.2.5) stems from the fact that when trying to prove Lipschitz bounds for solutions to equations as (2.1.2), the quantity $\mathcal{R}_a(x, Du(x))$ appears in all the crucial integral estimates. Examples of this will appear in (2.10.19) and (2.10.34). Assuming (2.2.3), and therefore the a priopri boundedness of $\mathcal{R}_a(x, Du(x))$ no matter how large the size of $|Du(x)|$ is, allows for a whole range of estimates and results. These must be then rediscussed in the nonuniformly elliptic case (2.2.6), up to the stage that a too fast growth of

$$z \mapsto \mathcal{R}_a(x, z) \tag{2.2.7}$$

with respect to $|z|$ implies the existence of non-Lipschitz, and even unbounded solutions [114, 170]. The whole point in the regularity theory of nonuniformly elliptic equations is then to find suitable growth assumptions on the map in (2.2.7) allowing to compensate the potential blow-up of the gradient of solutions, and therefore of $\mathcal{R}_a(x, Du(x))$, eventually leading to local Lipschitz regularity of solutions. Once this is known, the growth of (2.2.7) becomes irrelevant as Du remains bounded and so (typically) $\mathcal{R}_a(x, Du(x))$ does anyway. In this situation higher regularity of solutions can be often obtained adapting the methods developed for the uniformly elliptic case. Imposing a growth condition on (2.2.7) is a classical and natural approach in the literature [143, 159, 169–171, 202, 212] and we shall also follow it. Examples of this are given in Sect. 2.5 below.

2.2.2 Examples of Uniform Operators

Beside the Laplacean operator, obviously, and linear variations including coefficients, the next popular uniformly elliptic operator is

$$\begin{cases} -\text{div}\,(\mathfrak{c}(x)(|Du|^2 + s^2)^{(p-2)/2} Du) = 0 \\ 0 < \nu \leq \mathfrak{c}(x) \leq L, \quad s \in [0, 1], \quad p > 1 \end{cases} \tag{2.2.8}$$

which is the Euler-Lagrange equation of the functional

$$w \mapsto \int_\Omega (|Dw|^2 + s^2)^{p/2} \, dx \, .$$

In this case it is

$$\begin{cases} a(x, z) = \mathfrak{c}(x)(|z|^2 + s^2)^{(p-2)/2} z \\ \partial_z a(x, z) \approx_p \mathfrak{c}(x)(|z|^2 + s^2)^{(p-2)/2} \mathbb{I}_d \\ g_1(x, |z|) \approx g_2(x, |z|) \approx \mathfrak{c}(x)(|z|^2 + s^2)^{(p-2)/2} \end{cases} \quad (2.2.9)$$

and

$$\mathcal{R}_a(x, z) \lesssim \frac{L}{\nu} \frac{\max\{p-1, 1\}}{\min\{p-1, 1\}} \, .$$

The choice $s = 0$ and $\mathfrak{c}(\cdot) \equiv 1$ leads to

The p-Laplacean operator

$$-\Delta_p u = -\operatorname{div}(|Du|^{p-2} Du) = 0 \, . \quad (2.2.10)$$

See [90, 165, 166, 216, 217] for basic regularity in this last case; conditions on coefficients can be also relaxed, see for instance [8] and related references. The one in (2.2.8) belongs to the family of operators with standard polynomial p-growth. These have been treated at length in the literature where general equations of the type (2.1.2) have been considered under the assumptions

$$\begin{cases} |a(x, z)| + |\partial_z a(x, z)|(|z|^2 + s^2)^{1/2} \leq L(|z|^2 + s^2)^{(p-1)/2} \\ \nu(|z|^2 + s^2)^{(p-2)/2} |\xi|^2 \leq \partial_z a(x, z)\xi \cdot \xi \end{cases} \quad (2.2.11)$$

for every choice of $x \in \Omega$ and $z, \xi \in \mathbb{R}^n$, $|z| \neq 0$; as already specified in (2.1.6), it is $0 < \nu \leq L$ and $p > 1$. These assumptions are classical after the work of Ladyzhenskaya and Uraltseva [158] and are modelled on (2.2.9). In this case we have

$$\mathcal{R}_a(x, z) \lesssim_p \frac{L}{\nu} \, .$$

Sometimes conditions (2.2.11) are replaced by the weaker

$$v|z|^p - L \leq a(x,z) \cdot z, \quad |a(x,z)| \leq L|z|^{p-1} + L, \tag{2.2.12}$$

where $a(\cdot)$ is only supposed to be Carathéodory regular. The growth conditions $(2.2.11)_1$ fix the concept of (local) *energy solutions* as distributional solutions that belong to the Sobolev space $W^{1,p}_{\text{loc}}(\Omega)$, that is

Definition 2.2.1 A function u is an energy solution to (2.1.2) under assumptions (2.2.12) and $\mu \in W^{-1,p'}(\Omega)$ (the dual of $W^{1,p}_0(\Omega)$), iff $u \in W^{1,p}_{\text{loc}}(\Omega)$ and it is a distributional solution in the sense of Definition 2.1.2. In particular (2.1.4) holds for every $\varphi \in W^{1,p}(\Omega)$ which is compactly supported in Ω.

Notice that, as already mentioned, conditions (2.2.12) are implied by (2.2.11) (modulo changing the values of v, L), see also [3]. This is the type of solution which is usually considered in this setting. Uniformly elliptic problems are indeed often characterized by the fact of having a natural and robust notion of energy solutions, which is indeed fixed by the growth conditions of the vector field $a(\cdot)$, as in the present setting.

Uniformly elliptic equations are not necessarily of p-polynomial type as in (2.2.11), but growth conditions intrinsically determined by certain nonlinear functions can be considered too. We are thinking of operators of the type

$$-\operatorname{div}(\tilde{a}(|Du|)Du) = 0 \tag{2.2.13}$$

under assumptions

$$\begin{cases} -1 < i_a \leq \dfrac{\tilde{a}'(t)t}{\tilde{a}(t)} \leq s_a < \infty & \text{for every } t > 0 \\ \tilde{a} \colon (0, \infty) \to [0, \infty) \text{ is of class } C^1_{\text{loc}}(0, \infty). \end{cases} \tag{2.2.14}$$

In this case we find

$$\mathcal{R}_a(z) \lesssim \frac{\max\{s_a + 1, 1\}}{\min\{i_a + 1, 1\}}.$$

The one in (2.2.13) is the Euler-Lagrange equation of the functional

$$w \mapsto \int_\Omega A(|Dw|)\,dx, \quad A(t) := \int_0^t \tilde{a}(y)y\,dy. \tag{2.2.15}$$

Notice that (2.2.9) falls in the realm of (2.2.13) with $\tilde{a}(t) \equiv (t^2 + s^2)^{(p-2)/2}$ so that $i_a = s_a = p-2$ and $A(t) \equiv (t^2+s^2)^{p/2}/p$. These problems are naturally well-posed in the Orlicz space $W^{1,A(\cdot)}_{\text{loc}}(\Omega)$, defined as the set of $W^{1,1}_{\text{loc}}(\Omega)$-regular functions w such that $A(|Dw|) \in L^1_{\text{loc}}(\Omega)$. In turn, energy solutions to (2.2.13) are defined as distributional solutions belonging to $W^{1,A(\cdot)}_{\mathrm{loc}}(\Omega)$. An instance of a regularity result valid for energy solutions in this setting is presented in Theorem 2.8.4. A typical example of a functional as in (2.2.15) is provided by

$$w \mapsto \int_\Omega |Dw|^p \log(1+|Dw|)\,dx, \qquad p>1. \tag{2.2.16}$$

For such problems, and related regularity theory of solutions, references we recommend are for instance [55–59, 91, 137, 138, 161, 189]. Observe that in the limiting case $p=1$ the functional in (2.2.16) ceases to be uniformly elliptic. See Sect. 2.2.3 below.

2.2.3 Examples of Nonuniform Operators

Functionals as

$$w \mapsto \mathcal{F}_j(w,\Omega) := \int_\Omega F_j(x,Dw)\,dx$$

for $j \in \{1,\ldots,4\}$, where

$$\begin{cases} F_1(x,Dw) := |Dw|^p + \sum_{k=1}^m \mathfrak{a}_k(x)|D_kw|^{p_i}, \quad 1 < p \le p_k,\ 0 \le \mathfrak{a}_k(\cdot) \le L \\[1ex] F_2(x,Dw) := \exp(|Dw|^p), \quad p \ge 1 \\[1ex] F_3(x,Dw) := \exp(\exp((|Dw|^p))), \quad p \ge 1 \\[1ex] F_4(x,Dw) := \mathfrak{c}(x)|Dw|\log(1+|Dw|), \quad 0 < \nu \le \mathfrak{c}(x) \le L, \end{cases}$$

are all nonuniformly elliptic. In fact, the best upper bounds we can find on the ellipticity ratios, defined in (2.2.2), are

$$\begin{cases} \mathcal{R}_{F_1}(z) \lesssim_{\mathfrak{a}_i(\cdot)} |z|^{q-p}+1, \quad q := \max\{p_i\} \\ \mathcal{R}_{F_2}(z) \lesssim |z|^p + 1 \\ \mathcal{R}_{F_3}(z) \lesssim |z|^p \exp(|z|^p)+1 \\ \mathcal{R}_{F_4}(z) \lesssim_{L/\nu} \log(1+|z|)+1. \end{cases} \tag{2.2.17}$$

See also [14, 78] for the computations relevant to (2.2.17). $\mathcal{F}_1$ is a typical example of functional with so-called (p,q)-growth conditions. The ellipticity ratio grows as

$|Du|^{q-p}$, and therefore polynomially with respect to the gradient of a minimizer u. We shall come back on such functionals later on, in Sect. 2.5. Integrands where each partial derivative is penalized with its own exponent are called anisotropic integrands and are amongst the most intensively studied within the realm of nonuniform operators; see for instance [30, 169, 218]. Functionals $\mathcal{F}_2$ and $\mathcal{F}_3$ are relevant examples of functionals with fast, non-polynomial growth conditions, considered for instance in [14, 78, 92, 96, 103, 162, 172, 173]; see Sect. 2.8.3. In the case of $\mathcal{F}_3$ the ellipticity ratio still grows polynomially. Finally, $\mathcal{F}_4$ is an example of a functional with so-called nearly linear growth. See Sect. 2.9 for more.

Nonuniform ellipticity is a classical topic in the theory of PDE. In the polynomial growth case authors who worked on the topic are Ladyzhenskaya and Ural'tseva [158, 159], Hartman and Stampacchia [130], Trudinger [212, 213], Ivokina and A. P. Oskolkov [143], Oskolkov [191], Serrin [202], A. V. Ivanov [140–142], Leon Simon [204, 205], Ural'tseva and Urdaletova [218], Lieberman [160], just to mention a few. We shall describe more recent contributions later on, in Sect. 2.5.

2.2.4 Soft Nonuniform Ellipticity [78]

Uniform ellipticity is a way to express that the eigenvalues of a given operator, that at this stage depend on the solution itself as described in (2.2.1), self-rebalance. Ultimately, they mimic the situation of the classical Poisson equation. Of course substantiating this last assertion requires the full strength of the entire nonlinear regularity theory, starting from De Giorgi-Nash-Moser. Passing from the linear to the nonlinear case requires building a completely different world of ideas. Nevertheless, the final outcome is that things like

$$\begin{cases} \text{Schauder type estimates [116, 117, 161, 165, 166]} \\ \text{Calderón-Zygmund estimates (see [182] for an overview)} \\ \text{Potential estimates (see Sect. 2.4.2)} \end{cases} \qquad (2.2.18)$$

work more or less exactly as in the linear case. All in all, reducing the nonlinear theory to the linear one in terms of results is a gigantic achievement. These results are actually concerned with the case of equations involving vector fields satisfying (2.2.11) or (2.2.13) and (2.2.14). It is therefore reasonable to argue that, whenever condition (2.2.3) is satisfied, then such parts of regularity theory would still hold. Surprisingly enough, things turn out to be different. For this we consider the so-called

The Double Phase Functional

$$\begin{cases} w \mapsto \mathcal{D}(w, \Omega) := \int_\Omega \mathcal{H}(x, Dw)\,dx \\ \mathcal{H}(x, z) := |z|^p + \mathfrak{a}(x)|z|^q \\ 1 < p \leq q,\ 0 \leq \mathfrak{a}(\cdot) \in C^{0,\alpha}(\Omega),\ \alpha \in (0, 1] \end{cases} \qquad (2.2.19)$$

with corresponding Euler-Lagrange equation being

$$-\operatorname{div}(|Du|^{p-2}Du + (q/p)\mathfrak{a}(x)|Du|^{q-2}Du) = 0. \qquad (2.2.20)$$

Observe that in this case any local minimizer obviously belongs to $W^{1,p}_{\text{loc}}(\Omega)$, just because it makes the functional $\mathcal{D}$ locally finite. This kind of structure was introduced by Zhikov [220, 221] in the setting of Homogenization of strongly anisotropic materials [219, 222], where the geometry of the double composite, with hardening exponents p and q, is described by the zero set $\{\mathfrak{a}(\cdot) = 0\}$. A systematic study of qualitative properties of minima of the functional in (2.2.19) was made in [13, 60], after the first higher integrability results were obtained in [102]. Here we are mostly interested in the counterexamples developed in [102, 106]. Let us preliminary note that the functional in (2.2.19) is uniformly elliptic in the sense of Sect. 2.2.1. Indeed, a simple computation reveals that, with respect to the notation in (2.2.1), we have

$$\begin{cases} g_1(x, t) \approx \min\{p-1, 1\}t^{p-2} + \min\{q-1, 1\}\mathfrak{a}(x)t^{q-2} \\ g_2(x, t) \approx \max\{p-1, 1\}t^{p-2} + \max\{q-1, 1\}\mathfrak{a}(x)t^{q-2} \end{cases}$$

so that

$$\mathcal{R}_\mathcal{H}(x, z) \lesssim \frac{\max\{p-1, 1\}}{\min\{p-1, 1\}} + \frac{\max\{q-1, 1\}}{\min\{q-1, 1\}}. \qquad (2.2.21)$$

It follows that the equation in (2.2.20) is uniformly elliptic in the classical sense of Sect. 2.2.1, (2.2.3). Nevertheless the regularity features (2.2.18) generally fail for solutions to (2.2.20) and local minimizers of $\mathcal{D}$ in (2.2.19). It is in fact possible to find non-negative Hölder continuous functions $\mathfrak{a}(\cdot)$ leading to the existence of minimizers that do not even belong to $W^{1,q}_{\text{loc}}$ [102] and develop singularities on fractals with maximal Hausdorff dimension $n - p$ [7, 9, 106]; see also the two-dimensional constructions by Zhikov [220, 221]. This can be realised taking any choice of (p, q) such that

$$1 < p < n < n + \alpha < q, \qquad (2.2.22)$$

with a suitably constructed function $\mathfrak{a}(\cdot) \in C^{0,\alpha}(\Omega)$. Minima found in [102, 106] are globally bounded. Note that (2.2.22) implies

$$\frac{q}{p} > 1 + \frac{\alpha}{n} \quad \text{and} \quad q - p > \alpha. \qquad (2.2.23)$$

New and interesting counterexamples are in [6, 7, 9, 10]. All this happens in presence of classical uniform ellipticity (2.2.21), which is somehow counterintuitive according to the narrative drawn at the beginning of this section. The key to understand the failure of properties (2.2.18) is to introduce another type of ellipticity ratio, which is bound to detect milder forms of nonuniform ellipticity. Following [78], with $a(\cdot)$ being a vector field as considered in Sect. 2.2.1, we define

The Nonlocal Ellipticity Ratio

$$\mathcal{R}_a(z, B) := \frac{\sup_{x \in B} g_2(x, |z|)}{\inf_{x \in B} g_1(x, |z|)} \qquad \text{for any ball } B \subset \Omega \qquad (2.2.24)$$

or, equivalently, when $\partial_z a(\cdot)$ is symmetric

$$\mathcal{R}_a(z, B) := \frac{\sup_{x \in B} \text{ highest eigenvalue of } \partial_z a(x, z)}{\inf_{x \in B} \text{ lowest eigenvalue of } \partial_z a(x, z)}.$$

Of course, it is

$$\mathcal{R}_a(x, z) \leq \mathcal{R}_a(z, B) \quad \text{for every } x \in B.$$

Similar definitions can be given as in Sect. 2.2.1 concerning an integrand $F(\cdot)$ rather than a vector field. Needless to say, such different concepts coincide in the autonomous case, i.e., when the vector field and/or the integrand do not exhibit any direct dependence on x. We shall then say that a vector field $a: \Omega \times \mathbb{R}^n \to \mathbb{R}^n$ is *softly nonuniformly elliptic* if it is uniformly elliptic in the classical sense of (2.2.3) but, for at least one ball B, it happens that $\mathcal{R}_a(z, B)$ is unbounded, i.e.,

$$\sup_{x \in \Omega, |z| \neq 0} \mathcal{R}_a(x, z) < \infty \quad \text{and} \quad \sup_{|z| \neq 0} \mathcal{R}_a(z, B) = \infty. \qquad (2.2.25)$$

Softly nonuniformly elliptic integrands $F(\cdot)$ can be defined in a similar manner; these are such that $\partial_z F(\cdot)$ is softly nonuniformly elliptic. Back to the functional $\mathcal{D}$

in (2.2.19), we observe that

$$\mathcal{R}_{\mathcal{H}}(z, B) \approx \|\mathfrak{a}\|_{L^\infty(B)} |z|^{q-p} + 1 \lesssim |B|^{\alpha/n} |z|^{q-p} + 1 \qquad (2.2.26)$$

holds whenever $\{\mathfrak{a}(\cdot) \equiv 0\} \cap B$ is nonempty. This means that (2.2.25) occurs provided $\mathfrak{a}(\cdot)$ does not vanish almost everywhere. Using this second type of ellipticity ratio helps explaining the occurrence of irregularity phenomena in classical uniformly elliptic problems; these are evidently generated by what we now call soft nonuniform ellipticity. Moreover, being a nonlocal quantity, ratios of the type in (2.2.24) are more suitable to control estimates in nonautonomous problems, as these typically involve integral estimates on balls. In this respect, (2.2.23) reveals that the distance between q and p is too large with respect to α and n, therefore by (2.2.26) the ratio $\mathcal{R}_{\mathcal{H}}(z, B)$ grows fast enough to generate irregular minimizers according to the discussion made at the end of Sect. 2.2.1. Due to (2.2.23), the factor $|B|^{\alpha/n}$ in (2.2.26) is not sufficiently small to rebalance the growth with respect to $|z|$ on small balls B. Another instance of the same phenomena is provided by

The Variable Exponent Functional

$$w \mapsto \mathcal{V}(w, \Omega) := \int_\Omega |Dw|^{\mathfrak{p}(x)} \, dx \qquad (2.2.27)$$

where $\mathfrak{p} \colon \Omega \mapsto (1, \infty)$ is a continuous function. In this case we have that

$$\mathcal{R}(x, z) \lesssim \frac{\max\{q-1, 1\}}{\min\{p-1, 1\}}, \qquad q := \sup \mathfrak{p}(x), \quad p := \inf \mathfrak{p}(x),$$

and therefore the integrand $F(x, z) \equiv |z|^{\mathfrak{p}(x)}$ is uniformly elliptic in the classical, pointwise sense of Sect. 2.2.1. On the other hand, with $B \Subset \Omega$ being a ball, we note that

$$\begin{cases} \mathcal{R}(z, B) \approx |z|^{\mathfrak{p}_+(B) - \mathfrak{p}_-(B)} + 1 \\ \mathfrak{p}_+(B) := \sup_B \mathfrak{p}(x), \quad \mathfrak{p}_-(B) := \inf_B \mathfrak{p}(x). \end{cases} \qquad (2.2.28)$$

We therefore conclude that also the integrand in question is softly nonuniformly elliptic as long as $\mathfrak{p}(\cdot)$ is a non-constant function. Similarly to the double phase case, counterexamples to regularity of minima emerge when $\omega(\cdot)$, the modulus of continuity of $\mathfrak{p}(\cdot)$ i.e.,

$$|\mathfrak{p}(x) - \mathfrak{p}(y)| \leq \omega(|x - y|) \qquad \forall x, y \in \Omega,$$

fails to satisfy certain explicit decay properties. More precisely, if

$$\limsup_{\varrho \to 0} \omega(\varrho) \log \frac{1}{\varrho} = \infty, \qquad (2.2.29)$$

then examples of discontinuous minimizers are found by Zhikov [220, 221]; see also [10]. Indeed, notice that (2.2.28) implies

$$\mathcal{R}(z, B) \approx |z|^{\omega(c|B|^{1/n})} + 1, \qquad c \equiv c(n)$$

so that the oscillations of $\mathfrak{p}(\cdot)$ allow for a fast growth of the ellipticity ratio with respect to the size of the balls. The situation resembles the one for the double phase functional, where a balance between growth conditions with respect to the gradient and continuity rates of coefficients is necessary and sufficient for regularity of minima.

> **The Key Takeaway**
>
> There can be different notions of nonuniform ellipticity. Next to the classical, pointwise one, there is a softer notion aimed at explaining, for certain nonautonomous problems, lack of regularity of solutions although in presence of classical uniform ellipticity. Summarizing, we adopt the following classification:
>
> - Nonuniform ellipticity $\iff z \mapsto \mathcal{R}(x, z)$ is unbounded for at least one x.
> - Soft nonuniform ellipticity $\iff z \mapsto \mathcal{R}(x, z)$ is uniformly bounded (with respect to x) but $z \mapsto \mathcal{R}(z, B)$ is unbounded for at least one ball B.
> - Uniform ellipticity $\iff z \mapsto \mathcal{R}(z, B)$ is uniformly bounded for every ball B.

2.3 Regularity and Soft Nonuniform Ellipticity

Looking at the double phase functional $\mathcal{D}$ the twist is that when (2.2.23) are violated regularity of minima returns and the basic features of uniformly elliptic problems hold. More precisely we have the following two theorems:

Theorem 2.3.1 (Schauder Type [13, 60, 61]) *Let $u \in W^{1,1}_{\text{loc}}(\Omega)$ be a local minimizer of the functional $\mathcal{D}$ in (2.2.19). If either*

$$\frac{q}{p} \leq 1 + \frac{\alpha}{n} \tag{2.3.1}$$

or

$$u \in L^{\infty}_{\text{loc}}(\Omega) \text{ and } q \leq p + \alpha \tag{2.3.2}$$

holds, then Du is locally Hölder continuous in Ω.

Theorem 2.3.2 (Calderón-Zygmund Type [62, 74]) *Let $u \in W^{1,1}_{\text{loc}}(\Omega)$ be a distributional solution to*

$$\text{div}\,(|Du|^{p-2}Du + \mathfrak{a}(x)|Du|^{q-2}Du) = \text{div}\,(|F|^{p-2}F + \mathfrak{a}(x)|F|^{q-2}F)$$

such that $\mathcal{H}(\cdot, Du), \mathcal{H}(\cdot, F) \in L^1_{\text{loc}}(\Omega)$. If (2.3.1) holds, then

$$\mathcal{H}(\cdot, F) \in L^{\gamma}_{\text{loc}}(\Omega) \Longrightarrow \mathcal{H}(\cdot, Du) \in L^{\gamma}_{\text{loc}}(\Omega) \quad \text{for every } \gamma \geq 1. \tag{2.3.3}$$

Note that, when $\mathfrak{a}(\cdot) \equiv 0$, assertion (2.3.3) is a classical fact from Nonlinear Calderón-Zygmund theory; see [182] for a survey and a panorama of results. Furthermore, when $p = 2$ from (2.3.3) we recover a classical linear result of Calderón-Zygmund type. Note that the starting assumption $\mathcal{H}(\cdot, Du) \in L^1_{\text{loc}}(\Omega)$ is necessary already when $\mathfrak{a}(\cdot) \equiv 0$, as recently shown in [63]; see the comments after Theorem 2.6.2. Conditions (2.2.23) reveal that the occurrence of regularity/irregularity is linked to a subtle interaction between the so called gap q/p, the rate of Hölder of coefficients α and the ambient dimension n, so that a large α is able to compensate the growth of $\mathcal{R}_{\mathcal{H}}(z, B)$ with respect to $|z|$, as it is clear from (2.2.26). This time, $|B|^{\alpha/n}$ is small enough to rebalance the growth with respect to $|z|$ on shrinking balls B.

Similar considerations can be made for so-called multiphase variational integrals, that is, functionals of the type

$$\begin{cases} w \mapsto \int_{\Omega} \left(|Dw|^p + \sum_{k=1}^{m} \mathfrak{a}_k(x)|Dw|^{q_k}\right) dx \\ 0 \leq \mathfrak{a}_k(\cdot) \in C^{0,\alpha_k}_{\text{loc}}(\Omega), \quad 1 < p \leq q_1 \leq \ldots \leq q_m \end{cases} \tag{2.3.4}$$

for arbitrary number m of phases. In this case optimal conditions ensuring regularity are given by

$$\frac{q_k}{p} \leq 1 + \frac{\alpha_k}{n} \qquad (2.3.5)$$

for every $k \leq m$. Treated for the first time in [83], where this last fact was first proved, multiphase functionals have been already considered at length in the literature, see [4, 5] and related references. In particular, an optimal Calderón-Zygmund theory of the type reported in Theorem 2.3.2 has been established in [4, 71]. Finally, looking at the variable exponent functional $\mathcal{V}$ in (2.2.27), the condition

$$\limsup_{\varrho \to 0} \omega(\varrho) \log \frac{1}{\varrho} < \infty, \qquad (2.3.6)$$

which is complementary to (2.2.29), implies the local Hölder continuity of minima; moreover, assuming that $\omega(\varrho) \lesssim \varrho^\beta$ for some $\beta > 0$, implies the local gradient Hölder continuity of minima (maximal regularity and Schauder type result). Finally, assuming that the limit in (2.3.6) vanishes allows to construct a natural, intrinsic nonlinear Calderón-Zygmund theory. For an overview of regularity results on the variable exponent functional $\mathcal{V}$ in (2.2.27) we refer for instance to [180, Section 7].

Condition (2.3.6) plays a role similar to those that (2.3.1) and (2.3.5) have in the case of double and multiphase integrals, respectively. Actually, all these situations can be unified. This has been done in a recent series of papers by Hästo and Ok [137–139]. The authors consider general softly nonuniformly elliptic functionals and equations and prove local Hölder continuity of the gradients under a general condition on the regularity of the partial map $x \mapsto a(x, z)$. This condition gives back those known for the functionals $\mathcal{D}$ and $\mathcal{V}$.

The constructions implying the existence of irregular minimizers found in [102, 106, 220, 221] are linked to the occurrence of the so-called Lavrentiev phenomenon. This means that, for a suitable boundary datum $u_0 \in W^{1,\infty}(B_1(0))$

The Lavrentiev Phenomenon

$$\inf_{w \in u_0 + W_0^{1,p}(B_1(0))} \mathcal{D}(w, B_1(0))$$

$$< \inf_{w \in u_0 + W_0^{1,p}(B_1(0)) \cap W_{\text{loc}}^{1,q}(B_1(0))} \mathcal{D}(w, B_1(0)) \qquad (2.3.7)$$

holds, which is an obvious obstruction to regularity of minima. In fact, (2.3.7) implies that minima cannot belong to $W^{1,q}$ locally, so that even the basic bootstrap of regularity $W^{1,p}_{\text{loc}} \to W^{1,q}_{\text{loc}}$, fails. An interesting catch, pointed out in the approach originally introduced in [102], is that the very same condition on q/p implying regularity of minima, that is (2.3.1), implies the absence of Lavrentiev phenomenon for functionals

$$\begin{cases} w \mapsto \int_\Omega F(x, Dw)\,dx \\ F(x, z) \approx \mathcal{H}(x, z) + 1 \end{cases} \quad (2.3.8)$$

without any further assumption on $F(\cdot)$. Not even convexity $z \mapsto F(\cdot, z)$ is necessary assuming $(2.3.8)_2$. Specifically, in [102] it is proved that for every ball $B \Subset \Omega$ and $w \in W^{1,1}_{\text{loc}}(\Omega)$ such that $F(\cdot, Dw) \in L^1_{\text{loc}}(\Omega)$, it happens that

Approximation-in-Energy

$$\begin{cases} \text{there exists } \{w_k\}_k \subset C^\infty(\Omega) \text{ such that } w_k \to w \text{ strongly in } W^{1,p}(B) \\ \text{and } F(\cdot, Dw_k) \to F(\cdot, Dw) \text{ in } L^1(B) \end{cases} \quad (2.3.9)$$

provided (2.3.1) and (2.3.8) hold. In fact, somehow reversing the arrow, when considering a functional as in $(2.3.8)_1$, not necessarily satisfying $(2.3.8)_2$, assuming (2.3.9) allows to prove higher integrability of minima under suitable (p, q)-convexity assumptions on the integrand $F(\cdot)$ provided (2.3.1) takes place. See [102] for details.

After the initial approach introduced in [102], more connections between the occurrence of the approximation-in-energy property (2.3.9) and regularity of minima have been established, together with more results on the absence of Lavrentiev phenomenon [13, 35, 128, 129, 131, 137–139, 147, 148]. The analysis usually passes through the analysis of the Lavrentiev gap functional, see Remark 2.6.2. For such topics see also the recent survey [183]. For connections to related functions spaces and their abstract properties see the monograph [127].

In the last years there has been a large interest in regularity for soft uniformly ellptic problems, especially when considering double phase operators. Beyond the Hästo and Ok papers already cited above, we mention [36, 37] for different extensions. An interesting direction, which is certainly worth developing, concerns the case of manifold valued problems, considered for instance in [51, 69, 75]. In particular, in [75], the problem to minimize the double phase functional $\mathcal{D}$ amongst maps with values into the sphere has been considered. Partial regularity results and estimates of the size of the singular sets have been obtained using certain intrinsic Hausdorff measures. These measures have eventually proven to be relevant in various issues, as for instance removability of singularities [50] and in symmetry problems [22, 23]. Double phase degeneracies also appear in the setting of fully nonlinear problems, as first considered in [70] and then in [21, 66, 67, 72]. After the initial contribution in [84], nonlocal double phase problems were considered in [40, 41, 201].

> **The Key Takeaway**
> For softly uniformly elliptic problems with polynomial growth of the eigenvalues, a delicate balance between the growth with respect to the gradient and the modulus of continuity of coefficients is necessary and sufficient to guarantee regularity of solutions.

2.4 Nonlinear Potentials and A Priori Estimates

2.4.1 Potentials, Lorentz Spaces and Iterations

Let us recall a few basic definitions concerning linear and nonlinear potentials.

Definition 2.4.1 (Riesz and Havin-Mazya-Wolff Potentials) Let μ be a Borel measure with (locally) finite total mass defined on the open subset $\Omega \subset \mathbb{R}^n$, $n \geq 2$, and let $B_r(x_0) \Subset \Omega$ be a ball.

- The (truncated) Riesz potential $\mathbf{I}_\beta^\mu$ is defined by

$$\mathbf{I}_\beta^\mu(x_0, r) := \int_0^r \frac{|\mu|(B_\varrho(x_0))}{\varrho^{n-\beta}} \frac{d\varrho}{\varrho}, \qquad \beta > 0.$$

- The (nonlinear) Havin-Mazya-Wolff potential $\mathbf{W}_{\beta,p}^\mu$ is defined by

$$\mathbf{W}_{\beta,p}^\mu(x_0, r) := \int_0^r \left(\frac{|\mu|(B_\varrho(x_0))}{\varrho^{n-\beta p}} \right)^{1/(p-1)} \frac{d\varrho}{\varrho}, \qquad \beta > 0, \ p > 1.$$

The relation between truncated Riesz potentials and the classical ones is very simple, i.e.,

$$\mathbf{I}_\beta^\mu(x_0, r) \leq c(n) I_\beta(|\mu|)(x_0) = c(n) \int_{\mathbb{R}^n} \frac{d|\mu|(x)}{|x - x_0|^{n-\beta}} \quad \text{for every } r > 0. \tag{2.4.1}$$

Riesz potentials naturally occur when dealing with linear equations via fundamental solutions. On the other hand, starting by the fundamental work in [132, 133, 178], Havin-Mazya-Wolff potentials naturally intervene when passing to nonlinear equations of p-Laplacean type, that is, equations as (2.1.2) under assumptions (2.2.11), Eq. (2.2.10) being a chief model example. They can be used to control the pointwise behaviour of Sobolev functions. Moreover, a p-Laplacean version of the Wiener criterion can be derived using such potentials [145, 146, 164, 178]. The fine existence theory for non-homogeneous equations makes use of nonlinear potentials [193, 194]. Of course Riesz potentials can be realised as nonlinear potentials with a suitable choice of the parameters, i.e., $\mathbf{W}^{\mu}_{1,2} = \mathbf{I}^{\mu}_{2}$ and $\mathbf{W}^{\mu}_{1/2,2} = \mathbf{I}^{\mu}_{1}$.

Following [79], in the case μ is an integrable function we also have

Definition 2.4.2 With $t, \delta > 0$, $m, \theta \geq 0$, $\mu \in L^1(B_r(x_0))$ such that $|\mu|^m \in L^1(B_r(x_0))$, and with $B_r(x_0) \subset \mathbb{R}^n$, we define

$$\mathbf{P}^{m,\theta}_{t,\delta}(\mu; x_0, r) := \int_0^r \varrho^\delta \left(\fint_{B_\varrho(x_0)} |\mu|^m \, dx \right)^{\theta/t} \frac{d\varrho}{\varrho}.$$

Again, these potentials incorporate Havin-Mazya-Wolff potentials, and therefore Riesz potentials, in the sense

$$\mathbf{P}^{1,1}_{p-1, \frac{p\beta}{p-1}}(\mu; x_0, r) \equiv_{n,p} \mathbf{W}^{\mu}_{\beta, p}(x_0, r).$$

The behaviour of these nonlinear potentials with respect to Lorentz spaces is of interest here. The usual definition of Lorentz space $L(t, \gamma)(\Omega)$, with $t, \gamma \in (0, \infty)$, prescribes that, if $\mu \colon \Omega \to \mathbb{R}$ is a measurable function, then $\mu \in L(t, \gamma)(\Omega)$ iff

$$\|\mu\|_{t,\gamma,\Omega} := \left(t \int_0^\infty (\lambda^t |\{x \in \Omega : |\mu(x)| > \lambda\}|)^{\gamma/t} \frac{d\lambda}{\lambda} \right)^{1/\gamma} < \infty. \qquad (2.4.2)$$

See [126, 190, 209] for basic properties of such spaces. Lorentz spaces extend Lebesgue spaces and the second index tunes the first one in the sense that

$$\begin{cases} L(t_1, \gamma_1)(\Omega) \subset L(t_2, \gamma_2)(\Omega) \text{ for all } 0 < t_2 < t_1 < \infty, \ \gamma_1, \gamma_2 \in (0, \infty] \\ L(t, \gamma_1)(\Omega) \subset L(t, \gamma_2)(\Omega) \text{ for all } t \in (0, \infty), \ 0 < \gamma_1 \leq \gamma_2 \leq \infty \\ L(t, t)(\Omega) = L^t(\Omega) \text{ for all } t > 0 \end{cases}$$
(2.4.3)

hold with continuous inclusions. The catch with nonlinear potentials is now given by the following lemma, whose proof can be found in [79, Lemma 4.1].

Lemma 2.4.3 *Let $n \geq 2$, $t, \delta, \theta > 0$ be numbers such that*

$$\frac{n\theta}{t\delta} > 1. \qquad (2.4.4)$$

Let $B_{\tau_1} \Subset B_{\tau_1+r_0} \subset \mathbb{R}^n$ be two concentric balls with $\tau_1, r_0 \leq 1$, and let $\mu \in L^1(B_{\tau_1+r_0})$ be such that $|\mu|^m \in L^1(B_{\tau_0+r_0})$, where $m > 0$. Then

$$\|\mathbf{P}_{t,\delta}^{m,\theta}(\mu; \cdot, r_0)\|_{L^\infty(B_{\tau_1})} \leq \tilde{c}\|\mu\|_{L^{\frac{mn\theta}{t\delta}, \frac{m\theta}{t}}; B_{\tau_1+r_0}}^{\frac{m\theta}{t}}$$

$$\leq c(\varepsilon)\tilde{c}\|\mu\|_{L^{\frac{(1+\varepsilon)mn\theta}{t\delta}}(B_{\tau_1+r_0})}^{\frac{m\theta}{t}} \qquad (2.4.5)$$

holds for every $\varepsilon > 0$, with $\tilde{c} \equiv \tilde{c}(n, t, \delta, \theta)$.

By looking at (2.4.5) the reader might be slightly surprised by the fact that (2.4.5) covers the case $mn\theta/(t\delta) < 1$. This looks somehow in contradiction with the usual embedding properties of potentials, maximal operators and so forth. This is only apparent. In fact, in Lemma 2.4.3 one can always reduce to the case $m = 1$, so that, by (2.4.4), it is $mn\theta/(t\delta) = n\theta/(t\delta) > 1$. Indeed, notice that

$$\mathbf{P}_{t,\delta}^{m,\theta}(\mu; \cdot, r_0) = \mathbf{P}_{t,\delta}^{1,\theta}(|\mu|^m; \cdot, r_0).$$

Then, using Lemma 2.4.3 with $m = 1$, we find

$$\|\mathbf{P}_{t,\delta}^{m,\theta}(\mu; \cdot, r_0)\|_{L^\infty(B_{\tau_1})} = \|\mathbf{P}_{t,\delta}^{1,\theta}(|\mu|^m; \cdot, r_0)\|_{L^\infty(B_{\tau_1})}$$

$$\leq \tilde{c}\||\mu|^m\|_{L^{\frac{n\theta}{t\delta}, \frac{\theta}{t}}; B_{\tau_1+r_0}}^{\frac{\theta}{t}}$$

$$= \tilde{c}\|\mu\|_{L^{\frac{mn\theta}{t\delta}, \frac{m\theta}{t}}; B_{\tau_1+r_0}}^{\frac{m\theta}{t}}.$$

Note that in the last line we have used the very definition in (2.4.2). As we shall see in Sect. 2.10 below, nonlinear potentials can be used to get a priori estimates for solutions to nonlinear elliptic equations and minima of variational integrals. A key is the following lemma. It contains a pointwise version De Giorgi's geometric iteration [89]; its origins can be traced back in the seminal work of Kilpeläinen & Malý [146] and the proof can again be found in [79, Lemma 4.2].

Lemma 2.4.4 (Quantitative De Giorgi) Let $B_{r_0}(x_0) \subset \mathbb{R}^n$ be a ball, $n \geq 2$, and consider functions μ_i, $|\mu_i|^{m_i} \in L^1(B_{2r_0}(x_0))$, and constants

$$\chi > 1, \quad t \geq 1, \quad \delta_i, m_i, \theta_i > 0, \quad c_*, M_0 > 0, \quad \kappa_0, M_i \geq 0,$$

where $i \in \{1, \ldots, h\}$, $h \in \mathbb{N}$. Assume that $v \in L^t(B_{r_0}(x_0))$ is such that for all $\kappa \geq \kappa_0$, and for every ball $B_\varrho(x_0) \subset B_{r_0}(x_0)$, the inequality

$$\left(\fint_{B_{\varrho/2}(x_0)} (v-\kappa)_+^{t\chi} \, dx\right)^{1/\chi} \leq c_* M_0^t \fint_{B_\varrho(x_0)} (v-\kappa)_+^t \, dx$$

$$+ c_* \sum_{i=1}^h M_i^t \varrho^{t\delta_i} \left(\fint_{B_\varrho(x_0)} |\mu_i|^{m_i} \, dx\right)^{\theta_i} \quad (2.4.6)$$

holds, where we denote, as usual, $(v-\kappa)_+ := \max\{v-\kappa, 0\}$. If x_0 is a Lebesgue point of v in the sense that

$$\lim_{\varrho \to 0} (v)_{B_\varrho(x_0)} = v(x_0),$$

then

$$v(x_0) \leq \kappa_0 + cM_0^{\frac{\chi}{\chi-1}} \left(\fint_{B_{r_0}(x_0)} (v-\kappa_0)_+^t \, dx\right)^{1/t}$$

$$+ cM_0^{\frac{1}{\chi-1}} \sum_{i=1}^h M_i \mathbf{P}_{t,\delta_i}^{m_i,\theta_i}(\mu_i; x_0, 2r_0) \quad (2.4.7)$$

holds with $c \equiv c(n, \chi, \delta_i, \theta_i, c_*)$.

Remark 2.4.5 As we are going to see later on in Sect. 2.10, the crucial point in the above lemma, when passing from (2.4.6) to (2.4.7), is the quantified and explicit dependence on the constants M_i. We shall actually use Lemma 2.4.4 with $h = 1$, i.e., only one potential $\mathbf{P}_{t,\delta_i}^{m_i,\theta_i}$ will appear, but we prefer to report it in full generality.

Lemma 2.4.4 has a Moser iteration counterpart, which is the next one. This time potentials are not involved. The emphasis is again on the precise dependence on the multiple M_0.

Lemma 2.4.6 (Quantitative Moser) Let $B_{r_0} \subset \mathbb{R}^n$ be a ball and let $v \in L^{p/2}(B_{r_0})$, $p > 1$, be a non-negative function such that

$$\left(\fint_{B_{\varrho_1}} v^{(\gamma+p/2)\chi} \, dx\right)^{1/\chi} \leq \frac{M_0(1+\gamma)^t}{(\varrho_2 - \varrho_1)^{t_*}} \fint_{B_{\varrho_2}} v^{\gamma+p/2} \, dx$$

holds for every $\gamma \geq 0$, where M_0, t, t_*, χ are positive constants with $\chi > 1$, and where $B_{\tau_1} \Subset B_{\varrho_1} \Subset B_{\varrho_2} \Subset B_{\tau_2} \subset B_{r_0}$ are arbitrary concentric balls. Then it holds *that*

$$\|v\|_{L^\infty(B_{\tau_1})} \lesssim_{\chi,t,t_*} \left[\frac{M_0}{(\tau_2-\tau_1)^{t_*}}\right]^{\frac{2}{p}\frac{\chi}{\chi-1}} \|v\|_{L^{p/2}(B_{\tau_2})} .$$

The proof can be found in [76, Lemma 6.1] and it is nothing but the usual Moser iteration argument with a precise tracking of the dependence of the constants M_0.

2.4.2 Gradient Potential Estimates in the Uniformly Elliptic Case

In the past decade there has been an intensive research activity in the field of Nonlinear Potential Theory. In particular, pointwise gradient estimates via Riesz potentials for solutions to p-Laplacean type equations

$$-\operatorname{div} a(Du) = \mu \ (= \text{Borel measure with finite total mass})$$

under assumptions (2.2.11) have been discovered. It indeed holds

$$|Du(x_0)|^{p-1} \lesssim_{\text{data}} \mathbf{I}_1^\mu(x_0,r) + \left(\fint_{B_r(x_0)} (|Du|^2+s^2)^{t/2} \, dx\right)^{\frac{p-1}{t}} \qquad (2.4.8)$$

for a.e. $x_0 \in \Omega$ and $B_r(x_0) \Subset \Omega$, where t depends on p and $t=1$ for $p > 2-1/n$. Note that letting $r \to \infty$ in (2.4.8), recalling (2.4.1), and assuming suitable decay properties of Du at infinity yields

---

The Gradient Potential Estimate

$$-\Delta_p u = \mu \implies |Du(x_0)|^{p-1} \lesssim_{\text{data}} \int_{\mathbb{R}^n} \frac{d|\mu|(x)}{|x-x_0|^{n-1}} .$$

This is the same pointwise estimate that holds for the Poisson equation $-\Delta u = \mu$, apart from the scaling factor $p-1$. This result has been obtained first in [181] when $p=2$ and then in [98, 151, 186] for the case $p>1$. See [99] for earlier gradient estimates via Wolff potentials. Inequality in (2.4.8) holds once suitable notion of solutions to measure data problems are adopted as, in general, measure data problems do not support energy solutions (think of the fundamental solution of the Laplacean). The vectorial case for the p-Laplacean system and $p \geq 2$ is treated in [157], while parabolic cases are in [95, 150, 152, 153] after the initial results from [99] valid in the case $p = 2$. For an overview on nonlinear potential estimates we refer to [154]. Further results can be found in [185, 187]. For an interesting global counterpart of (2.4.8) involving rearrangements see [57]. Estimates employing Havin-Mazya-Wolff potentials, this time for u rather than Du, can be found in the groundbreaking work of Kilpeläinen and Malý [145, 146], see also [215] and again [99]. All such estimates can be also considered a part of what is called Nonlinear Calderón-Zygmund theory, for which we refer to the past CIME notes [182]. For an overview of the basic facts concerning Nonlinear Potential Theory and use of nonlinear potentials in the fine analysis of solutions to elliptic PDEs we refer to the classical treatises [2, 134, 164].

Size bounds often come along with companion continuity results. In this case, as it is shown in [151, 154, 157, 186, 187], it holds that

The Gradient Continuity Criterion

$$\lim_{r \to 0} \mathbf{I}_1^{\operatorname{div} a(Du)}(x, r) = 0 \text{ uniformly w.r.t. } x$$

$$\implies Du \text{ is continuous}. \qquad (2.4.9)$$

Estimate (2.4.8) immediately leads to a Lorentz space criterion for Lipschitz continuity of solutions. Indeed, notice that $\mathbf{P}_{1,1}^{1,1} \equiv \mathbf{I}_1^\mu$, so that (2.4.5) implies

$$\mu \in L(n, 1) \implies \mathbf{I}_1^\mu \in L^\infty \qquad (2.4.10)$$

locally. From this we find that Du is locally bounded via (2.4.8).

The Key Takeaway
Pointwise gradient bounds via Riesz potentials, typical of linear equations, actually hold for nonlinear problems too. No use of fundamental solutions is possible in this case and methods of proof are intrinsic and nonlinear in nature. Technical tools like De Giorgi type iterations can be framed in the setting of Nonlinear Potential Theory via the use of suitable nonlinear potentials.

2.5 Model Lipschitz Results in the Autonomous Case

Here we report a few sample yet significant regularity results for minima of nonuniformly, autonomous elliptic integrals of the type

$$w \mapsto \int_\Omega F(Dw)\,dx \qquad (2.5.1)$$

under the assumptions

$$\begin{cases} \nu[H_1(z)]^{p/2} \leq F(z) \leq L[H_1(z)]^{q/2} \\ \nu[H_1(z)]^{(p-2)/2}|\xi|^2 \leq \partial_{zz} F(z)\xi \cdot \xi & 1 < p \leq q \\ |\partial_{zz} F(z)| \leq L[H_1(z)]^{(q-2)/2}, \end{cases} \qquad (2.5.2)$$

satisfied for every $z, \xi \in \mathbb{R}^n$. The integrand $F: \mathbb{R}^n \to [0, \infty)$ is assumed to be C^2-regular. Integrands and functionals satisfying $(2.5.2)_1$ are nowadays known as functionals with (p, q)-growth conditions, a terminology introduced by Marcellini in [170]. Conditions (2.5.2) imply the following control or the ellipticity ratio of $F(\cdot)$:

$$\mathcal{R}_F(z) \lesssim_{\text{data}} |z|^{q-p} + 1. \qquad (2.5.3)$$

Similarly to what we have seen in Sect. 2.3, the crucial assumption to get regularity of minimizers is a bound of the type

$$\frac{q}{p} < 1 + \mathrm{o}(n), \qquad (2.5.4)$$

with

$$\mathrm{o}(n) \approx \frac{1}{n} \qquad (2.5.5)$$

for n large. As already explained, bounds as (2.5.4) serve to limitate the growth of the ellipticity ratio $\mathcal{R}_F(\cdot)$ with respect to the gradient variable when proving gradient boundedness of minima. Such bounds are necessary already in the autonomous case, as shown by explicit counterexamples [114, 170]. A model regularity result is the following:

Theorem 2.5.1 (Marcellini [170]) *Let* $u \in W^{1,1}_{\text{loc}}(\Omega)$ *be a minimizer of the functional in* (2.5.1) *under assumptions* (2.5.2) *and*

$$\frac{q}{p} < 1 + \frac{2}{n}. \tag{2.5.6}$$

Then Du *is locally bounded in* Ω.

While in nonautonomous case the optimal bound on q/p is known to be (2.3.1) by the counterexamples in [102, 106], for autonomous functionals (2.5.1) the optimal bound is still unknown. An example of progress in this direction is in the following:

Theorem 2.5.2 (Bella and Schäffner [17, 19, 199]) *Let* $u \in W^{1,1}_{\text{loc}}(\Omega)$ *be a minimizer of the functional in* (2.5.1) *under assumptions* (2.5.2) *and*

$$\frac{q}{p} < 1 + \frac{2}{n-1}. \tag{2.5.7}$$

Then Du *is locally bounded in* Ω.

The one in (2.5.7) is the best bound found on q/p up to now. Advances are in [46, 87], and are obtained under special structure conditions. The bound in (2.5.7) works in the vectorial case too, where it allows to prove partial regularity [196].

Bounds of the type in (2.5.4) also appear when proving lower order regularity. In this case the presence of coefficients plays no role. Indeed we have

Theorem 2.5.3 (Hirsch and Schäffner [135]) *Let* $u \in W^{1,1}_{\text{loc}}(\Omega)$ *be minimizer of the functional in* (2.1.1) *with* $\mu = 0$, *where* $F \colon \Omega \times \mathbb{R}^n \to \mathbb{R}$ *is Carathéodory regular and satisfies*

$$\begin{cases} \nu |z|^p \leq F(x,z) \leq L|z|^q + 1 \\ F(x, 2z) \leq LF(x,z) + L, \end{cases}$$

(continued)

Theorem 2.5.3 (continued)
for $(x, z) \in \Omega \times \mathbb{R}^n$, and where $1 < p \leq q$. Assume that

$$\frac{1}{p} - \frac{1}{q} \leq \frac{1}{n-1}. \tag{2.5.8}$$

Then u is locally bounded in Ω.

The bound in (2.5.8) is sharp for L^∞-regularity in view of the available counterexamples [114, 170]. As in the standard case $p = q$ [115], this time no regularity on $F(\cdot)$ is required. In particular, no convexity of $z \mapsto F(\cdot, z)$ is needed and not even continuity is required on $x \mapsto F(x, \cdot)$.

Finally, we mention that, as it already happened in (2.3.2), when starting by a solution which is already more regular, then bounds for Lipschitz continuity can be improved by using underlying interpolation effects. For instance, in perfect analogy with (2.3.2) we have

Theorem 2.5.4 (Choe [52]) *Let $u \in W^{1,1}_{\text{loc}}(\Omega) \cap L^\infty_{\text{loc}}(\Omega)$ be a minimizer of the functional in (2.5.1) under assumptions (2.5.2) and*

$$1 < p \leq q < p + 1. \tag{2.5.9}$$

Then Du is locally bounded in Ω.

Further results using non-dimensional bounds as (2.5.9) can be found for instance in [45].

Functionals and equations with (p, q)-growth nonuniform ellipticity have been the object of intensive investigation over the last decades. We refer to the surveys [174–176, 180, 183] for an overview. More recently, nonstandard growth conditions have been examined in the setting of nonlocal and mixed operators, see for instance [39, 40, 48, 49, 81, 84, 104, 188]. We forecast several further developments in this direction.

The Key Takeaway
Solutions to nonuniformly elliptic problems with polynomial (p, q)-growth of the eigenvalues are still regular provided the gap q/p is not to far from 1. This means that the ellipticity ratio cannot grow too fast with respect to

(continued)

the gradient, as framed in (2.5.3). While the asymptotic in (2.5.4)–(2.5.5) is known to be optimal, the sharp bound on q/p for Lipschitz regularity in the autonomous case remains unknown.

2.6 Nonuniformly Elliptic Schauder Estimates [79]

In treating Schauder type estimates we shall first present the results in [79], in this section. We shall eventually consider the more recent and improved ones in [82] in the next section. The reason for this is that we shall eventually give a sketch of the proofs in Sect. 2.10.3. This is easier in the case of the results in [79] but becomes less straightforward in the case of those from [82], which features more technically involved proofs.

2.6.1 Hopf, Caccioppoli and Schauder, Reloaded

So-called Schauder estimates for linear elliptic equations

$$-\operatorname{div}(\mathsf{A}(x)Du) = 0, \qquad \mathsf{A}(\cdot) \approx \mathbb{I}_d \qquad (2.6.1)$$

are a basic tool in PDE theory and they play a role in a variety of situations. For instance, they are fundamental ingredient in the proof of higher regularity of solutions to nonlinear elliptic equations. When referring to (2.6.1), the basic assertion is

$$\mathsf{A}(\cdot) \in C^{0,\alpha} \implies Du \in C^{0,\alpha} \qquad (2.6.2)$$

in local and/or global fashion, eventually depending on boundary conditions. In other words, the gradient of solutions inherits the regularity of coefficients, as long as the operator without coefficients allows (this is the case of "frozen coefficients"). Several variants of (2.6.2) are possible, for instance when considering linear equations in non-divergence form, which is in fact the most classical case. Schauder estimates are actually a classic achievement of Hopf [136], Caccioppoli [42] and Schauder [197, 198]; see also the work of Giraud [119, 120]. Modern proofs can be found in [44, 118, 206, 214].

After the classical, linear era, nonlinear versions of Schauder estimates, i.e., for equations of the type (2.1.2), were established in the uniformly elliptic case—see Giaquinta and Giusti's [116, 117], Manfredi's [165, 166] and Lieberman's [160] papers for full generality. A model result can be obtained by looking at solutions to

$$-\operatorname{div}(\mathfrak{c}(x)|Du|^{p-2}Du) = 0, \quad 0 < \nu \leq \mathfrak{c}(\cdot) \in C^{0,\alpha}$$

for which (2.6.2) holds locally as long as $\alpha < \alpha_0$ and solutions to (2.2.10) are C^{1,α_0} (see for instance [154, 165] for the precise meaning of this assertion).

All the known proofs of Schauder estimates rely on a two-step argument based on comparing the solution to the original problem with the solutions of problems with constant coefficients, that typically enjoy good a priori estimates. Combining these with suitable comparison estimates, and iterating, gradient Hölder continuity follows. The crucial point in this procedure is that all the estimates involved are homogeneous, and this allows to combine and iterate them. In turn, this is a direct consequence the uniform ellipticity of the equations considered. Observe that this scheme is common to both the linear and the nonlinear case.

On the other hand, when turning to the nonuniformly elliptic case, estimates are typically affected by a lack of homogeneity. An example is obviously given by functionals with (p, q)-growth, where two different exponents occur, generating an obvious lack of scaling. This is essentially the reason for which the problem of establishing nonlinear Schauder estimates has remained open for a long time, finally getting an answer in [79]. Here we present a few facts from this last paper. We shall actually include two model results for the sake of readability; the interested reader will find more in [79]. We start by the simplest possible variational case, that nevertheless already contains all the core difficulties and peculiarities of more general ones. We consider

$$\begin{cases} w \mapsto \int_\Omega \mathfrak{c}(x)F(Dw)\,dx, & 0 < \nu \leq \mathfrak{c}(\cdot) \leq L \\ |\mathfrak{c}(x_1) - \mathfrak{c}(x_2)| \leq L|x_1 - x_2|^\alpha, & \alpha \in (0, 1], \end{cases} \quad (2.6.3)$$

for every choice of $x_1, x_2 \in \Omega$, where $F(\cdot)$ satisfies

$$\begin{cases} F(\cdot) \in C^1(\mathbb{R}^n) \cap C^2(\mathbb{R}^n \setminus \{0\}) \\ \nu[H_s(z)]^{p/2} \leq F(z) \leq L[H_s(z)]^{q/2} + L[H_s(z)]^{p/2} \\ \nu[H_s(z)]^{(p-2)/2}|\xi|^2 \leq \partial_{zz}F(z)\xi \cdot \xi \\ |\partial_{zz}F(z)| \leq L[H_s(z)]^{(q-2)/2} + L[H_s(z)]^{(p-2)/2} \end{cases} \quad (2.6.4)$$

for all $z, \xi \in \mathbb{R}^n$, $|z| \neq 0$, $\xi \in \mathbb{R}^n$, and $H_s(\cdot)$ is defined in (2.1.7).

Theorem 2.6.1 *Let $u \in W^{1,1}_{\text{loc}}(\Omega)$ be a minimizer of the functional in (2.6.3), under assumptions (2.6.4). If*

$$\frac{q}{p} \leq 1 + \frac{1}{5}\left(\frac{\alpha}{n}\right)^2, \tag{2.6.5}$$

then Du is locally Hölder continuous in Ω. Moreover

$$\|Du\|_{L^\infty(B_{r/2})} \lesssim_{\text{data},\alpha} \left(\fint_{B_r} F(Dw)\,dx\right)^\kappa + 1 \tag{2.6.6}$$

holds whenever $B_r \Subset \Omega$ is a ball with $r \leq 1$, where $\kappa \equiv \kappa(n, p, q, \alpha) \geq 1$.

When the problem becomes non-degenerate, that is, when $s > 0$, we can then recover a result as in (2.6.2).

Theorem 2.6.2 *In the setting of Theorem 2.6.1 with $\alpha \in (0, 1)$, assume also that $p \geq 2$, $s > 0$, and that $\partial_{zz}F(\cdot)$ is continuous. Then $u \in C^{1,\alpha}_{\text{loc}}(\Omega)$.*

The variational case is particularly natural to start with as it provides a natural setting for nonuniformly elliptic problems, allowing to put all the emphasis on a priori regularity estimates. In fact, it dispenses us to distinguish from various definitions of solutions. Minimality automatically selects the right solutions as it involves a definition that works against all possible competitors (variations). The situation changes when considering equations, and already in the uniformly elliptic case. Indeed, by considering the distributional version of (2.2.10)

$$\int_\Omega |Du|^{p-2} Du \cdot D\varphi \, dx = 0 \quad \forall \, \varphi \in C_0^\infty(\Omega) \tag{2.6.7}$$

we notice that this makes actually sense whenever $u \in W^{1,p-1}_{\text{loc}}(\Omega)$; we are therefore in the realm of Definition 2.1.2. Such solutions, not necessarily belonging to the natural space $u \in W^{1,p}_{\text{loc}}(\Omega)$ as the standard energy ones from Definition 2.2.1, are called very weak solutions. Although some authors firmly believed of the contrary [144], in general very weak solutions to (2.2.10) do not belong $u \in W^{1,p}_{\text{loc}}(\Omega)$ [63]. Note that energy solutions are those allowing to test (2.6.7) with functions φ that are proportional to the solution itself. This is crucial in order to get regularity estimates, even basic ones such as Caccioppoli inequalities.

This very basic issue naturally reappears in the nonuniformly elliptic setting. As an example, we consider an equation of the type

$$-\operatorname{div} a(x, Du) = 0 \quad \text{in } \Omega \subset \mathbb{R}^n. \tag{2.6.8}$$

Here the vector field $a \colon \Omega \times \mathbb{R}^n \to \mathbb{R}^n$ is of class $C^1(\mathbb{R}^n \setminus \{0_{\mathbb{R}^n}\})$ with respect to gradient variable, and satisfies

$$\begin{cases} |a(x,z)| + |\partial_z a(x,z)|[H_s(z)]^{1/2} \\ \quad \leq L[H_s(z)]^{(q-1)/2} + L[H_s(z)]^{(p-1)/2} \\ \nu[H_s(z)]^{(p-2)/2}|\xi|^2 \leq \partial_z a(x,z)\xi \cdot \xi \\ |a(x_1,z) - a(x_2,z)| \\ \quad \leq L|x_1-x_2|^\alpha ([H_s(z)]^{(q-1)/2} + [H_s(z)]^{(p-1)/2}) \end{cases} \tag{2.6.9}$$

whenever $x_1, x_2 \in \Omega$ and $z, \xi \in \mathbb{R}^n$, $|z| \neq 0$. A natural ambiguity on the notion of weak solution now arises. Is this assumed to belong to $W^{1,p}$, as in the case of minima of functionals with (p,q)-growth? Or it is rather to be taken from $W^{1,q}$? In this last case, in the distributional form of (2.6.8), that is

$$\int_\Omega a(x, Du) \cdot D\varphi \, dx = 0, \quad \forall \varphi \in C_0^\infty(\Omega)$$

we can take test functions φ that are proportional to the solution u and again derive basic energy estimates. Therefore considering $W^{1,q}$-solutions corresponds to take energy solutions in this setting. Issues concerning the various notions of weak solutions to problems with (p,q)-growth conditions are examined in depth in Zhikov's papers [219–222]. In view of this discussion, there are now two possible approaches usually pursued in the literature

- Prove a priori estimates for more regular, i.e. $W^{1,q}$-solutions [158, 170, 206], thereby starting by energy solutions.
- Simultaneously proving both the existence and regularity for solutions to assigned boundary value problems, say for instance Dirichlet problems [14, 142, 170].

No distinction between the two approaches takes place when $p = q$. The second point from the above couple is the one we are now proposing, as already done in [79] (and in most of the literature). For this we consider the Dirichlet problem

$$\begin{cases} -\operatorname{div} a(x, Du) = 0 & \text{in } \Omega \\ u \equiv u_0 & \text{on } \partial\Omega, \end{cases} \quad u_0 \in W^{1, \frac{p(q-1)}{p-1}}(\Omega), \tag{2.6.10}$$

where $\Omega \subset \mathbb{R}^n$ is a bounded and Lipschitz domain.

Theorem 2.6.3 *Assume that the vector field $a(\cdot)$ satisfies (2.6.9). If*

$$\frac{q}{p} \leq 1 + \frac{p-1}{10p}\left(\frac{\alpha}{n}\right)^2, \quad (2.6.11)$$

then there exists a solution $u \in W^{1,p}(\Omega)$ to the Dirichlet problem (2.6.10), such that Du is locally Hölder continuous in Ω. Moreover, the estimate

$$\|Du\|_{L^\infty(\Omega_0)} \lesssim_{\text{data},\alpha} \frac{1}{[\text{dist}(\Omega_0, \partial\Omega)]^\kappa} \left(\int_\Omega (|Du_0|+1)^{\frac{p(q-1)}{p-1}} dx + 1\right)^\kappa$$

holds whenever $\Omega_0 \Subset \Omega$ is an open subset, with $\kappa \equiv \kappa(n, p, q, \alpha) \geq 1$.

Theorem 2.6.4 *In the setting of Theorem 2.6.3 with $\alpha \in (0, 1)$, assume also that $p \geq 2$, $s > 0$, and that $\partial_z A(\cdot)$ is continuous on $\Omega \times \mathbb{R}^n$. Then $u \in C^{1,\alpha}_{\text{loc}}(\Omega)$.*

In the case $p = q = 2$, Theorem 2.6.4 is a classical result from the uniformly elliptic theory [42, 116, 117, 136, 197]. The main point in the proof of Theorems 2.6.1–2.6.4 is that plain perturbation arguments, of the type used in the nonuniformly elliptic setting, do not work due to the aforementioned lack of homogeneous estimates (estimates that are not invariant under scaling). To overcome this point, in [79] a new approach via renormalized Caccioppoli inequalities in fractional Sobolev spaces is considered. After this, a fractional version of De Giorgi's geometric iteration technique, and the use of nonlinear potentials via Lemma 2.4.4, leads to establish gradient boundedness. Once this is achieved, one can then adapt more standard perturbation techniques. We shall expand on this in Sect. 2.10.3 below.

Remark 2.6.5 On the contrary of (2.6.5), the bound in (2.6.11) deteriorates as $p \to 1$, forcing $q/p \to 1$. In both cases the factors $1/5$ and $1/10$ have been introduced in order to simplify the presentation and can be replaced by slightly larger quantities. For instance, (2.6.5) can be replaced by

$$\frac{q}{p} < 1 + \mathrm{k}(n, p, q, \alpha)\left(\frac{\alpha}{n}\right)^2, \quad \frac{1}{5} < \mathrm{k}(n, p, q, \alpha) < 1$$

where $\mathrm{k}(n, p, q, \alpha)$ is an involved function of its parameters. See [79, Proposition 7.1].

2.6.2 Relaxation

For general functionals of the type

$$w \mapsto \mathcal{F}_{\mathbf{x}}(w, \Omega) := \int_{\Omega} F(x, Dw)\,dx \qquad (2.6.12)$$

Theorem 2.6.1 continues to hold provided

$$\begin{cases} z \mapsto F(x, z) \text{ satisfies } (2.6.4) \text{ uniformly with respect to } x \in \Omega \\ |\partial_z F(x_1, z) - \partial_z F(x_2, z)| \\ \leq L|x_1 - x_2|^{\alpha}([H_\mu(z)]^{(q-1)/2} + [H_\mu(z)]^{(p-1)/2}) \end{cases} \qquad (2.6.13)$$

and the approximation-in-energy property (2.3.9) holds for the minimizer in question u. This is a in a sense a necessary and natural assumption in view of the potential occurrence of the Lavrentiev phenomenon (2.3.7). In fact, more can be said. The first step is to define the so-called relaxed functional [29, 105, 107, 167–169, 200]

$$\overline{\mathcal{F}}_{\mathbf{x}}(w, U) := \inf_{\{w_k\} \subset W^{1,q}(U)} \left\{ \liminf_k \mathcal{F}_{\mathbf{x}}(w_k, U) : w_k \rightharpoonup w \text{ in } W^{1,p}(U) \right\} \qquad (2.6.14)$$

for every open subset $U \subset \Omega$ and $w \in W^{1,1}_{\text{loc}}(\Omega)$. Considering this type of lower semicontinuous envelope goes back to Lebesgue, Caccioppoli, Serrin and De Giorgi. In the nonuniformly elliptic setting, it appears for the first time in the work of Marcellini [167, 168]. Accordingly, the Lavrentiev gap functional is defined by

$$\mathcal{L}_{\mathcal{F}_{\mathbf{x}}}(w, U) := \begin{cases} \overline{\mathcal{F}}_{\mathbf{x}}(w, U) - \mathcal{F}_{\mathbf{x}}(w, U) & \text{if } \mathcal{F}_{\mathbf{x}}(w, U) < \infty \\ 0 & \text{if } \mathcal{F}_{\mathbf{x}}(w, U) = \infty. \end{cases} \qquad (2.6.15)$$

for every $w \in W^{1,1}(U)$. We have

- By $W^{1,p}$-weak lower semicontinuity of $\mathcal{F}_{\mathbf{x}}(\cdot, U)$ we have $\mathcal{F}_{\mathbf{x}}(\cdot, U) \leq \overline{\mathcal{F}}_{\mathbf{x}}(\cdot, U)$ and therefore $\mathcal{L}_{\mathcal{F}_{\mathbf{x}}}(\cdot, U) \geq 0$. It follows that $w \in W^{1,p}(U)$ whenever $\overline{\mathcal{F}}_{\mathbf{x}}(w, U)$ is finite.
- A function $u \in W^{1,p}(\Omega)$ is a minimizer of $\overline{\mathcal{F}}_{\mathbf{x}}(\cdot, \Omega)$ iff $\overline{\mathcal{F}}_{\mathbf{x}}(u, \Omega)$ is finite and $\overline{\mathcal{F}}_{\mathbf{x}}(u, \Omega) \leq \overline{\mathcal{F}}_{\mathbf{x}}(w, \Omega)$ holds whenever $w \in u + W_0^{1,1}(\Omega)$.
- Assume that $u \in W^{1,1}_{\text{loc}}(\Omega)$ is a minimizer of the original functional $\mathcal{F}_{\mathbf{x}}$ in (2.6.12) such that $\mathcal{L}_{\mathcal{F}_{\mathbf{x}}}(u, B) \equiv 0$ for every ball $B \Subset \Omega$. Then

$$\overline{\mathcal{F}}_{\mathbf{x}}(u, B) = \mathcal{F}_{\mathbf{x}}(u, B) \leq \mathcal{F}_{\mathbf{x}}(w, B) \leq \overline{\mathcal{F}}_{\mathbf{x}}(w, B) \qquad (2.6.16)$$

holds whenever $w \in u + W_0^{1,1}(B)$. Therefore u also minimize $\overline{\mathcal{F}_{\mathbf{x}}}(\cdot, B)$ for every ball $B \Subset \Omega$.

This last point allows to follow a natural strategy, devised in [102]: first, proving regularity of minima of the relaxed functional, and, from this, deducing the regularity of minima when the Lavrentiev gap vanishes. This is in the next results, still taken from [79].

Theorem 2.6.6 *Let $u \in W^{1,p}(\Omega)$ be a minimizer of the functional $\overline{\mathcal{F}_{\mathbf{x}}}(\cdot, \Omega)$, where Ω is a Lipschitz regular domain, and under assumptions* (2.6.5) *and* (2.6.13). *Then Du is locally Hölder continuous in Ω. Moreover*

$$\|Du\|_{L^\infty(\Omega_0)} \lesssim_{\text{data},\alpha} \frac{1}{[\text{dist}(\Omega_0, \partial\Omega)]^\kappa} \left[\overline{\mathcal{F}_{\mathbf{x}}}(u, \Omega) + 1\right]^\kappa \quad (2.6.17)$$

holds whenever $\Omega_0 \Subset \Omega$ is an open subset, where $\kappa \equiv \kappa(n, p, q, \alpha) \geq 1$.

Corollary 2.6.7 *Let $u \in W^{1,1}_{\text{loc}}(\Omega)$ be a minimizer of the functional $\mathcal{F}_{\mathbf{x}}$ in* (2.6.12), *under assumptions* (2.6.5) *and* (2.6.13). *Assume that*

$$\mathcal{L}_{\mathcal{F}_{\mathbf{x}}}(u, B) = 0 \quad \text{holds for every ball } B \Subset \Omega. \quad (2.6.18)$$

Then Du is locally Hölder continuous in Ω. Moreover,

$$\|Du\|_{L^\infty(B_t)} \lesssim_{\text{data},\alpha} \frac{1}{(r-t)^\kappa} \left[\mathcal{F}_{\mathbf{x}}(u, B_r) + 1\right]^\kappa \quad (2.6.19)$$

holds whenever $B_t \Subset B_r \Subset \Omega$ are concentric balls with $r \leq 1$, where $\kappa \equiv \kappa(n, p, q, \alpha) \geq 1$.

To make the Theorem 2.6.6 and Corollary 2.6.7 really effective, it remains to understand when (2.6.18) holds. This is in fact equivalent to require that the approximation-in-energy (2.3.9) takes place for u on every ball $B \Subset \Omega$. This in fact happens in several common situations. For instance, via a simple convolution argument it is easy to see [102] that if there exists a convex function $G \colon \mathbb{R}^n \to [0, \infty)$ such that

$$G(z) \lesssim F(x, z) \lesssim G(z) + 1,$$

then $\mathcal{L}_{\mathcal{F}_x}(\cdot, B) \equiv 0$ holds for every ball $B \Subset \Omega$. This is the case of Theorem 2.6.1, that in fact follows from Corollary 2.6.7. The property is in a sense self-reproductive, as if $F(\cdot)$ is an integrand such that (2.3.9) holds, then (2.3.9) also holds for any other Carathéodory integrand $F_1(\cdot)$ for which

$$F(x, z) \lesssim F_1(x, z) \lesssim F(x, z) + 1$$

holds true. For instance (2.3.9) holds whenever

$$|z|^p + \mathfrak{a}(x)|z|^q \lesssim F(x, z) \lesssim |z|^p + \mathfrak{a}(x)|z|^q + 1$$

is satisfied, as described before (2.3.8). See Sect. 2.3 and [102] for further cases in which (2.3.9) holds.

2.6.3 Non-Differentiable Functionals

A classical series of results in the Calculus of Variations are concerned with functionals of the type

$$w \mapsto \int_\Omega [F(Dw) + \mathrm{h}(x, w)] \, dx \qquad (2.6.20)$$

treated for instance in [47, 116, 130, 149, 192, 207]. The situation is as follows. When the Carathéodory function $\mathrm{h}(x, y)$ is differentiable with respect to the second variable, regularity of minima can be proved using the Euler-Lagrange equation, that is

$$\mathrm{div}\, \partial_z F(Du) = \partial_y \mathrm{h}(x, u) \,. \qquad (2.6.21)$$

Things change when $\mathrm{h}(\cdot)$ is not differentiable, let's say it only satisfies

$$|\mathrm{h}(x, y_1) - \mathrm{h}(x, y_2)| \leq L|y_1 - y_2|^\alpha, \quad \alpha \in (0, 1] \qquad (2.6.22)$$

and Eq. (2.6.21) cannot be used simply because the right-hand side does not exist (unless $\alpha = 1$). In the uniformly elliptic case a breakthrough was achieved in [116] where it was proved that, if $\partial_z F(\cdot)$ satisfies (2.2.11) with $p = 2$ and $\mathrm{h}(\cdot)$ is a bounded and Hölder continuous functions, then Du is locally Hölder continuous in Ω. Later on, the literature reports several extensions of this kind of result in the case $p \neq 2$, starting by Manfredi [165, 166]. All such methods fail when $F(\cdot)$ is not uniformly elliptic, for the same reasons explained in Sect. 2.6 (lack of homogeneous estimates). Here we report the first gradient regularity result for minima of non-differentiable functionals of the type in (2.6.20), obtained in [79].

Theorem 2.6.8 *Let* $u \in W^{1,1}_{\text{loc}}(\Omega)$ *be a minimizer of the functional in* (2.6.20), *under assumptions* (2.6.4) *and* (2.6.22) *and*

$$\frac{q}{p} \leq 1 + \frac{1}{5}\left(1 - \frac{\alpha}{p}\right)\frac{\alpha}{n}. \qquad (2.6.23)$$

Then Du is locally Hölder continuous in Ω. Moreover,

$$\|Du\|_{L^\infty(B_t)} \lesssim_{\text{data},\alpha} \frac{1}{(r-t)^\kappa} \left[\|F(Du)\|_{L^1(B_r)} + \|\text{h}(\cdot, u)\|_{L^1(B_r)} + 1\right]^\kappa$$

holds whenever $B_t \Subset B_r \Subset \Omega$ are concentric balls with $r \leq 1$, where $\kappa \equiv \kappa(n, p, q, \alpha) \geq 1$.

The Key Takeaway
For uniformly elliptic problems Schauder estimates follow via perturbation methods. This is not possible in the nonuniformly elliptic case, due to basic lack of homogeneous estimates. They actually fail when not assuming a bound of the type $q/p \leq 1 + \text{o}(\alpha/n)$, encoding a direct and delicate interaction between growth of the ellipticity ratio and Hölder continuity of coefficients (see Sect. 2.2.4). On the positive side, Schauder estimates can be proved in the nonuniformly elliptic case by means of fractional spaces based techniques (see Sect. 2.10.3).

2.7 The Sharp Growth Rate in Nonuniformly Elliptic Schauder Theory [82]

When looking at Theorem 2.6.1, and in particular at the gap bound (2.6.5), it is natural to wonder whether this can be improved in at least something of the form

$$\frac{q}{p} < 1 + \text{o}(n, \alpha), \qquad \text{o}(\alpha, n) \approx \frac{\alpha}{n} \qquad (2.7.1)$$

that would be in line with (2.5.6) and (2.5.7). In other words, in something that approaches 1 linearly, rather that quadratically, with respect to α/n. In fact, we have

Theorem 2.7.1 *Let $u \in W^{1,p}(\Omega)$ be a minimizer of the functional $\overline{\mathcal{F}_x}(\cdot, B)$ in (2.6.14) for every ball $B \Subset \Omega$ under assumptions (2.6.13). If*

$$\frac{q}{p} < 1 + \frac{\alpha}{n}, \qquad (2.7.2)$$

then with $B_\varrho \Subset \Omega$ being a ball, $\varrho \in (0, 1]$, and $\theta \in (0, 1)$,

- *The local Lipschitz estimate*

$$\|Du\|_{L^\infty(B_{\varrho/2})} \lesssim_{\mathrm{data},\alpha} \varrho^{-n\kappa} \overline{\mathcal{F}_x}(u, B_\varrho)^\kappa + 1$$

holds with $\kappa \equiv \kappa(n, p, q, \alpha) \geq 1$.
- *The local Hölder estimate*

$$[Du]_{0,\alpha_*; B_{\theta\varrho}} \leq c \qquad (2.7.3)$$

holds with both $\alpha_* \in (0, 1)$ and $c \geq 1$ depending on $n, p, q, \alpha, L, \theta$ and $\overline{\mathcal{F}_x}(u, B_\varrho)$.
- *If, in addition, $s > 0$ and $\partial_{zz} F(\cdot)$ is continuous, then $u \in C^{1,\alpha}_{\mathrm{loc}}(\Omega)$ when $\alpha < 1$. Specifically,*

$$[Du]_{0,\alpha; B_{\theta\varrho}} \leq c_*$$

holds as in (2.7.3), where c_* depends also on the modulus of continuity of $\partial_{zz} F(\cdot)$ and on s.

Needless to say, in the case the Lavrentiev gap functional vanishes, we recover the result for the original functional.

Corollary 2.7.2 *Theorem 2.7.1 extends to any minimizer u of the original functional in (2.6.12) provided $\mathcal{L}_{\mathcal{F}_x}(u, B) = 0$ holds for every ball $B \Subset \Omega$.*

Theorem 2.7.1 is sharp, in particular with respect to (2.7.2), in view of the counterexamples in [102, 106]. Comments on the proof of Theorem 2.7.1 can be found in Sect. 2.10.5.

Theorem 2.6.8 *Let* $u \in W^{1,1}_{\text{loc}}(\Omega)$ *be a minimizer of the functional in* (2.6.20), *under assumptions* (2.6.4) *and* (2.6.22) *and*

$$\frac{q}{p} \leq 1 + \frac{1}{5}\left(1 - \frac{\alpha}{p}\right)\frac{\alpha}{n}. \qquad (2.6.23)$$

Then Du is locally Hölder continuous in Ω. Moreover,

$$\|Du\|_{L^\infty(B_t)} \lesssim_{\text{data},\alpha} \frac{1}{(r-t)^\kappa}\left[\|F(Du)\|_{L^1(B_r)} + \|\mathrm{h}(\cdot,u)\|_{L^1(B_r)} + 1\right]^\kappa$$

holds whenever $B_t \Subset B_r \Subset \Omega$ are concentric balls with $r \leq 1$, where $\kappa \equiv \kappa(n,p,q,\alpha) \geq 1$.

The Key Takeaway
For uniformly elliptic problems Schauder estimates follow via perturbation methods. This is not possible in the nonuniformly elliptic case, due to basic lack of homogeneous estimates. They actually fail when not assuming a bound of the type $q/p \leq 1 + \mathrm{o}(\alpha/n)$, encoding a direct and delicate interaction between growth of the ellipticity ratio and Hölder continuity of coefficients (see Sect. 2.2.4). On the positive side, Schauder estimates can be proved in the nonuniformly elliptic case by means of fractional spaces based techniques (see Sect. 2.10.3).

2.7 The Sharp Growth Rate in Nonuniformly Elliptic Schauder Theory [82]

When looking at Theorem 2.6.1, and in particular at the gap bound (2.6.5), it is natural to wonder whether this can be improved in at least something of the form

$$\frac{q}{p} < 1 + \mathrm{o}(n,\alpha), \qquad \mathrm{o}(\alpha,n) \approx \frac{\alpha}{n} \qquad (2.7.1)$$

that would be in line with (2.5.6) and (2.5.7). In other words, in something that approaches 1 linearly, rather that quadratically, with respect to α/n. In fact, we have

Theorem 2.7.1 *Let $u \in W^{1,p}(\Omega)$ be a minimizer of the functional $\overline{\mathcal{F}}_\mathbf{x}(\cdot, B)$ in (2.6.14) for every ball $B \Subset \Omega$ under assumptions (2.6.13). If*

$$\frac{q}{p} < 1 + \frac{\alpha}{n}, \qquad (2.7.2)$$

then with $B_\varrho \Subset \Omega$ being a ball, $\varrho \in (0, 1]$, and $\theta \in (0, 1)$,

- *The local Lipschitz estimate*

$$\|Du\|_{L^\infty(B_{\varrho/2})} \lesssim_{\text{data},\alpha} \varrho^{-n\kappa} \overline{\mathcal{F}}_\mathbf{x}(u, B_\varrho)^\kappa + 1$$

holds with $\kappa \equiv \kappa(n, p, q, \alpha) \geq 1$.
- *The local Hölder estimate*

$$[Du]_{0,\alpha_*; B_{\theta\varrho}} \leq c \qquad (2.7.3)$$

holds with both $\alpha_* \in (0, 1)$ and $c \geq 1$ depending on $n, p, q, \alpha, L, \theta$ and $\overline{\mathcal{F}}_\mathbf{x}(u, B_\varrho)$.
- *If, in addition, $s > 0$ and $\partial_{zz} F(\cdot)$ is continuous, then $u \in C^{1,\alpha}_{\text{loc}}(\Omega)$ when $\alpha < 1$. Specifically,*

$$[Du]_{0,\alpha; B_{\theta\varrho}} \leq c_*$$

holds as in (2.7.3), where c_* depends also on the modulus of continuity of $\partial_{zz} F(\cdot)$ and on s.

Needless to say, in the case the Lavrentiev gap functional vanishes, we recover the result for the original functional.

Corollary 2.7.2 *Theorem 2.7.1 extends to any minimizer u of the original functional in (2.6.12) provided $\mathcal{L}_{\mathcal{F}_\mathbf{x}}(u, B) = 0$ holds for every ball $B \Subset \Omega$.*

Theorem 2.7.1 is sharp, in particular with respect to (2.7.2), in view of the counterexamples in [102, 106]. Comments on the proof of Theorem 2.7.1 can be found in Sect. 2.10.5.

2.8 Universal Lorentz Conditions for Lipschitz Regularity

2.8.1 Stein Type Theorems

An essentially equivalent formulation of a classical theorem of Stein [208] asserts

Stein's Theorem

$$\triangle u \in L(n, 1) \Longrightarrow Du \text{ is continuous}. \tag{2.8.1}$$

This is sharp [53]. Surprisingly enough, this condition is in a sense universal, and fo instance holds for energy solutions of the p-Laplacean system

Nonlinear Stein Theorem [155]

$$\triangle_p u \in L(n, 1) \Longrightarrow Du \text{ is continuous}. \tag{2.8.2}$$

More in general, once the proper functional setting and the definition of solution is settled, the result continues to hold—at C^1 or $C^{0,1}$ regularity level—when replacing Eq. (2.8.1) with (2.1.2) in the case of

- A p-Laplacean type operator as in (2.2.11) with Dini-continuous coefficients $x \mapsto a(x, \cdot)$ [97, 151, 153, 154].
- The p-Laplacean system (again with Dini coefficients) [97, 155, 157] and degenerate systems involving differential forms [203]. See also [33, 34].
- General uniformly elliptic operators of the type in (2.2.13)–(2.2.14), [12].
- General nonuniformly elliptic operators, also with fast exponential growth, especially in the variational case [14, 19, 78, 85]. See Sects. 2.8.3 and 2.10.2.
- Fully nonlinear operators [68, 195] and renormalized p-Laplacean operators (via viscosity solutions) [11].
- General nonlinear elliptic systems, without Uhlenbeck type structure, and in the context of partial regularity [38, 156]. This continues to hold in the case of minimizers of quasiconvex functionals [73, 86].

Moreover

- Global Lipschitz regularity results for solutions to Dirichlet and Neumann problems, under suitable regularity on the boundary, can be obtained [55–57].
- Many of the results above extend to parabolic equations; see [150, 152, 153].

The gradient continuity of solutions is essentially linked to (2.4.10). In fact, it holds that

$$\mu \in L(n,1) \implies \lim_{r \to 0} \mathbf{I}_1^\mu(x,r) = 0 \text{ uniformly w.r.t. } x$$

and therefore the continuity of Du follows by the criterion displayed in (2.4.9). This is the general scheme of the various nonlinear versions of the original Stein's theorem, from degenerate to fully nonlinear elliptic problems. We refer to [154] for an overview.

2.8.2 Non-differentiable Stein [79]

Of course solutions to (2.8.1) locally minimize the functional in (2.6.20) when $F(Dw) \equiv |Dw|^2/2$ and $h(x,w) = -\mu(x)w$. It is therefore natural to wonder what happens in more general cases. For this we consider a general integrand $F(\cdot)$ and

$$\begin{cases} |h(x,y_1) - h(x,y_2)| \leq \mu(x)|y_1 - y_2|^\alpha, & \alpha \in (0,1] \\ \mu \in L\left(\dfrac{n}{\alpha}, \dfrac{1}{2-\alpha}\right). \end{cases} \quad (2.8.3)$$

Theorem 2.8.1 *Let $u \in W_{\mathrm{loc}}^{1,1}(\Omega)$ be a minimizer of the functional in (2.6.20), under assumptions (2.6.4) with $p = 2$, (2.6.23) and (2.8.3). Then $Du \in L_{\mathrm{loc}}^\infty(\Omega, \mathbb{R}^n)$ and moreover*

$$\|Du\|_{L^\infty(B_t)}$$
$$\leq \frac{c}{(r-t)^\kappa}\left[\|F(Du)\|_{L^1(B_r)} + \|h(\cdot,u)\|_{L^1(B_r)} + \|\mu\|_{n/\alpha,1/(2-\alpha);B_r} + 1\right]^\kappa$$

holds whenever $B_t \Subset B_r \Subset \Omega$ are concentric balls with $r \leq 1$, where $c \equiv c(\mathrm{data},\alpha) \geq 1$ and $\kappa \equiv \kappa(n,p,q,\alpha) \geq 1$.

Notice that, when $\alpha = 1$, the Lorentz condition $(2.8.3)_2$ gives, as expected, $\mu \in L(n,1)$. Moreover, following the monotonicity properties in $(2.4.3)_{1,2}$, condition $(2.8.3)_2$ gets weaker when α increases, again as expected. A more general result can be found in [79].

2.8.3 Stein Type Theorems in the Fast Growth Case [14, 78]

There are integrands whose nonuniform ellipticity cannot be controlled via polynomial growth of the eigenvalues. These can be of the type

$$w \mapsto \int_\Omega \exp(\exp(\ldots \exp(|Dw|^p)\ldots))\,dx\,, \qquad p \geq 1\,, \tag{2.8.4}$$

treated for instance in [96, 103, 162, 172, 173]. Following the computations in [14, (6.13)], in the case of (2.8.4) we find the following optimal upper bound on the ellipticity ratio

$$\mathcal{R}(z) \lesssim t^{p-1} \exp(\exp(\ldots \exp(|z|^p)\ldots)) + 1 \tag{2.8.5}$$

where, if $k \geq 1$ is the number of nested exponentials involved in (2.8.4), the number in (2.8.5) is $k - 1$ (it is zero when (2.8.4) involves one exp only and in this case $\mathcal{R}(z)$ grows polynomially in $|z|$). We report a sample result, concerning a full nonautonomous version of (2.8.4). A main point here is the dependence on coefficients, that are assumed to be Sobolev functions, as first considered in the (p, q)-case in [100]. For this, we fix functions $\{p_k(\cdot)\}$ and $\{c_k(\cdot)\}$, all defined on the open subset $\Omega \subset \mathbb{R}^n$, such that

$$\begin{cases} 1 < p_{\mathrm{m}} \leq p_0(\cdot) \leq p_M\,, & 0 < p_{\mathrm{m}} \leq p_k(\cdot) \leq p_M\,, \quad \text{for } k \geq 1 \\ 0 < \nu \leq c_k(\cdot) \leq L\,, & p_k(\cdot), c_k(\cdot) \in W^{1,d}(\Omega)\,, \ d > n\,, \quad \text{for } k \geq 0\,. \end{cases} \tag{2.8.6}$$

For every $k \in \mathbb{N}$, we next inductively define $\mathbf{e}_k : \Omega \times [0, \infty) \to \mathbb{R}$ by

$$\begin{cases} \mathbf{e}_{k+1}(x, t) := \exp\left(c_{k+1}(x)\,[\mathbf{e}_k(t)]^{p_{k+1}(x)}\right) \\ \mathbf{e}_0(x, t) := \exp\left(c_0(x)t^{p_0(x)}\right)\,, \end{cases}$$

and consider functionals

$$w \mapsto \mathcal{E}_k(w, \Omega) := \int_\Omega [\mathbf{e}_k(x, |Dw|) - w\mu]\,dx\,. \tag{2.8.7}$$

For this we have the following result, taken from [78], and valid in the vectorial case too $N \geq 1$:

Theorem 2.8.2 *Let $u \in W^{1,1}_{\mathrm{loc}}(\Omega, \mathbb{R}^N)$ be a minimizer of the functional $\mathcal{E}_k$ in (2.8.7) for some $k \in \mathbb{N}$, under assumptions (2.8.6) and such that $\mu \in L(n, 1)$ with $n \geq 3$. Then $Du \in L^\infty_{\mathrm{loc}}(\Omega, \mathbb{R}^{N \times n})$.*

For more general cases and the two-dimensional situation $n = 2$, we refer to [14, 78]. Amongst the various applications of the methods in [78] we mention the possibility of proving sharp conditions for Lipschitz regularity of solutions to obstacle problems involving nonuniformly elliptic functionals. For this we consider a measurable function $\psi: \Omega \to \mathbb{R}$ and the convex set

$$\mathcal{K}_\psi(\Omega) := \{w \in W^{1,1}_{\text{loc}}(\Omega): w(x) \geq \psi(x) \text{ for a.e. } x \in \Omega\}.$$

We then say that a function $u \in W^{1,1}_{\text{loc}}(\Omega) \cap \mathcal{K}_\psi(\Omega)$ is a constrained local minimizer of $\mathcal{E}_k$ if, for every open subset $\tilde{\Omega} \Subset \Omega$, we have $\mathcal{E}_k(u; \tilde{\Omega}) < \infty$ and if $\mathcal{E}_k(u; \tilde{\Omega}) \leq \mathcal{E}_k(w; \tilde{\Omega})$ holds for every competitor $w \in u + W^{1,1}_0(\tilde{\Omega})$ such that $w \in \mathcal{K}_\psi(\tilde{\Omega})$.

Theorem 2.8.3 *Let $u \in W^{1,1}_{\text{loc}}(\Omega) \cap \mathcal{K}_\psi(\Omega)$ be a constrained local minimizer of $\mathcal{E}_k$ in (2.8.7) for some $k \in \mathbb{N}$, with $\mu \equiv 0$ and $n \geq 3$. If $\psi \in W^{2,1}_{\text{loc}}(\Omega)$ and $|D^2\psi| \in L(n, 1)$, then $Du \in L^\infty_{\text{loc}}(\Omega, \mathbb{R}^n)$.*

Theorem 2.8.3 continues to hold when, instead of functionals as in (2.8.7), we consider general functionals with (p, q)-growth as for instance the one in Theorem 2.5.1. Details are again in [78].

Commenting on the growth of the ellipticity ratio (2.8.5) is probably useful here. In the case of functionals with (p, q)-growth conditions a quantified rate of Hölder continuity of coefficients

$$x \mapsto \frac{\partial_z F(x, z)}{(|z|^2 + s^2)^{(q-1)/2}}, \qquad |z| + s \neq 0 \tag{2.8.8}$$

is necessary to guarantee Lipschitz continuity of minima; see Section 2.3 and the gap bound in (2.3.1). The same holds even assuming Sobolev differentiability on coefficients, that is

$$\frac{|\partial_{zx} F(x, z)|}{(|z|^2 + s^2)^{(q-1)/2}} \lesssim g(x) \in L^d(\Omega), \qquad d > n. \tag{2.8.9}$$

This is for instance the approach from [78, 100], where (2.8.9) is used with d large enough to satisfy

$$\frac{q}{p} < 1 + \frac{1}{n} - \frac{1}{d} \iff d > \frac{np}{p - (q - p)n}. \tag{2.8.10}$$

Note that Sobolev-Morrey embedding theorem implies that the vector field in (2.8.8) belongs to $C^{0,\alpha}$ with $\alpha := 1 - n/d$, so that (2.8.10) and (2.3.1) actually coincide. This means that a quantified rate of integrability d is needed anyway. On the other hand, in Theorem 2.8.2 no lower bound on d is needed. This is in striking contrast with the fact that, while in the case of (2.8.7) the ellipticity ratio grows exponentially, it just grows polynomially fast in (p, q)-growth functionals. This apparently counterintuitive situation is explained as follows. A relevant quantity implicitly appearing in the estimates is in fact a renormalized ratio

$$\sup_{x \in B} \frac{\mathcal{R}_F(x, z)}{F(x, z)}$$

for a fixed ball $B \subset \Omega$. While in the (p, q) case the above quantity decays to zero (as $|z| \to \infty$) polynomially, in the exponential case the convergence to zero is exponentially fast (compare (2.8.5)). This explains the stronger requirement on the integrability of coefficients in the polynomial growth case and also reflects the fact that exponential type functionals force Lipschitz estimates more strongly due to their fast growth (although they are more delicate to treat from the technical view point). The case of functionals with exponential growth therefore reveals to be closer to that of uniformly elliptic functionals and operators, for which we have the following borderline result, once again from [78], that deals with elliptic systems:

Theorem 2.8.4 *Let $u \in W^{1,1}_{\mathrm{loc}}(\Omega, \mathbb{R}^N)$ be a distributional solution to*

$$-\operatorname{div}(\mathfrak{c}(x)\tilde{a}(|Du|)Du) = \mu, \quad 0 < \nu \leq \mathfrak{c}(\cdot) \leq L,$$

such that $A(\cdot, |Du|) \in L^1(\Omega)$, where $A(\cdot)$ is defined in (2.2.15), and under assumptions (2.2.14). If $\mu, |D\mathfrak{c}| \in L(n, 1)$ with $n \geq 3$, then $Du \in L^\infty_{\mathrm{loc}}(\Omega, \mathbb{R}^{N \times n})$.

The borderline nature of the Theorem 2.8.4 with respect to Theorem 2.8.2 relies in that $L^d \subset L(n, 1) \subset L^n$ holds for every $d > n$; see (2.4.3).

The Key Takeaway
Condition $\mu \in L(n, 1)$ in (2.1.1) and (2.1.2) is universal for Lipschitz and/or C^1-regularity of minima and solutions, as it works both in the uniformly and in the nonuniformly elliptic case. It does not depend on the specific operator/integrand considered but it is ultimately an embedding type effect linked to any form of ellipticity. New Lorentz conditions appear when considering non-differentiable functionals.

2.9 Schauder Estimates at Nearly Linear Growth [80]

The functionals of the type (2.1.1) considered in the previous sections have superlinear polynomial growth in the sense that

$$|z|^p \lesssim F(x,z), \qquad p > 1.$$

Therefore integrands with so-called nearly linear growth are excluded. In fact, these are characterized by

$$\lim_{|z| \to \infty} \frac{F(x,z)}{|z|} = \infty, \qquad \lim_{|z| \to \infty} \frac{F(x,z)}{|z|^p} = 0 \quad \forall \, p > 1.$$

This happens for integrals of the type

$$w \mapsto \int_\Omega \mathfrak{c}(x)|Dw|\log(1+|Dw|)\,dx, \qquad 0 < \nu \leq \mathfrak{c}(\cdot) \leq L \tag{2.9.1}$$

that we already encountered in Sect. 2.2.3. Going to the realm of (p,q)-growth problems when $p=1$, a borderline case of the double phase functional in (2.2.19) is

$$w \mapsto \int_\Omega \left[\mathfrak{c}(x)|Dw|\log(1+|Dw|) + \mathfrak{a}(x)|Dw|^q \right] dx. \tag{2.9.2}$$

This one was first considered in [76] under Sobolev differentiability assumption on the coefficients $\mathfrak{c}(\cdot)$, $\mathfrak{a}(\cdot)$. A main point here is that, although the double phase integral in (2.2.19) is softly nonuniformly elliptic in the sense of (2.2.25), the functional in (2.9.2) is nonuniformly elliptic in the classical sense of (2.2.6). Indeed, the best upper bound we can find for the ellipticity ratio is

$$\mathcal{R}(x,z) \lesssim \log(1+|z|) + 1.$$

For this very reason, the extension to the functional (2.9.2) of the techniques from [13, 60, 61, 137, 138], devised to treat softly nonuniformly elliptic integrals, is impossible. A different approach has been introduced in [80] and gives

Theorem 2.9.1 *Let* $u \in W^{1,1}_{\text{loc}}(\Omega)$ *be a local minimizer of the functional in* (2.9.2), *with*

$$\begin{cases} 0 \leq \mathfrak{a}(\cdot) \in C^{0,\alpha}(\Omega), & 1 < q < 1 + \alpha/n \\ \mathfrak{c}(\cdot) \in C^{0,\alpha_0}_{\text{loc}}(\Omega), & 1/\Lambda \leq \mathfrak{c}(\cdot) \leq \Lambda, \end{cases} \tag{2.9.3}$$

(continued)

Theorem 2.9.1 (continued)

where $\alpha, \alpha_0 \in (0, 1)$ and $\Lambda \geq 1$. Then Du is locally Hölder continuous in Ω and moreover, for every ball $B_r \Subset \Omega$, $r \leq 1$, there holds

$$\|Du\|_{L^\infty(B_{r/2})} \leq c \left(\fint_{B_r} \left[|Du| \log(1 + |Du|) + \mathfrak{a}(x)|Du|^q \right] dx \right)^\kappa + c$$

with

$$\begin{cases} c \equiv c(n, q, \Lambda, \alpha, \alpha_0, \|a\|_{C^{0,\alpha}}, \|\mathfrak{c}\|_{C^{0,\alpha_0}}) \geq 1 \\ \kappa \equiv \kappa(n, q, \alpha, \alpha_0) \geq 1. \end{cases}$$

The main assumption in Theorem 2.9.1 is

$$q < 1 + \frac{\alpha}{n}, \tag{2.9.4}$$

which is the obvious borderline version of (2.3.1) (let $p \to 1$). The equality case in (2.2.1), which is lacking in (2.9.4), is based on Gehring type results [74] that typically miss in the nearly linear growth case. In the non-degenerate case, the gradient Hölder continuity exponent can be quantified and we have an analog of Theorems 2.6.2 and 2.6.4, that is

Theorem 2.9.2 *Under assumptions (2.9.3), local minimizers of the functional*

$$w \mapsto \int_\Omega \left[\mathfrak{c}(x)|Dw| \log(1 + |Dw|) + \mathfrak{a}(x)(|Dw|^2 + 1)^{q/2} \right] dx$$

are locally $C^{1,\tilde{\alpha}/2}$-regular in Ω, where $\tilde{\alpha} := \min\{\alpha_0, \alpha\}$. In particular, local minimizers of the functional in (2.9.1) are locally $C^{1,\alpha_0/2}$-regular provided (2.9.3)$_2$ is assumed.

One might of course wonder whether or not it is possible to approach linear growth conditions from below more closely, for instance considering arbitrary compositions of logarithms as follows:

$$w \mapsto \int_\Omega \left[F(x, Dw) + \mathfrak{a}(x)(|Dw|^2 + s^2)^{q/2} \right] dx, \tag{2.9.5}$$

where $s \in [0, 1]$ and

$$\begin{cases} F(x, z) \equiv \mathfrak{c}(x)|z|L_{k+1}(|z|) & \text{for } k \geq 0 \\ L_{k+1}(|z|) = \log(1 + L_k(|z|)) & \text{for } k \geq 0, \quad L_0(|z|) = |z|. \end{cases} \quad (2.9.6)$$

Here $\mathfrak{c}(\cdot)$ is as in $(2.9.3)_2$. For this we have to cook up a more technical setting. We consider continuous integrands $F: \Omega \times \mathbb{R}^n \to [0, \infty)$ such that $z \mapsto F(x, z) \in C^2(\mathbb{R}^n)$ for every $x \in \Omega$ and satisfying

$$\begin{cases} \nu|z|g(z) \leq F(x, z) \leq L(|z|g(z) + 1) \\ \dfrac{\nu|\xi|^2}{(|z|^2+1)^{\gamma/2}} \leq \partial_{zz}F(x,z)\xi \cdot \xi, \quad |\partial_{zz}F(x,z)| \leq \dfrac{Lg(|z|)}{(|z|^2+1)^{1/2}} \\ |\partial_z F(x,z) - \partial_z F(y,z)| \leq L|x-y|^{\alpha_0} g(|z|), \end{cases} \quad (2.9.7)$$

for every choice of $x, y \in \Omega$, $z, \xi \in \mathbb{R}^n$, where $\alpha_0 \in (0, 1)$ and $1 \leq \gamma < 3/2$ being fixed constants. Here $g: [0, \infty) \to [1, \infty)$ is a non-decreasing, concave and unbounded function such that $t \mapsto tg(t)$ is convex. Moreover, we assume that for every $\varepsilon > 0$ there exists a constant $c_g(\varepsilon)$ such that

$$g(t) \leq c_g(\varepsilon)t^\varepsilon \quad \text{holds for every } t \geq 1. \quad (2.9.8)$$

Conditions (2.9.7) and (2.9.8) imply the local Lipschitz continuity of minima. Local Hölder continuity needs one more condition, that is

$$|\partial_z F(x, z)| \leq L|z| \quad \text{holds for every } |z| \leq 1. \quad (2.9.9)$$

We then have

> **Theorem 2.9.3** *Let $u \in W^{1,1}_{\text{loc}}(\Omega)$ be a local minimizer of the functional in (2.9.5) under assumptions $(2.9.3)_1$ and (2.9.7) and (2.9.8). There exists $\gamma_m \in (1, 2)$, depending only n, q, α_0, α, such that, if $1 \leq \gamma < \gamma_m$, then*
>
> $$\|Du\|_{L^\infty(B_{r/2})} \leq c \left(\fint_{B_r} [F(x, Du) + \mathfrak{a}(x)(|Du|^2 + s^2)^{q/2}] \, dx \right)^\kappa + c$$
>
> *holds whenever $B_r \Subset \Omega$, $r \leq 1$. Here it is $c \equiv c(n, q, \nu, L, \Lambda, \alpha, \alpha_0)$, $\|\mathfrak{a}\|_{C^{0,\alpha}}, \|\mathfrak{c}\|_{C^{0,\alpha_0}} \geq 1$ and $\kappa \equiv \kappa(n, q, \alpha_0, \alpha) \geq 1$. Moreover, assuming also (2.9.9) implies that Du is locally Hölder continuous is Ω. Finally, if also $s > 0$ and $\partial_{zz}F(\cdot)$ is continuous, then $u \in C^{1,\tilde{\alpha}/2}_{\text{loc}}(\Omega)$-regular with $\tilde{\alpha} = \min\{\alpha_0, \alpha\}$.*

As a matter of fact, Theorems 2.9.1 and 2.9.2 follow from Theorem 2.9.3 with $\gamma = 1$. In order to treat functionals in (2.9.5) with the choice in (2.9.6) it is sufficient to apply Theorem 2.9.3 with any $\gamma > 1$. For more details the reader is referred to [80]. The proof of Theorem 2.9.3 is again based on the use of fractional energy estimates. This time a careful use of the double phase structure of the integral (2.9.5) is necessary, and involves an ad hoc anisotropic version of the Bernstein technique. The final outcome is the Lipschitz regularity of minima under the optimal bound (2.9.4). Once again, once local Lipschitz regularity is gained, one can (carfully) adapt more standard perturbation methods.

The methods introduced in [80] are general enough to provide a blueprint to treat more general situations. One of these is in [88], where the authors considered multi-phase models with nearly linear growth. In this case (three phases case) a model is provided by the functional

$$w \mapsto \int_\Omega \left[\mathfrak{c}(x)|Dw|\log(1+|Dw|) + \mathfrak{a}(x)|Dw|^q + \mathfrak{b}(x)|Dw|^\gamma \right] dx \qquad (2.9.10)$$

where

$$\begin{cases} 1 < q < 1 + \dfrac{\alpha}{n}, & 1 < \gamma < 1 + \dfrac{\beta}{n} \\ \\ 0 \leq \mathfrak{a}(\cdot) \in C^{0,\alpha}(\Omega), & 0 \leq \mathfrak{b}(\cdot) \in C^{0,\beta}(\Omega) \end{cases}$$

with $\alpha, \beta \in (0, 1]$ and $\mathfrak{c}(\cdot)$ as in $(2.9.3)_2$. In this situation, as it has been proved in [88], minimizers of the functional in (2.9.10) have locally Hölder continuous first derivatives. Needless to say, the result extends to functionals of the type

$$w \mapsto \int_\Omega \left[\mathfrak{c}(x)|Dw|\log(1+|Dw|) + \sum_{k=1}^m \mathfrak{a}_k(x)|Dw|^{q_k} \right] dx$$

with

$$1 < q_k < 1 + \frac{\alpha_k}{n}, \quad 0 \leq \mathfrak{a}_k(\cdot) \in C^{0,\alpha_k}(\Omega), \quad \alpha_k \in (0, 1]. \qquad (2.9.11)$$

The case of multi-phase functionals with super-linear growth, i.e., the one displayed in (2.3.4), was originally considered in [83]. Observe that conditions (2.9.11) are the sharp analogue of those in (2.3.5) when approaching nearly linear growth conditions, that is, when formally letting $p \to 1$ in (2.3.4) and (2.3.5).

Functionals with nearly linear growth have been the object of intensive investigation over the years. For this we mention, for instance, [24–26, 101, 109, 111, 177]. Their nonuniform ellipticity stems from their closeness to linear growth functionals, like for instance the minimal surface one, that are always nonuniformly elliptic; for this see [15, 16, 122–125] and related references for recent developments. Beside their theoretical interest, nearly linear growth integrals also appeared in relation to applications in Plasticity and Non-Newtonian Fluid-dynamics [108, 110–112].

> **The Key Takeaway**
> Schauder type estimates still hold for minima of nearly linear growth integrals. These are in general nonuniformly elliptic in the classical, pointwise sense (2.2.6) and therefore perturbation methods are ruled out. Again, a direct proof of Lipschitz continuity is needed and goes via fractional Caccioppoli inequalities and nonlinear potential theoretic methods.

2.10 Renormalized Caccioppoli Inequalities and Nonlinear Potentials

In this section we shall try to present and synthesize some of the methods from [14, 78–80, 82], where the use of nonlinear potential theoretic methods in the setting of nonuniformly elliptic problems was introduced. As mentioned in Sect. 2.1, in order to give a unified treatment and to emphasize the main ideas, we shall pick a few special cases, often not the most general ones. These can be anyway treated via more specific variants of the same leading principles, sometimes discussed in the commentaries. We shall demonstrate four different applications of the general ideas in order to emphasize the flexibility of the approach presented and provide a gradual, gentle introduction to the most technical cases. We shall start by a toy model in Sect. 2.10.1, that actually deals with classical linear uniformly elliptic case, in order to highlight the connections between uniformly and nonuniformly elliptic problems. Of course we shall confine ourselves to give sketches of the proofs as a complete treatment would just (more than) double the length of these notes.

2.10.1 Linear Elliptic Equations

We consider a linear elliptic equation

$$-\operatorname{div}(\mathbf{A}(x)Dv) = \mu, \qquad v\mathbb{I}_d \leq \mathbf{A}(\cdot) \leq L\mathbb{I}_d \qquad (2.10.1)$$

where $\mathbf{A}(\cdot)$ has measurable entries. We want to prove that

$$\mu \in L(n/2, 1) \implies v \in L^\infty_{\text{loc}}(\Omega) \qquad (2.10.2)$$

provided $n > 4$. Of course we start assuming that $v \in W^{1,2}_{\mathrm{loc}}(\Omega)$, i.e., it is a standard energy solution (according to Definition 2.2.1). Notice that assuming $\mu \in L(n/2, 1)$ implies that $\mu \in L^2$ holds provided $n \geq 4$ by (2.4.3)$_2$.

The Classical Caccioppoli Inequality

$$\int_{B_{\varrho/2}} |D(v-\kappa)_+|^2 \, dx \lesssim_{\nu,L} \frac{1}{\varrho^2} \int_{B_\varrho} (v-\kappa)_+^2 \, dx + \varrho^2 \int_{B_\varrho} |\mu|^2 \, dx \qquad (2.10.3)$$

now holds for every ball $B_\varrho \Subset \Omega$. As usual, here it is $(v-\kappa)_+ := \max\{v-\kappa, 0\}$ and we are taking any $\kappa \geq 0$. The proof is very standard and can be obtained testing (2.10.1) by $\eta^2(v-\kappa)_+$, where $\eta \in C_0^1(B_\varrho)$ is such that $\eta \equiv 1$ on $B_{\varrho/2}$ and $|D\eta| \lesssim 1/\varrho$ (see for instance [121, Chapter 7]). Sobolev embedding theorem now implies the following:

Reverse Type Inequality with Remainder

$$\left(\fint_{B_{\varrho/2}} (v-\kappa)_+^{2\chi} \, dx \right)^{1/\chi} \lesssim_{n,\nu,L} \fint_{B_\varrho} (v-\kappa)_+^2 \, dx + \varrho^4 \fint_{B_\varrho} |\mu|^2 \, dx$$

$$(2.10.4)$$

where $\chi := 2^*/2 > 1$ and 2^* has been defined in (2.1.5). Applying Lemma 2.4.4 with the choices $h = 1, m_1 = t = \delta_1 = 2, \theta_1 = 1, M_0 = M_1 \approx 1, \kappa_0 = 0$, yields

$$v(x_0) \lesssim_{n,\nu,L} \left(\fint_{B_{r_0}(x_0)} v^2 \, dx \right)^{1/2} + \mathbf{P}^{2,1}_{2,2}(\mu; x_0, 2r_0) \, .$$

This holds whenever $B_{2r_0}(x_0) \Subset \Omega$ such that x_0 is Lebesgue point of v. By observing that $-v$ solves (2.10.1) with μ replaced by $-\mu$, we arrive at the pointwise nonlinear potential estimate

$$|v(x_0)| \lesssim_{n,\nu,L} \left(\fint_{B_{r_0}(x_0)} v^2 \, dx \right)^{1/2} + \mathbf{P}^{2,1}_{2,2}(\mu; x_0, 2r_0) \qquad (2.10.5)$$

that again holds whenever $B_{2r_0}(x_0) \Subset \Omega$, for a.e. x_0. We now consider a ball $B_r \Subset \Omega$; applying (2.10.5) to $B_{r_0}(x_0)$, $r_0 = r/4$, for all $x_0 \in B_{r/2}$ for which (2.10.5) holds, we get

$$\|v\|_{L^\infty(B_{r/2})} \lesssim_{n,v,L} \left(\fint_{B_r} v^2 \, dx\right)^{1/2} + \|\mathbf{P}^{2,1}_{2,2}(\mu;\cdot,r/2)\|_{L^\infty(B_{r/2})}.$$

It is now time to use Lemma 2.4.3; this is possible since (2.4.4) is verified by our assumption $n > 4$. We therefore conclude with

$$\|v\|_{L^\infty(B_{r/2})} \lesssim_{n,v,L} \left(\fint_{B_r} v^2 \, dx\right)^{1/2} + \|\mu\|_{n/2,1;B_r} \qquad (2.10.6)$$

that holds for every ball $B_r \Subset \Omega$.

Commentary 2.10.1 Both the result in (2.10.2) and estimate (2.10.6) are sharp with respect to the function space $L(n/2, 1)$ considered for the right-hand side. The result continues to hold in the cases $n = 3, 4$, that are missing here due to the use of the nonlinear potential $\mathbf{P}^{2,1}_{2,2}$. This is in fact a consequence of another, more basic nonlinear potential estimate, due to Kilpelainen & Maly [146] and valid for general quasilinear elliptic equations with measure data right-hand side, that in the case $p = 2$ reduces to

$$|v(x_0)| \lesssim_{n,v,L} \mathbf{I}^\mu_2(x_0,r) + \fint_{B_r(x_0)} v \, dx \approx_n \mathbf{P}^{1,1}_{1,2}(\mu;x_0,r) + \fint_{B_r(x_0)} v \, dx.$$

See also [154]. This again implies (2.10.2) by Lemma 2.4.3 as this time (2.4.4) is satisfied for $n > 2$ (see also [54]). In fact, more general results are available for equations of the type (2.1.2) under assumptions (2.2.12). In this case, for suitably defined solutions to (2.1.2), and restricting for simplicity to the case $p \geq 2$, we have the

---

Wolff Potential Estimate [146, 154]

$$|v(x_0)| \lesssim_{\text{data}} \mathbf{W}^\mu_{1,p}(x_0,r) + \left(\fint_{B_r(x_0)} |v|^{p-1} \, dx\right)^{1/(p-1)}$$

$$\approx_{n,p} \mathbf{P}^{1,1}_{p-1,p/(p-1)}(\mu;x_0,r) + \left(\fint_{B_r(x_0)} |v|^{p-1} \, dx\right)^{1/(p-1)} \qquad (2.10.7)$$

for a.e. $x_0 \in \Omega$ and balls $B_r(x_0) \Subset \Omega$.

In this case, again as a consequence of Lemma 2.4.3, applied with $m = \theta = 1$, $t = p-1$ and $\delta = p/(p-1)$, we find

$$\mu \in L\left(\frac{n}{p}, \frac{1}{p-1}\right) \Longrightarrow v \in L^\infty_{\mathrm{loc}}(\Omega) \qquad (2.10.8)$$

provided

$$\frac{n}{t\delta} = \frac{n}{p} > 1, \qquad (2.10.9)$$

with the a priori estimate

$$\|v\|_{L^\infty(B_{r/2})} \lesssim_{\mathrm{data}} \left(\fint_{B_r} |v|^{p-1}\, dx\right)^{1/(p-1)} + \|\mu\|^{1/(p-1)}_{n/p, 1/(p-1); B_r} \qquad (2.10.10)$$

that holds for every ball $B_r \Subset \Omega$. In fact, this last one reduces to (2.10.6) when $p = 2$. Let us briefly see how to recover (2.10.10) without appealing to (2.10.7), but rather using more standard energy estimates as done for (2.10.6). For equations of the type in (2.1.2), under assumptions (2.2.12), the following Caccioppoli type inequality holds:

$$\int_{B_{\varrho/2}} |D(v-\kappa)_+|^p\, dx \lesssim_{v,L,p} \frac{1}{\varrho^p} \int_{B_\varrho} (v-\kappa)_+^p\, dx + \varrho^{\frac{p}{p-1}} \int_{B_\varrho} |\mu|^{\frac{p}{p-1}}\, dx \qquad (2.10.11)$$

for every ball $B_\varrho \Subset \Omega$; see [121, Chapter 7]. Then, again via Sobolev embedding, we come to the analog of (2.10.4), that is

$$\left(\fint_{B_{\varrho/2}} (v-\kappa)_+^{p\chi}\, dx\right)^{1/\chi} \lesssim_{\mathrm{data}} \fint_{B_\varrho} (v-\kappa)_+^p\, dx + \varrho^{\frac{p^2}{p-1}} \fint_{B_\varrho} |\mu|^{\frac{p}{p-1}}\, dx \qquad (2.10.12)$$

for some $\chi \equiv \chi(n, p) > 1$. We can now proceed as after (2.10.4), this time the relevant nonlinear potential being $\mathbf{P}^{p/(p-1),1}_{p,p/(p-1)}$ (Lemma 2.4.4). This yields (2.10.8) upon checking that (Lemma 2.4.3)

$$\frac{n}{t\delta} > 1 \iff n > \frac{p^2}{p-1}. \qquad (2.10.13)$$

Therefore we again come to (2.10.10), and again with a restricted range of dimensions n. The difference between (2.10.9) and (2.10.13) stems from the kind of *underlying energy estimates* they rely on. In the case of (2.10.11) one is using estimates at the natural energy level $W^{1,p}$. These are natural estimates for energy solutions, but yield weaker results. In order to get estimates *below the natural growth exponent* like (2.10.7), more sophisticated techniques are needed [146, 154]. They relate to so-called measure data problems [20, 27, 28, 65, 163] and usually work at the energy level fixed by the space $W^{1,p-1}$.

> **The Key Takeaway**
> Energy inequalities of the type in (2.10.3) and (2.10.11), with related reverse Hölder type inequalities (2.10.4) and (2.10.12), automatically encode L^∞-bounds on solutions under optimal regularity assumptions on μ, via nonlinear potentials. Similar De Giorgi type iteration schemes work below the natural growth exponent allowing for slightly larger range of parameters [146, 154]. The crucial point here is that, as the proof in [14, 79] of Lemma 2.4.4 reveals, the basic estimates involved (2.10.4) and (2.10.12) must be homogeneous with respect to the function v. This happens here as a consequence of the assumed uniform ellipticity in (2.10.1) and (2.2.12).

2.10.2 Nonuniformly Elliptic Problems [14, 78]

Here we switch to gradient estimates and present a few basic methods from [14, 78]. For the sake of simplicity we shall confine ourselves to the case $n \geq 3$, again referring to [14, 78] for the two dimensional case $n = 2$. In this section we consider the simplified case of functionals of the type

$$w \mapsto \int_\Omega [F(Dw) - \mu w]\, dx\,, \quad (2.10.14)$$

where $F(\cdot)$ is a C^2-regular integrand satisfying (2.5.2) as in Sect. 2.5. Moreover, we assume that $\mu \in L(n, 1)$. Here a disclaimer is necessary: [14, 78] are long papers and reporting here a self-contained proof of the results would be impossible. In particular, a lengthy and nonstandard procedure is necessary in [14, 78] in order to approximate the original nonuniformly elliptic problem with uniformly elliptic ones and this is aimed at making all the computations necessary for the a priori estimates legal. The complexity of such a procedure stems from the fact that this is tailored to cover elliptic functionals whose growth of the eigenvalues is faster than polynomial, therefore going much beyond the (p, q) case considered here. We shall therefore follow the classical approach of delivering the core a priori estimates for more regular solutions and data (see (2.10.15)). Of course the shape of the estimates will not quantitatively incorporate such a priori assumption that can be eventually removed by the aforementioned approximation procedure. We now fix a local minimizer u of the functional in (2.10.14) and assume that

$$u \in W^{1,\infty}_{\mathrm{loc}}(\Omega) \cap W^{2,2}_{\mathrm{loc}}(\Omega), \qquad \mu \in L^{\infty}_{\mathrm{loc}}(\Omega). \tag{2.10.15}$$

In particular, this will guarantee that all the forthcoming integrals will be finite. We now proceed via a variant of the so-called Bernstein technique. In the uniformly elliptic case this classical method prescribes that, if u solves an equation of the type $\mathrm{div}\, a(Du) = 0$, then, under appropriate differentiability assumptions, functions of the type $|Du|^{\gamma}$ are subsolutions of certain linear elliptic equations, for suitable powers $\gamma > 0$. As such, they are locally bounded functions and therefore original solutions u are locally Lipschitz regular. In turn, upper bounds for subsolutions typically goes via Caccioppoli type inequalities on level sets and De Giorgi type iterations. In our setting we are using the Euler-Lagrange equation

$$-\mathrm{div}\, \partial_z F(Du) = \mu \tag{2.10.16}$$

of the functional in (2.10.14), and this poses two problems. First, it is nonuniformly elliptic. Second, the right-hand side does not vanish. Therefore the subsolution type scheme described above falls short of deriving estimates. Nevertheless, by differentiating (2.10.16) we can use a raw version of Bernstein's method. We derive a Caccioppoli type inequality for a certain convex function of $|Du|$, as at this stage we do not have any information concerning any type of subsolution property due to the presence of a general right-hand side μ. From this Caccioppoli inequality we directly derive local gradient bounds with the aid of nonlinear potentials. Specifically, we start defining

$$\begin{cases} G(t) := \int_0^t (y^2 + s^2)^{(p-2)/2} y\, dy = \dfrac{1}{p}\left[(t^2 + s^2)^{p/2} - s^p\right] \\ v := G(|Du|). \end{cases} \tag{2.10.17}$$

We next consider concentric balls

$$B_{\varrho} \equiv B_{\varrho}(x_0) \subset B_{r_0}(x_0) \subset B_{2r_0}(x_0) \Subset \Omega. \tag{2.10.18}$$

Following [14, Lemma 4.5], the Caccioppoli type inequality

$$\int_{B_{\varrho/2}} |D(v-\kappa)_+|^2\, dx \lesssim_{\mathrm{data}} \frac{1}{\varrho^2} \int_{B_{\varrho}} [\mathcal{R}_F(Du) + 1](v-\kappa)_+^2\, dx$$

$$+ \int_{B_{\varrho}} (|Du|^2 + 1)|\mu|^2\, dx \tag{2.10.19}$$

holds for every ball $B_{\varrho} \Subset \Omega$ and number $\kappa \geq 0$, where $\mathcal{R}_F(\cdot)$ denotes the ellipticity ratio of the integrand $F(\cdot)$. Note that this is a point where (2.10.15) is needed. In fact (2.10.19) is obtained differentiating the Euler-Lagrange equation (2.10.16) so that (2.10.15) guarantees that this is possible and that all the terms generated by such a procedure make sense and are finite. By (2.5.2) the best upper bound we can get on $\mathcal{R}_F(Du)$ is

$$\mathcal{R}_F(Du) \lesssim_{L/\nu} |Du|^{q-p} + 1. \tag{2.10.20}$$

In comparison to (2.10.3), inequality (2.10.19) fails to be homogeneous with respect to v due to the appearance of $\mathcal{R}_F(Du)$. This makes proceeding as after (2.10.3) impossible. Using (2.10.20) we renomalize (2.10.19) getting the following

Renormalized Caccioppoli Inequality

$$\begin{cases} \int_{B_{\varrho/2}} |D(v-\kappa)_+|^2 \, dx \\ \qquad \lesssim_{\text{data}} \dfrac{M^{q-p}}{\varrho^2} \int_{B_\varrho} (v-\kappa)_+^2 \, dx + M^2 \int_{B_\varrho} |\mu|^2 \, dx \\ \|Du\|_{L^\infty(B_\varrho)} + 1 \leq M, \quad v \approx |Du|^p. \end{cases} \tag{2.10.21}$$

By paying the prize of the appearance of M, we now have an inequality which is homogeneous with respect to v. Recalling (2.10.18), the number M is from on chosen to be such that

$$\|Du\|_{L^\infty(B_{r_0}(x_0))} + 1 \leq M \tag{2.10.22}$$

so that (2.10.21) holds for every choice of balls B_ϱ in (2.10.18). Sobolev embedding theorem now yields

Renormalized Reverse Type Inequality with Remainder

$$\left(\fint_{B_{\varrho/2}} (v-\kappa)_+^{2\chi} \, dx \right)^{1/\chi} \lesssim_{\text{data}} M^{q-p} \fint_{B_\varrho} (v-\kappa)_+^2 \, dx + M^2 \varrho^2 \fint_{B_\varrho} |\mu|^2 \, dx \tag{2.10.23}$$

where (recall here it is $n > 2$)

$$\chi = \frac{n}{n-2} > 1. \tag{2.10.24}$$

By (2.10.22) inequality (2.10.23) holds for every choice of balls as in (2.10.18) and therefore we are allowed to apply Lemma 2.4.4 with $h = 1$, $m_1 = t = 2$, $\delta_1 = \theta_1 = 1$, $M_0 = M^{(q-p)/2}$, $M_1 = M$ and $\kappa_0 = 0$. This yields

$$v(x_0) \lesssim_{\text{data}} M^{\frac{n(q-p)}{4}} \left(\fint_{B_{r_0}(x_0)} v^2 \, dx \right)^{1/2} + M^{\frac{(q-p)(n-2)}{4}+1} \mathbf{P}_{2,1}^{2,1}(\mu; x_0, 2r_0).$$

Finally, recalling (2.10.17), we further estimate

$$|Du(x_0)| \lesssim_{\text{data}} M^{\frac{n}{4}\left(\frac{q}{p}-1\right)+\frac{1}{2}} \left(\fint_{B_{r_0}(x_0)} [F(Du) + 1] \, dx \right)^{\frac{1}{2p}}$$

$$+ M^{\frac{n-2}{4}\left(\frac{q}{p}-1\right)+\frac{1}{p}} \left[\mathbf{P}_{2,1}^{2,1}(\mu; x_0, 2r_0) \right]^{\frac{1}{p}} + 1. \quad (2.10.25)$$

All in all, this last estimate holds whenever $B_{2r_0}(x_0) \Subset \Omega$ and M satisfies (2.10.22).

The Pointwise-to-Global Trick
A general scheme to turn the pointwise estimate (2.10.25) into a global one on $B_{r/2}$ getting rid of the presence on $\|Du\|_{L^\infty}$ on the right-hand side, which is implicit in the choice of M in (2.10.22).

We consider a ball $B_r \Subset \Omega$, $r \le 1$, and concentric balls

$$B_{r/2} \Subset B_{\tau_1} \Subset B_{\tau_2} \Subset B_r, \quad (2.10.26)$$

and apply (2.10.27) at the generic point $x_0 \in B_{\tau_1}$, with $r_0 := (\tau_2 - \tau_1)/2$, and with the choice $M := \|Du\|_{L^\infty(B_{\tau_2})} + 1$, so that (2.10.22) is met as $B_{2r_0}(x_0) \subset B_{\tau_2}$. We gain

$$\mathrm{h}(\tau_1) \lesssim_{\text{data}} \frac{[\mathrm{h}(\tau_2)]^{\frac{n}{4}\left(\frac{q}{p}-1\right)+\frac{1}{2}}}{(\tau_2 - \tau_1)^{\frac{n}{2p}}} \left(\int_{B_r} [F(Du) + 1] \, dx \right)^{\frac{1}{2p}}$$

$$+ [\mathrm{h}(\tau_2)]^{\frac{n-2}{4}\left(\frac{q}{p}-1\right)+\frac{1}{p}} \|\mathbf{P}_{2,1}^{2,1}(\mu; \cdot, \tau_2 - \tau_1)\|_{L^\infty(B_{\tau_1})}^{1/p} + 1, \quad (2.10.27)$$

where we have set

$$\mathrm{h}(t) := \|Du\|_{L^\infty(B_t)} + 1, \quad r/2 \le t \le r. \quad (2.10.28)$$

We now require that

$$\frac{n}{4}\left(\frac{q}{p}-1\right)+\frac{1}{2}<1 \iff \frac{q}{p}<1+\frac{2}{n} \qquad (2.10.29)$$

and

$$\frac{n-2}{4}\left(\frac{q}{p}-1\right)+\frac{1}{p}<1 \iff \frac{q}{p}<1+\frac{4(p-1)}{p(n-2)}. \qquad (2.10.30)$$

These allow to use Young's inequality in (2.10.27) thereby getting

$$h(\tau_1) \leq \frac{h(\tau_2)}{2} + \frac{c}{(\tau_2-\tau_1)^{\frac{2n}{(n+2)p-nq}}} \left(\int_{B_r} [F(Du)+1]\,dx\right)^{\frac{2}{(n+2)p-nq}}$$
$$+ c\|\mathbf{P}_{2,1}^{2,1}(\mu;\cdot,\tau_2-\tau_1)\|_{L^\infty(B_{\tau_1})}^{\frac{4}{4(p-1)-(n-2)(q-p)}}$$

with $c \equiv c(\text{data})$. Using Lemma 2.4.3 (recall we are assuming $n>2$ and this allows to verify (2.4.4)) we estimate

$$\|\mathbf{P}_{2,1}^{2,1}(\mu;\cdot,\tau_2-\tau_1)\|_{L^\infty(B_{\tau_1})} \lesssim_n \|\mu\|_{n,1;B_{\tau_2}} \leq \|\mu\|_{n,1;B_r}$$

so that we gain

$$h(\tau_1) \leq \frac{h(\tau_2)}{2} + \frac{c}{(\tau_2-\tau_1)^{\frac{2n}{(n+2)p-nq}}} \left(\int_{B_r} [F(Du)+1]\,dx\right)^{\frac{2}{(n+2)p-nq}}$$
$$+ c\|\mu\|_{n,1;B_r}^{\frac{4}{4(p-1)-(n-2)(q-p)}}.$$

Applying Lemma 2.10.2 below we finally arrive at

The Lorentz a Priori Estimate

$$\|Du\|_{L^\infty(B_{r/2})} \lesssim_{\text{data}} \left(\fint_{B_r} F(Du)\,dx\right)^{\frac{2}{(n+2)p-nq}}$$
$$+ \|\mu\|_{n,1;B_r}^{\frac{4}{4(p-1)-(n-2)(q-p)}} + 1. \qquad (2.10.31)$$

Summarising, by (2.10.29) and (2.10.30) estimate (2.10.31) holds provided

$$\frac{q}{p} < 1 + \min\left\{\frac{2}{n}, \frac{4(p-1)}{p(n-2)}\right\}. \qquad (2.10.32)$$

The one in (2.10.31) is a Lipschitz a priori bound that does not depend on the regularity of u assumed in (2.10.15) but just of the energy and the right-hand datum μ. It is therefore suitable for a match with approximation arguments bound to remove (2.10.15). The final outcome is that the implication

$$\mu \in L(n,1) \implies Du \in L^{\infty}_{\text{loc}}(\Omega) \qquad (2.10.33)$$

and estimate (2.10.31) hold for any general local minimizer u.

Lemma 2.10.2 ([115, Lemma 1.1]) *Let* $\mathrm{h}\colon [r/2, r] \to \mathbb{R}$ *be a non-negative and bounded function, and let a, b, γ be non-negative numbers. Assume that the inequality in*

$$\mathrm{h}(\tau_1) \leq \frac{\mathrm{h}(\tau_2)}{2} + \frac{a}{(\tau_2 - \tau_1)^{\gamma}} + b,$$

holds whenever $r/2 \leq \tau_1 < \tau_2 \leq r$. *Then*

$$\mathrm{h}(r/2) \leq \frac{c(\gamma)a}{r^{\gamma}} + c(\gamma)b$$

holds too.

Commentary 2.10.3 The method exposed applies to general nonuniformly elliptic problems, including those with fast growth conditions as in (2.8.4), see [14, 78]. We have chosen the (p, q)-setting in order to give a streamlined presentation of some of the main ideas, especially those concerned with a priori estimates. We also stress that, although we confined ourselves to the scalar case, the approach presented here works in the vectorial case too when considering functionals with radial structure, i.e., $F(z) \equiv \tilde{F}(|z|)$. This last assumption is necessary already in the uniformly elliptic case, otherwise counterexamples to Lipschitz regularity emerge; see [180, 210] and related references. For this and more general situations we again refer to [14, 78]. Here the essence of the matter is better understood when $\mu \equiv 0$. In this case (2.10.19) becomes

$$\int_{B_{\varrho/2}} |D(v-\kappa)_+|^2 \, dx \lesssim_{\text{data}} \frac{1}{\varrho^2} \int_{B_\varrho} \mathcal{R}_F(Du)(v-\kappa)_+^2 \, dx. \qquad (2.10.34)$$

See again [14, Lemma 4.5]. In the uniformly elliptic regime $\mathcal{R}_F(Du) \lesssim 1$ this last inequality reduces to (2.10.3) (with $\mu \equiv 0$) and the standard De Giorgi's iteration applies. In the nonuniform case we pass to the renormalized Caccioppoli inequality

$$\int_{B_{\varrho/2}} |D(v-\kappa)_+|^2 \, dx \lesssim_{\text{data}} \frac{M^{q-p}}{\varrho^2} \int_{B_\varrho} (v-\kappa)_+^2 \, dx. \tag{2.10.35}$$

The nonuniform ellipticity has in a sense disappeared and is now encoded in the presence of M. From (2.10.35) we arrive at the following version of (2.10.31):

Reference Lipschitz Estimate

$$\|Du\|_{L^\infty(B_{r/2})} \lesssim_{\text{data}} \left(\fint_{B_r} F(Du) \, dx \right)^{\mathfrak{s}/q} + 1 \tag{2.10.36}$$

that holds provided

$$\frac{q}{p} < 1 + \frac{2}{n}. \tag{2.10.37}$$

In (2.10.36) we find

The Exponent $\mathfrak{s}$

$$\mathfrak{s} := \begin{cases} \dfrac{2q}{(n+2)p - nq} & \text{if } n \geq 3 \\[1ex] \text{any number } > \dfrac{q}{2p-q} & \text{if } n = 2 \text{ and } q > p \\[1ex] 1 & \text{if } n = 2 \text{ and } q = p. \end{cases} \tag{2.10.38}$$

This is a very streamlined and short proof of Theorem 2.5.1, and (2.10.36) is exactly the a priori estimate found by Marcellini in his classic paper [170] (our proof works in this case also when $n > 2$ as $\mu \equiv 0$). In fact, by passing to renormalized energy inequalities like (2.10.35), we are in position to use uniformly elliptic iteration procedures. Notice that, at this stage, it is crucial to get a precise control of the constants M_i in Lemma 2.4.4 in order to conclude, an information that is not needed in the uniformly elliptic case (2.10.3) and (2.10.4), as seen in Sect. 2.10.1. Surprisingly, this single fact allows to reduce the analysis of nonuniformly elliptic problems to that of uniformly elliptic ones. This idea is general and can be implemented also via Moser's iteration method; see Sect. 2.11. Moreover, the method above is general enough to provide another proof of Theorem 2.5.2. This has been done by Bella & Schäffer [19], who cleverly combined the methods of [17, 18] and those from [14] to prove (2.10.33) under the improved bound

$$\frac{q}{p} < 1 + \min\left\{\frac{2}{n-1}, \frac{4(p-1)}{p(n-3)}\right\}$$

for $n > 3$; this improves (2.10.32). When $\mu = 0$ this bound reduces to (2.5.7). We finally remark that in the case $p = q$ (uniform ellipticity) estimate (2.10.31) reduces to

$$\|Du\|_{L^\infty(B_{r/2})} \lesssim_{\text{data}} \left(\fint_{B_r} |Du|^p \, dx\right)^{1/p} + \|\mu\|_{n,1;B_r}^{1/(p-1)} + 1$$

which the standard a priori estimate for p-Laplacean type problems [97, 155, 157].

The Key Takeaway
The analysis of nonuniformly elliptic problems can be reduced to that of uniform ones via suitable renormalized Caccioppoli inequalities (2.10.21). These are aimed at overcoming the lack of homogeneous estimates which are typical of the nonuniform case. The mechanism works thanks to a precise quantitative control of the constants along the relevant iterative schemes (Lemma 2.4.4). In this setting nonlinear potentials naturally appear along the iterations and encode in a sharp way the regularity properties of the data.

2.10.3 Fractional Approach to Nonuniform Ellipticity [79]

As already mentioned, Lipschitz regularity is the focal point in nonuniformly elliptic problems. The classical approach to $C^{0,1}$-regularity is to start differentiating the Euler-Lagrange equation (2.10.16) (in the x_s direction, $1 \leq s \leq n$). The resulting differentiated equation formally looks like (2.10.1), with $v \equiv D_s u$, $A(x) \equiv \partial_{zz} F(Du(x))$ and μ replaced by $\partial_{x_s} \mu$ (in the distributional sense). This leads to the basic energy inequality (2.10.19), which is the starting result for the subsequent developments. Here we want to address what happens in the case of functionals with Hölder continuous coefficients, for simplicity looking at the model case (2.6.3) and Theorem 2.6.1, whose proof we are indeed going to sketch. Therefore from now on we shall assume that

$$\frac{q}{p} \leq 1 + \frac{1}{5}\left(\frac{\alpha}{n}\right)^2, \tag{2.10.39}$$

holds. Moreover, as explained in [79], we can confine ourselves on the case

$$0 < \alpha < 1. \tag{2.10.40}$$

On the other hand, the differentiable case $\alpha = 1$—Lipschitz continuous coefficients—can also be dealt with by more standard methods, see also Sect. 2.11. In the case of Theorem 2.6.1 the Euler-Lagrange equation of the functional is therefore

$$-\operatorname{div}\left[\mathfrak{c}(x)\partial_z F(Du)\right] = \mu. \tag{2.10.41}$$

This time we cannot differentiate (2.10.41) as $\mathfrak{c}(\cdot)$ is only Hölder continuous, so that the previous approach fails immediately unless $\mathfrak{c}(\cdot)$ is differentiable, i.e., $\alpha = 1$. This is in fact the point leading to consider indirect, perturbation methods to prove gradient regularity, and this is the classical basic strategy pursued in the proof of Schauder estimates. As mentioned before, this approach is succesful in the uniformly elliptic case and falls short of working in nonuniform problems. We therefore go back to the direct approach, but we perform a "fractional differentiation" of (2.10.41). In other words, while Lipschitz continuous coefficients lead to energy estimates involving second order derivatives of solutions as (2.10.19), Hölder coefficients lead to energy estimates featuring higher order fractional derivatives of solutions. This approach, developed in [79], is really too long to reproduce here and we confine ourselves to give a sketch of it; moreover, we shall confine for simplicity to the case $p \geq 2$. For this, we start recalling the basic definition of fractional Sobolev spaces. In this case, we use so-called Sobolev-Slobodevsky spaces, equipped with Gagliardo norms [1, 93].

Definition 2.10.4 Let $\beta \in (0, 1)$, $N \in \mathbb{N}$, $n \geq 2$, and let $\Omega \subset \mathbb{R}^n$ be an open subset. The space $W^{\beta,2}(\Omega, \mathbb{R}^N)$ consists of maps $w \colon \Omega \to \mathbb{R}^N$ such that

$$\|w\|_{W^{\beta,2}(\Omega)} := \|w\|_{L^2(\Omega)} + \left(\int_\Omega \int_\Omega \frac{|w(x) - w(y)|^2}{|x - y|^{n+2\beta}}\, dx\, dy\right)^{1/2}.$$

The local variant $W^{\beta,2}_{\mathrm{loc}}(\Omega, \mathbb{R}^N)$ is defined by requiring that $w \in W^{\beta,2}_{\mathrm{loc}}(\Omega, \mathbb{R}^N)$ iff $w \in W^{\beta,2}(\tilde{\Omega}, \mathbb{R}^N)$ for every open subset $\tilde{\Omega} \Subset \Omega$.

Fractional spaces come along with their own embedding; see for instance [93, Theorem 6.7]. We shall use it on balls $B_\varrho \subset \mathbb{R}^n$ in the form

$$\left(\fint_{B_\varrho} |w|^{\frac{2n}{n-2\beta}} \, dx \right)^{\frac{n-2\beta}{2n}} \lesssim_{n,\beta} \varrho^\beta \left(\int_{B_\varrho} \int_{B_\varrho} \frac{|w(x) - w(y)|^2}{|x-y|^{n+2\beta}} \, dx \, dy \right)^{1/2}$$
$$+ \left(\fint_{B_\varrho} |w|^2 \, dx \right)^{1/2}, \quad (2.10.42)$$

that holds whenever $w \in W^{\beta,2}(B_\varrho)$. This follows via a simple scaling argument.

Proceeding as in the scheme of a priori estimates for more regular minimizers, we shall now consider a local minimizer u of the functional which is such that

$$u \in C^1_{\text{loc}}(\Omega). \quad (2.10.43)$$

As usual, such an additional regularity assumption can be removed by an approximation procedure (see [79]). With the function v being defined as in (2.10.17), we have

Renormalized Fractional Caccioppoli Inequality

$$\int_{B_{\varrho/2}} \int_{B_{\varrho/2}} \frac{|(v-\kappa)_+(x) - (v-\kappa)_+(y)|^2}{|x-y|^{n+2\beta}} \, dx \, dy$$
$$\lesssim_{\text{data},\alpha,\beta} \frac{M^{\mathfrak{s}(q-p)}}{\varrho^{2\beta}} \int_{B_\varrho} (v-\kappa)_+^2 \, dx$$
$$+ \frac{M^{\mathfrak{s}q+p-\mathfrak{b}}}{\varrho^{2\beta}} \varrho^{2\alpha} \int_{B_\varrho} (|Du|+1)^{2q-2p+\mathfrak{b}} \, dx. \quad (2.10.44)$$

This holds whenever $p \geq 2$, $B_\varrho \Subset \Omega$, $\kappa \geq 0$, and for any choice of numbers $M, \mathfrak{b}, \beta$ obeying

$$\|Du\|_{L^\infty(B_\varrho)} + 1 \leq M, \quad 0 \leq \mathfrak{b} \leq p, \quad 0 < \beta < \frac{\alpha}{1+\alpha}, \quad (2.10.45)$$

where $\mathfrak{s}$ has been defined in (2.10.38).

Remark 2.10.5 (The Exponent $\mathfrak{s}$ in (2.10.38)) Note that definition of $\mathfrak{s}$ makes sense as $(n+2)p - nq > 0$, and this is a consequence of (2.10.37) (which is of course implied by (2.10.39)). As one could guess from its shape, the derivation of (2.10.44), as an a priori estimate, does not yet require the bound in (2.10.39). At this stage we only need to assume (2.10.37). Also note that

$$\mathfrak{s} \geq 1 \iff 1 \leq \frac{q}{p} < 1 + \frac{2}{n} \tag{2.10.46}$$

and that $\mathfrak{s} = 1$ iff $p = q$. The exponent $\mathfrak{s}$ plays a special role here, as, in fact, estimate (2.10.36) is a basic building block in the proof of (2.10.44). See the Commentary 2.10.6 below.

Inequality (2.10.44) is an analog of (2.10.21). In both cases we have a Caccioppoli type estimate where the integral of a higher derivative of $(v - \kappa)_+$ is controlled by the integral of $(v - \kappa)_+$. In the case of (2.10.21) we have a first derivative of $(v - \kappa)_+$, while in (2.10.44) we have a fractional derivative. The last terms are concerned wit the so-called *external ingredients*: right-hand side data μ when looking at (2.10.21), presence of coefficients when dealing with (2.10.44). Both inequalities are homogeneous with respect to $(v - \kappa)_+$, i.e., they have been renormalized hiding the nonuniform ellipticity using the constant M. A main difference between the two is the presence of the free renormalization parameter b in (2.10.44), that will be used later on. We can now apply (2.10.42) with $w \equiv (v - \kappa)_+$ in (2.10.44) and the result is

Renormalized Fractional Reverse Type Inequality with Remainder

$$\left(\fint_{B_{\varrho/2}} (v - \kappa)_+^{2\chi} \, dx\right)^{1/\chi} \lesssim_{\text{data},\alpha,\beta} M^{\mathfrak{s}(q-p)} \fint_{B_\varrho} (v - \kappa)_+^2 \, dx$$
$$+ M^{\mathfrak{s}q+p-\mathrm{b}} \varrho^{2\alpha} \fint_{B_\varrho} (|Du| + 1)^{2q-2p+\mathrm{b}} \, dx$$

$$\tag{2.10.47}$$

where

$$\chi \equiv \chi(\beta) := \frac{n}{n - 2\beta} > 1. \tag{2.10.48}$$

Again, (2.10.47) parallels (2.10.23), with a different value of χ (compare (2.10.24) with (2.10.48)). To proceed, we fix balls as in (2.10.26), we take

$$M := \|Du\|_{L^\infty(B_{\tau_2})} + 1$$

and use (2.10.27) at the generic point $x_0 \in B_{\tau_1}$, with $r_0 := (\tau_2 - \tau_1)/2$, so that $B_{2r_0}(x_0) \subset B_{\tau_2}$. This leads us to apply Lemma 2.4.4, with $h = 1, m_1 \equiv 2q - 2p + \mathrm{b}$,

$\theta_1 \equiv 1, t \equiv 2, \delta_1 \equiv \alpha, M_0 \equiv M^{\mathfrak{s}(q-p)/2}, M_1 \equiv M^{(\mathfrak{s}q+p-b)/2}, \kappa_0 \equiv 0$, thereby getting

$$v(x_0) \lesssim_{\text{data},\alpha,\beta} M^{\frac{\chi\mathfrak{s}(q-p)}{2(\chi-1)}} \left(\fint_{B_{r_0}(x_0)} v^2 \, dx \right)^{1/2}$$
$$+ M^{\frac{\mathfrak{s}(q-p)}{2(\chi-1)} + \frac{\mathfrak{s}q+p-b}{2}} \mathbf{P}_{2,\alpha}^{2q-2p+b,1}(|Du|+1; x_0, 2r_0)$$

so that, recalling (2.10.17)

$$|Du(x_0)| \lesssim_{\text{data},\alpha,\beta} M^{\frac{\chi\mathfrak{s}}{2(\chi-1)}\left(\frac{q}{p}-1\right)+\frac{1}{2}} \left(\fint_{B_{r_0}(x_0)} (|Du|^p + 1) \, dx \right)^{\frac{1}{2p}}$$
$$+ M^{\frac{\mathfrak{s}}{2(\chi-1)}\left(\frac{q}{p}-1\right)+\frac{\mathfrak{s}q+p-b}{2p}} \left[\mathbf{P}_{2,\alpha}^{2q-2p+b,1}(|Du|+1; x_0, 2r_0) \right]^{1/p} + 1.$$

The inequalities in the last two displays hold for every ball $B_{2r_0}(x_0) \Subset \Omega$ as we are assuming that Du is continuous (2.10.43). We can now use the covering argument in concentric balls (2.10.26) as after (2.10.27), thereby obtaining

$$h(\tau_1) \tag{2.10.49}$$
$$\lesssim_{\text{data},\alpha,\beta} \frac{[h(\tau_2)]^{\frac{\chi\mathfrak{s}}{2(\chi-1)}\left(\frac{q}{p}-1\right)+\frac{1}{2}}}{(\tau_2-\tau_1)^{\frac{n}{2p}}} \left(\int_{B_r} [F(Du)+1] \, dx \right)^{\frac{1}{2p}}$$
$$+[h(\tau_2)]^{\frac{\mathfrak{s}}{2(\chi-1)}\left(\frac{q}{p}-1\right)+\frac{\mathfrak{s}q+p-b}{2p}} \|\mathbf{P}_{2,\alpha}^{2q-2p+b,1}(|Du|+1;\cdot,\tau_2-\tau_1)\|_{L^\infty(B_{\tau_1})}^{1/p} + 1$$

where $h(\cdot)$ is again defined as in (2.10.28) and remark that b is still a free parameter. Now we want to proceed as after (2.10.27). For this, we first want to check that the L^∞-norm of $\mathbf{P}_{2,\alpha}^{2q-2p+b,1}$ is finite. This means applying Lemma 2.4.3 and therefore verifying that

$$\frac{n\theta}{t\delta} \equiv \frac{n}{2\alpha} > 1 \quad \text{and} \quad \frac{mn\theta}{t\delta} \equiv \frac{n(2q-2p+b)}{2\alpha} < p. \tag{2.10.50}$$

Note that the first inequality in the above display is automatically verified by (2.10.40) and that the second one is equivalent to require

$$\frac{q}{p} < 1 + \frac{\alpha}{n} - \frac{b}{2p}. \tag{2.10.51}$$

These ensure that

$$\|\mathbf{P}_{2,\alpha}^{2q-2p+b,1}(|Du|+1;\cdot,\tau_2-\tau_1)\|_{L^\infty(B_{\tau_1})} \le c\|Du\|_{L^p(B_r)}^{q-p+b/2} + c$$

$$\le c\left(\int_{B_r} \mathfrak{c}(x)F(Du)\,dx\right)^{\frac{q}{p}-1+\frac{b}{2p}} + c$$

$$\le c\left(\int_{B_r} F(Du)\,dx\right)^{\frac{q}{p}-1+\frac{b}{2p}} + c \qquad (2.10.52)$$

holds with $c \equiv c(\mathtt{data}, b)$ as a consequence of (2.4.5). Next, we have to make sure that the exponents of $h(\tau_2)$ in (2.10.49) are smaller than one, i.e.,

$$\begin{cases} \dfrac{\chi(\beta)\mathfrak{s}}{2(\chi(\beta)-1)}\left(\dfrac{q}{p}-1\right) + \dfrac{1}{2} < 1 \\ \dfrac{\mathfrak{s}}{2(\chi(\beta)-1)}\left(\dfrac{q}{p}-1\right) + \dfrac{\mathfrak{s}q+p-b}{2p} < 1. \end{cases} \qquad (2.10.53)$$

In fact, the second inequality in the above display implies the first and reduces to

$$\left(\dfrac{q}{p}-1\right)\dfrac{\mathfrak{s}n}{4\beta} + \dfrac{\mathfrak{s}}{2} - \dfrac{b}{2p} < \dfrac{1}{2}. \qquad (2.10.54)$$

In terms of q/p, condition (2.10.54) translates into

$$\dfrac{q}{p} < 1 + \left(1-\mathfrak{s}+\dfrac{b}{p}\right)\dfrac{2\beta}{\mathfrak{s}n}. \qquad (2.10.55)$$

Notice that in (2.10.51) and (2.10.55) we can still choose the values of β and b appearing in (2.10.45). To do this, note that while the right-hand side of (2.10.51) is a decreasing function of b, the right-hand side of (2.10.55) is increasing. We then optimize the choice of b by equalizing the two expressions and this leads to choose

$$b \equiv b(\beta) = \dfrac{2p[\mathfrak{s}\alpha + 2\beta(\mathfrak{s}-1)]}{\mathfrak{s}n + 4\beta} > 0. \qquad (2.10.56)$$

Notice that this is an admissible value provided $b \le p$. Recalling the definition of $\mathfrak{s}$ in (2.10.38), we note that

$$b < \dfrac{2p\alpha}{n} \iff \mathfrak{s} < 1 + \dfrac{2\alpha}{n}. \qquad (2.10.57)$$

On the other hand we have

$$(2.10.39) \implies \mathfrak{s} < 1 + \dfrac{\alpha}{2n} \qquad (2.10.58)$$

and this proves the admissibility of b in (2.10.57). Notice that condition (2.10.57) is independent of β. In fact,

$$t \in [0, p] \mapsto b(t) \text{ is decreasing} \iff \mathfrak{s} < 1 + \frac{2\alpha}{n} \qquad (2.10.59)$$

and therefore $p \geq b(0) = 2p\alpha/n > b(\beta)$, which is again (2.10.57). By inserting the value of b found in (2.10.56) in the right-hand side of (2.10.51) (which is the same that inserting b in (2.10.55)), gives the unified condition

$$\frac{q}{p} < 1 + \frac{2\beta}{n}\left[\frac{2\alpha - n(\mathfrak{s} - 1)}{\mathfrak{s} n + 4\beta}\right]. \qquad (2.10.60)$$

Notice that the right-hand side is an increasing function of β as consequence of (2.10.59). This leads us to consider the (forbidden) limiting case

$$\beta = \frac{\alpha}{1+\alpha}$$

in (2.10.60), thereby getting

$$\frac{q}{p} < 1 + \frac{2\alpha}{(1+\alpha)n}\left[\frac{2\alpha - n(\mathfrak{s}-1)}{\mathfrak{s} n + 4\alpha/(1+\alpha)}\right] =: 1 + \mathcal{R}_1(n, p, q, \alpha)\frac{\alpha^2}{n^2}$$

with

$$\mathcal{R}_1(n, p, q, \alpha) = \frac{2}{1+\alpha}\left[\frac{2 - n(\mathfrak{s}-1)/\alpha}{\mathfrak{s} + \frac{4\alpha}{n(1+\alpha)}}\right].$$

Using (2.10.58) we observe that $\mathcal{R}_1 > 2/3$. It follows that we can choose $\beta < \alpha/(1+\alpha)$ close enough to $\alpha/(1+\alpha)$ to guarantee that (2.10.60) holds so that both (2.10.51) and (2.10.53) hold too, with the corresponding choice of $b \equiv b(\beta)$ in (2.10.56). We can apply Young's inequality in (2.10.49), that, together with (2.10.52), yields

$$\mathsf{h}(\tau_1) \leq \frac{\mathsf{h}(\tau_2)}{2} + \frac{c}{(\tau_2 - \tau_1)^{n\kappa}}\left(\int_{B_r}[F(Du) + 1]\,dx\right)^\kappa$$

with $c \equiv c(\text{data}, \alpha)$ and $\kappa \equiv \kappa(n, p, q, \alpha) > 1$, and with $\mathsf{h}(\cdot)$ being defined as in (2.10.28). Applying Lemma 2.10.2 we conclude with

$$\|Du\|_{L^\infty(B_{r/2})} \lesssim_{\text{data},\alpha} \left(\fint_{B_r} F(Du)\,dx\right)^\kappa + 1$$

that is (2.6.6) under the additional a priori regularity assumption (2.10.43). As already mentioned, this can be removed via an approximation argument [79, Corollary 3] so that we obtain the full statement of Theorem 2.6.1. Full proofs, including the case $1 < p < 2$, can be found in [79].

Commentary 2.10.6 A crucial point in the proof of Theorem 2.6.1 is the fractional energy estimate (2.10.44), that, as already mentioned, cannot be obtained via direct differentiation of (2.10.41), on the contrary of (2.10.19). Inequality (2.10.44) is obtained via a sort of nonlinear dyadic/atomic decomposition technique, finding its roots in [149], and that resembles the one used for Besov spaces [2, 211] in the setting of Littlewood-Paley theory. This roughly goes as follows (see [79] for the full details). First we rescale the problem to $B_1(0)$ by passing to

$$x \to \frac{u(x_0 + \varrho x)}{\varrho}. \qquad (2.10.61)$$

For simplicity we shall still denote this function by u. We select a scale of differentiation $0 < |h| \ll 1$, where $h \in \mathbb{R}^n$, and consider small cubes and balls with radius $|h|^{1/(1+\alpha)} \gg |h|$. The idea is to cover $B_{1/2}$ with a lattice of disjoint dyadic cubes $\{Q_k \equiv Q_h(x_k)\}_{k \leq \mathfrak{n}}$ with (equal) sidelength $\approx |h|^{1/(1+\alpha)}$, centred at points $x_k \in B_{1/2}$ and with sides parallel to the coordinate axes. The number $\mathfrak{n}$ of such cubes is comparable to $|h|^{-n/(1+\alpha)}$. It follows that any dilated family $\{tQ_k\}$, $t \geq 1$ has the finite intersection property. This means that each cube of the type tQ_k does not touch others of the same type more that a finite number $\mathfrak{n} \equiv \mathfrak{n}(n, t)$ of times. The same happens for the outer balls $\{B_k \equiv B_h(x_k)\}$, defined as the smallest concentric balls to Q_k such that $Q_k \subset B_k$ (for this it is sufficient to consider the corresponding outer cubes, i.e., the smallest concentric cubes containing B_k; these are still of the type tQ_k, with $t = \sqrt{n}$). We take $|h|$ small enough to guarantee that $8B_k \Subset B_1$ (recall that B_k is centred in $B_{1/2}(0)$). All in all, we have

$$\left| B_{1/2}(0) \setminus \bigcup_{k \leq \mathfrak{n}} Q_k \right| = 0, \qquad Q_i \cap Q_j = \emptyset \Leftrightarrow i \neq j$$

and

$$\lambda(B_{1/2}(0)) \leq \sum_{k \leq \mathfrak{n}} \lambda(B_k) \leq \sum_{k \leq \mathfrak{n}} \lambda(8B_k) \lesssim_n \lambda(B_1(0)), \qquad (2.10.62)$$

that holds for any Borel measure $\lambda(\cdot)$ with finite total mass defined on $B_1(0)$. Next, on each ball $8B_k$, one defines $v_k \equiv v(B_k) \in W^{1,q}(8B_k)$ solving

$$\begin{cases} -\operatorname{div}[\mathfrak{c}(x_k)\partial_z F(Dv_k)] = 0 \\ v_k \equiv u \text{ on } \partial 8B_k . \end{cases} \qquad (2.10.63)$$

The function v_k plays the role of an "atom" in the sense of the dyadic decompositions used to decompose functions Besov spaces [2, 211]. In a very rough sense, we shall see that the atomic decomposition in question looks like

$$u(x) \approx \sum_k v_k(x) 1_{B_k}(x) + o(|h|)$$

see (2.10.70) below, where $o(|h|)$ is measured in a suitable sense. This time we employ atoms that are more special functions than those considered in the usual Besov spaces decompositions (see for instance [2, Definition 4.6.1]). Indeed, they themselves are solutions to nonlinear equations (2.10.63). This allows to such a decomposition to adapt more closely to the original problem, which is nonlinear. To proceed, we apply to v_k a variant of the classical Bernstein technique, in the sense that we get the following energy inequality

$$\int_{2B_k} |D(G(|Dv_k|) - \kappa)_+|^2 \, dx$$

$$\lesssim \frac{1}{|h|^{\frac{2}{1+\alpha}}} \int_{4B_k} \mathcal{R}_F(Dv_k)(G(|Dv_k|) - \kappa)_+^2 \, dx \qquad (2.10.64)$$

for every $\kappa \geq 0$, which is in fact (2.10.34); see [79, Lemma 5.1] and recall that $G(\cdot)$ is defined in (2.10.17). Here and until the end of this section all the constants implied in the symbol $\lesssim$ will depend at most on data and α. By standard properties of finite differences we get

$$\int_{B_k} |\tau_h(G(|Dv_k|) - \kappa)_+|^2 \, dx \lesssim |h|^2 \int_{2B_k} |D(G(|Dv_k|) - \kappa)_+|^2 \, dx \qquad (2.10.65)$$

where, as usual, $\tau_h w(x) := w(x+h) - w(x)$ denotes the standard finite difference operator of a function w in direction $h \in \mathbb{R}^n$. Note that here we have used that

$$|h| \leq |h|^{1/(1+\alpha)} \implies B_k + h \subset 2B_k \,. \qquad (2.10.66)$$

Combining (2.10.65) and (2.10.64) we gain

$$\int_{B_k} |\tau_h(G(|Dv_k|) - \kappa)_+|^2 \, dx$$

$$\lesssim |h|^{\frac{2\alpha}{1+\alpha}} \int_{4B_k} \mathcal{R}_F(Dv_k)(G(|Dv_k|) - \kappa)_+^2 \, dx \,. \qquad (2.10.67)$$

Using additional a priori estimates for solutions to (2.10.63), that is, applying (2.10.36) to v_k, then gives

$$\|Dv_k\|_{L^\infty(4B_k)} \lesssim \|Du\|_{L^\infty(8B_k)}^5 + 1 \lesssim M^5$$

where 𝔰 is in (2.10.54) and provided M is as in (2.10.45), [79, Proposition 5.3]. Recalling (2.10.20), and using the last inequality in (2.10.67), we find

$$\int_{B_k} |\tau_h(G(|Dv_k|) - \kappa)_+|^2 \, dx$$
$$\lesssim |h|^{\frac{2\alpha}{1+\alpha}} M^{(q-p)\mathfrak{s}} \int_{4B_k} (G(|Dv_k|) - \kappa)_+^2 \, dx \,. \tag{2.10.68}$$

The one in (2.10.68) is a sort of Nikolski space estimate at the microlocal level B_k. Indeed, recall that a function $w \in L^2(\Omega)$ belongs to Nikolski space $\mathcal{N}^{s,2}(\Omega)$ iff

$$\int_{\Omega_{|h|}} |\tau_h w|^2 \, dx \lesssim |h|^{2s}, \qquad \forall \, h \in \mathbb{R}^n, \tag{2.10.69}$$

where $\Omega_h := \{x \in \Omega : \text{dist}(x, \partial\Omega) > |h|\}$. Such spaces relate well with Sobolev-Slobodevski spaces (see Lemma 2.10.7 below and [3] for more details); all of them are special cases of a larger family of fractional spaces called Besov spaces [211]. Here the word microlocal refers to the fact that in (2.10.68) the differentiability gain $|h|^{2\alpha/(1+\alpha)}$ is obtained on a domain B_k whose diameter is proportional to $|h|^{1/(1+\alpha)}$, and therefore depends on $|h|$, on the contrary of the classical definition of Nikolski space in (2.10.69). A point worth remarking here is that the gain of the factor $|h|^{2\alpha/(1+\alpha)}$ comes from the combination of (2.10.64) and (2.10.65); in this respect choosing the size of the sidelength of Q_k to be comparable to $|h|^{1/(1+\alpha)}$, and therefore larger to the differentiation scale $|h|$, is crucial. The special exponent $1/(1+\alpha)$ optimizes the procedure.

Estimate (2.10.68) is eventually transferred to u via the comparison estimate

$$\int_{4B_k} |(G(|Du|) - \kappa)_+ - (G(|Dv_k|) - \kappa)_+|^2 \, dx$$
$$\lesssim |h|^{\frac{2\alpha}{1+\alpha}} M^{\mathfrak{s}p+p-\mathfrak{b}} \varrho^{2\alpha} \int_{8B_k} (|Du| + 1)^{2q-2p+\mathfrak{b}} \, dx \,. \tag{2.10.70}$$

We transfer the microlocal information in (2.10.68) from v_k to u on B_k, as follows

$$\int_{B_k} |\tau_h(G(Du) - \kappa)_+|^2 \, dx \lesssim \int_{B_k} |\tau_h(G(Dv_k) - \kappa)_+|^2 \, dx$$
$$+ \int_{B_k} |(G(Du(\cdot + h)) - \kappa)_+ - (G(Dv_k(\cdot + h)) - \kappa)_+|^2 \, dx$$
$$+ \int_{B_k} |(G(Du) - \kappa)_+ - (G(Dv_k) - \kappa)_+|^2 \, dx$$

$$\lesssim \int_{B_k} |\tau_h(G(Dv_k) - \kappa)_+|^2 \, dx$$
$$+ \int_{2B_k} |(G(Du) - \kappa)_+ - (G(Dv_k) - \kappa)_+|^2 \, dx \,.$$

Note that here we have again used (2.10.66). Using (2.10.68) and (2.10.70) we then find

$$\int_{B_k} |\tau_h(G(Du) - \kappa)_+|^2 \, dx \lesssim |h|^{\frac{2\alpha}{1+\alpha}} M^{(q-p)\mathsf{s}} \int_{4B_k} (G(|Dv_k|) - \kappa)_+^2 \, dx$$
$$+ |h|^{\frac{2\alpha}{1+\alpha}} M^{\mathsf{s}p+p-\mathsf{b}} \varrho^{2\alpha} \int_{8B_k} (|Du| + 1)^{2q-2p+\mathsf{b}} \, dx$$

and finally applying (2.10.70) once again, we conclude with

$$\int_{B_k} |\tau_h(G(Du) - \kappa)_+|^2 \, dx \lesssim |h|^{\frac{2\alpha}{1+\alpha}} M^{\mathsf{s}(q-p)} \int_{8B_k} (G(Du) - \kappa)_+^2 \, dx$$
$$+ |h|^{\frac{2\alpha}{1+\alpha}} M^{\mathsf{s}q+p-\mathsf{b}} \varrho^{2\alpha} \int_{8B_k} (|Du| + 1)^{2q-2p+\mathsf{b}} \, dx \,.$$

Adding up the above inequalities over the decomposition $\{B_k\}$, and using (2.10.62), we obtain

$$\int_{B_{1/2}(0)} |\tau_h(G(Du) - \kappa)_+|^2 \, dx \lesssim |h|^{\frac{2\alpha}{1+\alpha}} M^{\mathsf{s}(q-p)} \int_{B_1(0)} (G(Du) - \kappa)_+^2 \, dx$$
$$+ |h|^{\frac{2\alpha}{1+\alpha}} M^{\mathsf{s}q+p-\mathsf{b}} \varrho^{2\alpha} \int_{B_1(0)} (|Du| + 1)^{2q-2p+\mathsf{b}} \, dx \,.$$

We are now in the position to use Lemma 2.10.7 below, that gives

$$\int_{B_{1/2}(0)} \int_{B_{1/2}(0)} \frac{|(G(Du) - \kappa)_+(x) - (G(Du) - \kappa)_+(y)|^2}{|x-y|^{n+2\beta}} \, dx \, dy$$
$$\lesssim M^{\mathsf{s}(q-p)} \int_{B_1(0)} (G(Du) - \kappa)_+^2 \, dx$$
$$+ M^{\mathsf{s}q+p-\mathsf{b}} \varrho^{2\alpha} \int_{B_1(0)} (|Du| + 1)^{2q-2p+\mathsf{b}} \, dx \,.$$

This holds whenever $0 < \beta < \alpha/(1+\alpha)$. From this (2.10.44) follows rescaling as in (2.10.61).

Lemma 2.10.7 ([77, Lemma 1]) *Let $B_\varrho \Subset B_r \subset \mathbb{R}^n$ be concentric balls with $r \leq 1$, $w \in L^2(B_r, \mathbb{R}^k)$, $s \geq 1$ and assume that, for $\beta \in (0, 1]$, $\mathcal{H} \geq 1$, there holds*

$$\|\tau_h w\|_{L^2(B_\varrho)} \leq \mathcal{H}|h|^\beta \quad \text{for every } h \in \mathbb{R}^n \text{ with } 0 < |h| \leq \frac{r - \varrho}{K}, \text{ where } K \geq 1.$$

Then it holds that

$$\|w\|_{W^{\alpha_0, 2}(B_\varrho)}$$
$$\lesssim_{n,s} \frac{1}{(\beta - \alpha_0)^{1/2}} \left(\frac{r - \varrho}{K}\right)^{\beta - \alpha_0} \mathcal{H} + \left(\frac{K}{r - \varrho}\right)^{n/2 + \alpha_0} \|w\|_{L^2(B_\varrho)}.$$

The approach outlined here is robust and flexible enough to allow for several applications. The point is to choose the right form of the gradient function v playing the role of subsolution in the setting of Bernstein technique. An example is given in [80], where, in order to prove Theorems 2.9.1 and 2.9.3, a fractional Caccioppoli inequality has been derived for the function

$$v = |Du|^{2-\gamma} + \mathfrak{a}(\cdot)|Du|^q, \tag{2.10.71}$$

where $\gamma \in [1, 2)$ is suitably close to 1 as described in Theorem 2.9.3. One takes $\gamma = 1$ in the case of Theorem 2.9.1 and of the model functional in (2.9.2). The choice (2.10.71) reflects the structure of the functional considered in (2.9.5). Specifically, the scheme outlined above works with an inequality of the type

$$\int_{B_{\varrho/2}} \int_{B_{\varrho/2}} \frac{|(v - \kappa)_+(x) - (v - \kappa)_+(y)|^2}{|x - y|^{n+2\beta}} \, dx \, dy$$
$$\lesssim \frac{M^{2\mathfrak{s}_1}}{\varrho^{2\beta}} \int_{B_\varrho} (v - \kappa)_+^2 \, dx$$
$$+ \frac{M^{2\mathfrak{s}_2}}{\varrho^{2\beta}} \varrho^{2\alpha} \int_{B_\varrho} (|Du| + 1)^{2(q-1+\delta_2)} \, dx$$
$$+ \frac{M^{2\mathfrak{s}_3}}{\varrho^{2\beta}} \varrho^\alpha \int_{B_\varrho} (|Du| + 1)^{q-1+\delta_2} \, dx$$
$$+ \frac{M^{2\mathfrak{s}_3}}{\varrho^{2\beta}} \varrho^{\alpha_0} \int_{B_\varrho} (|Du| + 1)^{3\delta_2} \, dx \tag{2.10.72}$$

where v is as in (2.10.71) and this time M satisfies the intrinsic relation

$$\sup_{x \in B_\varrho} |Du(x)|^{2-\gamma} + \mathfrak{a}(x)|Du(x)|^q + 1 \leq M.$$

Here the free parameters $\delta_{1,2,3}$ can be chosen small at will and the parameters $\mathfrak{s}_{1,2,3}$ are suitable numbers related to $\delta_{1,2,3}$ and γ, q. Inequality (2.10.72) provides useful estimates provided the sharp bound (2.9.4) is satisfied and γ is suitably close to 1. The crucial point in [80] is the intrinsic approach fully using the structure in (2.9.5) for which the choice in (2.10.71) becomes effective.

We finally observe that the approach to L^∞-bounds via fractional De Giorgi's iterations has been introduced, independently, in [43] and [181]. In both cases the starting point is a fractional Caccioppoli inequality. In case of [43] this is natural as the problem considered involves a nonlocal operator (like a fractional power of the Laplacean) and therefore the starting energy inequality comes from simple testing. In [181] the equation considered is of the type in (2.1.2), with $p = 2$ and μ being a Borel measure with finite total mass, and the decomposition argument explained in this section is required. In this last case, fractional Caccioppoli's inequalities are obtained for the function $v = |Du|$. An approach using fractional De Giorgi's classes has been used later in [64, 184]. See [179, 180] for fractional estimates for nonlinear elliptic systems and their applications to singular sets estimates.

> **The Key Takeaway**
> When full higher derivatives of solutions are not available, try fractional ones. The scheme [renormalized energy estimates $\oplus$ De Giorgi type iterations $\oplus$ nonlinear potentials] is robust enough to resist the lack of both full second order derivatives and uniform ellipticity. This allows to replace differentiability of coefficients with Hölder continuity and to give a unified approach to gradient boundedness of solutions, avoiding perturbation methods already in classical cases.

2.10.4 Stein Type Theorems [79]

Here we briefly sketch the proof of the a priori estimate of Theorem 2.8.1, and again in the case $p \geq 2$; we keep the notation introduced in the previous sections. Full details of the proofs can be found in [79]. As described in Sect. 2.6.3, this time we cannot even use (2.6.21) since $y \mapsto h(\cdot, y)$ is only Hölder continuous. On the other hand, it is still possible to get a fractional Caccioppoli inequality [79, Proposition 5.2], that now looks like

$$\int_{B_{\varrho/2}} \fint_{B_{\varrho/2}} \frac{|(v-\kappa)_+(x) - (v-\kappa)_+(y)|^2}{|x-y|^{n+2\beta}} \, dx \, dy$$

$$\lesssim_{\text{data},\alpha,\beta} \frac{M^{\mathfrak{s}(q-2)}}{\varrho^{2\beta}} \fint_{B_\varrho} (v-\kappa)_+^2 \, dx + \frac{M^{\mathfrak{s}q}}{\varrho^{2\beta}} \varrho^{\frac{2\alpha}{2-\alpha}} \fint_{B_\varrho} \mu^{\frac{2}{2-\alpha}} \, dx \, . \quad (2.10.73)$$

This holds whenever $B_\varrho \Subset \Omega$ and $\kappa \geq 0$, where M is as in (2.10.21), $\mathfrak{s}$ as in (2.10.38), v as in (2.10.17) and this time β satisfies

$$0 < \beta < \frac{\alpha}{2+\alpha}. \tag{2.10.74}$$

Of course, we still assume the a propri regularity in (2.10.43); this additional property can be removed via a (delicate) approximation procedure as detailed in [79]. Using (2.10.42) with $w \equiv (v - \kappa)_+$ in (2.10.73) yields

$$\left(\fint_{B_{\varrho/2}} (v-\kappa)_+^{2\chi} \, dx \right)^{1/\chi}$$

$$\lesssim_{\text{data},\alpha,\beta} M^{\mathfrak{s}(q-2)} \fint_{B_\varrho} (v-\kappa)_+^2 \, dx + M^{\mathfrak{s}q} \varrho^{\frac{2\alpha}{2-\alpha}} \fint_{B_\varrho} \mu^{\frac{2}{2-\alpha}} \, dx$$

where $\chi \equiv \chi(\beta) > 1$ is as in (2.10.48). Again, compare this last inequality with (2.10.4) and with (2.10.23). In all cases the last integrals account for the external ingredients: coefficients or data. We are now in business to apply Lemma 2.4.4, that gives

$$v(x_0) \lesssim_{\text{data},\alpha,\beta} M^{\frac{\chi \mathfrak{s}(q-2)}{2(\chi-1)}} \left(\fint_{B_{r_0}(x_0)} v^2 \, dx \right)^{1/2}$$

$$+ M^{\frac{\mathfrak{s}(q-2)}{2(\chi-1)} + \frac{\mathfrak{s}q}{2}} \mathbf{P}_{2,\alpha/(2-\alpha)}^{2/(2-\alpha),1}(\mu; x_0, 2r_0)$$

so that, after a few elementary manipulations, we get

$$|Du(x_0)| \lesssim_{\text{data},\alpha,\beta} M^{\frac{\chi \mathfrak{s}}{2(\chi-1)}(\frac{q}{2}-1)+\frac{1}{2}} \left(\fint_{B_{r_0}(x_0)} (|Du|^2 + 1) \, dx \right)^{1/4}$$

$$+ M^{\frac{\mathfrak{s}}{2(\chi-1)}(\frac{q}{2}-1)+\frac{\mathfrak{s}q}{4}} \left[\mathbf{P}_{2,\alpha/(2-\alpha)}^{2/(2-\alpha),1}(\mu; x_0, 2r_0) \right]^{1/2}.$$

As after (2.10.26), we find

$$h(\tau_1) \lesssim_{\text{data},\alpha,\beta} \frac{[h(\tau_2)]^{\frac{\chi \mathfrak{s}}{2(\chi-1)}(\frac{q}{2}-1)+\frac{1}{2}}}{(\tau_2 - \tau_1)^{\frac{n}{2p}}} \left(\int_{B_r} [F(Du) + 1] \, dx \right)^{1/4}$$

$$+ [h(\tau_2)]^{\frac{\mathfrak{s}}{2(\chi-1)}(\frac{q}{2}-1)+\frac{\mathfrak{s}q}{4}} \| \mathbf{P}_{2,\alpha/(2-\alpha)}^{2/(2-\alpha),1}(\mu; \cdot, \tau_2 - \tau_1) \|_{L^\infty(B_{\tau_1})}^{1/2} + 1.$$

From this inequality we can now proceed as after (2.10.27). Lemma 2.4.3 now gives

$$\| \mathbf{P}_{2,\alpha/(2-\alpha)}^{2/(2-\alpha),1}(\mu; \cdot, \tau_2 - \tau_1) \|_{L^\infty(B_{\tau_1})} \lesssim_{n,\alpha} \| f \|_{n/\alpha, 1/(2-\alpha); B_r}^{1/(2-\alpha)}.$$

Then we require

$$\begin{cases} \dfrac{\chi(\beta)\mathsf{s}}{2(\chi(\beta)-1)}\left(\dfrac{q}{2}-1\right)+\dfrac{1}{2}<1 \\ \dfrac{\mathsf{s}}{2(\chi(\beta)-1)}\left(\dfrac{q}{2}-1\right)+\dfrac{\mathsf{s}q}{4}<1. \end{cases}$$

Also in this case the second inequality implies the first and therefore we finally ask that

$$\left(\dfrac{q}{2}-1\right)\dfrac{\mathsf{s}(n-2\beta)}{4\beta}+\dfrac{\mathsf{s}q}{4}<1.$$

In turn, by using (2.6.23), we can find β within the range (2.10.74) such that the above inequality is true. The rest of the proof, leading to the a priori estimate of Theorem 2.8.1, follows as in Sect. 2.10.3.

2.10.5 Optimal Gap Bounds [82]

A key to the proof of Theorem 2.7.1 is a preliminary higher integrability result, i.e.,

$$Du \in L^q_{\mathrm{loc}}(\Omega), \quad \text{for every } q < \infty. \tag{2.10.75}$$

This is based on a careful use of delicate Besov spaces techniques. Besov spaces methods are sometimes employed in uniformly elliptic problems with a certain degree of lack of ellipticity like for instance operators in the Heisenberg group [94], or in nonlocal problems [31, 32, 113]. The proof of (2.10.75) relies on estimates of the type

$$\left\|\dfrac{\tau_h(\tau_h u)}{|h|^{s_i}}\right\|_{L^{q_i}} \leq c_i \tag{2.10.76}$$

for $h \in \mathbb{R}^n$ and $|h|$ suitably small and with sequences $q_i \to \infty$, $1 < s_i \to 1$ an $c_i \to \infty$. Such estimates eventually imply (2.10.75). The proof of (2.10.75) also yields an intermediate result, connected to Theorem 2.3.1, (2.3.2), and Theorem 2.5.4, in the sense that the local boundedness of minima makes the gap bounds independent of n.

Theorem 2.10.8 (Non-dimensional Gap Bound) *Let* $u \in W^{1,p}_{\mathrm{loc}}(\Omega) \cap L^{\infty}_{\mathrm{loc}}(\Omega)$ *be a minimizer of the functional* $\overline{\mathcal{F}_x}(\cdot, B)$ *in (2.6.14) for every ball* $B \Subset \Omega$, *under assumptions (2.6.13) with* $p \leq n$ *and*

$$q < p + \alpha \min\{1, p/2\}. \qquad (2.10.77)$$

Then (2.10.75) holds. Moreover,

$$\|Du\|^{\mathfrak{q}}_{L^{\mathfrak{q}}(B_{\varrho/2})} \lesssim_{\mathrm{data}} \left(\frac{\|u - (u)_{B_{\varrho}}\|_{L^{\infty}(B_{\varrho})}}{\varrho} + 1 \right)^{\mathfrak{b}_{\mathfrak{q}}} \left[\overline{\mathcal{F}_x}(u, B_{\varrho}) + \varrho^n \right]$$

holds for every $\mathfrak{q} < \infty$, *and for every ball* $B_{\varrho} \Subset \Omega$, $\varrho \leq 1$, *where* $\mathfrak{b}_{\mathfrak{q}} \equiv \mathfrak{b}_{\mathfrak{q}}(n, p, q, \alpha, \mathfrak{q})$. *Finally, if* $\mathcal{L}_{\mathcal{F}_x}(u, B) = 0$ *holds for every ball* $B \Subset \Omega$, *then the same result holds for* $W^{1,p}_{\mathrm{loc}}(\Omega) \cap L^{\infty}_{\mathrm{loc}}(\Omega)$-*regular minimizers* u *of* $\mathcal{F}_x$ *in (2.6.12).*

When $p \geq 2$ condition (2.10.77) is sharp, again by the counterexamples in [102, 106]. Note that in Theorem 2.10.8 we are restricting to the case $p \leq n$ otherwise $u \in L^{\infty}_{\mathrm{loc}}(\Omega)$ is automatically verified and indeed condition (2.10.77) is more restrictive than $q/p < 1 + \alpha/n$.

Going back to the proof of Theorem 2.7.1, the use of higher integrability (2.10.75) allows to upgrade the fractional technique explained in Sect. 2.10.3 and improve the exponents via a new fractional Caccioppoli type inequality of the type in (2.10.44). A crucial role in obtaining the sharp condition $q/p < 1 + \alpha/n$ is at this stage played by the use of Theorem 2.5.2 (and the related a priori estimate). In other words, estimate (2.10.36) holds with

The New Exponent $\mathfrak{s}$

$$\mathfrak{s} := \begin{cases} \dfrac{2q}{(n+1)p - (n-1)q} & \text{if either } n \geq 4 \text{ or } n = 2 \\[2mm] \text{any number} > \dfrac{q}{2p-q} & \text{if } n = 3 \text{ and } q > p \\[2mm] 1 & \text{if } n = 3 \text{ and } q = p. \end{cases} \qquad (2.10.78)$$

This follows using the results in [17, 19, 199]. This new exponent is smaller than the one appearing in (2.10.38). Summarizing, passing from the first gap bound in (2.6.5) to the sharp one $q/p < 1 + \alpha/n$ goes in two steps:

- The structural improvement. Using (2.10.75) allows to improve the quadratic decay gap bound in (2.6.5) to a linear one, that is something of the form (2.7.1).
- The fine improvement. The use of estimate (2.10.36) with the new exponent $\mathfrak{s}$ in (2.10.78), allowing to finally reach $q/p < 1 + \alpha/n$. In this respect note that the use of the results in [17, 19, 199] is really essential in that the use of the previous exponent in (2.10.38) would at this stage lead to yet the suboptimal bound

$$\frac{q}{p} < 1 + \frac{\alpha}{n+1}.$$

2.11 Moser Strikes Again [76]

In the above section we have seen the general three-step scheme

$$\begin{cases} \text{Renormalized Caccioppoli type inequalities} \\ \text{De Giorgi type iterations} \\ \text{Use of Nonlinear potentials.} \end{cases} \quad (2.11.1)$$

As observed in [76] a similar, parallel approach using the first two points in (2.11.1) can be employed via Moser iteration. We indeed sketch how to obtain an a priori estimate for minima of functionals of the type (2.1.1) with $\mu \equiv 0$. The assumptions we are considering here are

$$\begin{cases} [H_1(z)]^{p/2} \leq F(x,z) \leq L[H_1(z)]^{q/2} \\ [H_1(z)]^{(p-2)/2}|\xi|^2 \leq \partial_{zz}F(x,z)\xi \cdot \xi \\ |\partial_{zz}F(x,z)|[H_1(z)]^{1/2} + |\partial_{zx}F(x,z)| \leq L[H_1(z)]^{(q-1)/2}, \end{cases} \quad (2.11.2)$$

for every $x \in \Omega$, $z, \xi \in \mathbb{R}^n$, where $L \geq 1$; the integrand $F\colon \Omega \times \mathbb{R}^n \to [0,\infty)$ is assumed to be locally C^2-regular with respect to the gradient variable and is continuous; finally its derivatives are Carathéodory-regular. We recall that, according to (2.1.7), it is $H_1(z) = |z|^2 + 1$. Assumptions (2.11.2) guarantee that minimizers u satisfy the a priori estimate

$$\|Du\|_{L^\infty(B/2)} \lesssim_{\text{data}} \left(\fint_B F(x, Du)\,dx\right)^{\frac{1}{(nb+1)p - nbq}} + 1, \quad (2.11.3)$$

whenever $B \equiv B_\varrho \Subset \Omega$ is such that $\varrho \leq 1$, where data $= (n, p, q, L)$ and

$$\mathfrak{b} := \begin{cases} 1 \text{ if } n > 2 \\ > 1 \text{ such that } 2\mathfrak{b}(q - p) < p \text{ if } n = 2, \end{cases} \tag{2.11.4}$$

provided

$$\frac{q}{p} < 1 + \frac{1}{n} \tag{2.11.5}$$

is satisfied and the approximation in energy property (2.3.9) holds for $w \equiv u$ (absence of Lavrentiev phenomenon), i.e.,

$$\begin{cases} \text{there exists } \{u_k\}_k \subset C^\infty(B) \text{ such that } u_k \to u \text{ strongly in } W^{1,p}(B) \\ \text{and } F(\cdot, Du_k) \to F(\cdot, Du) \text{ in } L^1(B). \end{cases} \tag{2.11.6}$$

Notice that condition (2.11.5) is optimal by the counterexamples in [102, 106]; see Sect. 2.2.4. Notice also that in (2.11.6) it is equivalent to require $\{u_k\}_k \subset W^{1,q}(B)$ due to the growth assumptions in $(2.11.2)_1$.

We shall give a rapid proof of (2.11.3) as an a priori estimate, that is, assuming the a priori higher integrability $u \in W^{1,\infty}(B)$; the general case can then be treated via an approximation argument, based on (2.11.6), as for instance shown in [102]. The point we want to stress now is the following: similarly to the method demonstrated in Sect. 2.10, renormalizing the relevant energy inequalities, and taking advantage of the precise dependence of the constants along the iterations (in this case Lemma 2.4.6) rapidly leads to (2.11.3). The outcome is a very streamlined proof that in fact literally reduces estimates for (p, q)-growth functionals to those valid when $p = q$. For the proof of (2.11.3) we preliminary observe that we can reduce to the case $B \equiv B_1 := B_1(0)$ by using the same scaling argument in (2.10.61) (in the following lines we denote $\mathcal{B}_\tau \equiv B_\tau(0)$). The function in (2.10.61) locally minimizes the same functional endowed with new (rescaled) integrand $(x, z) \mapsto F(x_0 + \varrho x, z)$, which satisfies (2.11.2) as we are assuming that $\varrho \leq 1$. We still denote by u the blown-up function in (2.10.61). Following [76], we first recall that in the uniformly elliptic case $p = q$ the inequality

$$\left(\int_{\mathcal{B}_{\varrho_1}} v^{(\gamma + p/2)\chi} \, dx \right)^{1/\chi} \lesssim_{n,p} \frac{L^2 (1 + \gamma)^2}{(\varrho_2 - \varrho_1)^2} \int_{\mathcal{B}_{\varrho_2}} v^{\gamma + p/2} \, dx, \quad \forall \gamma \geq 0 \tag{2.11.7}$$

with

$$v := H_1(Du) = |Du|^2 + 1$$

holds whenever $\mathcal{B}_{\tau_1} \Subset \mathcal{B}_{\varrho_1} \Subset \mathcal{B}_{\varrho_2} \Subset \mathcal{B}_{\tau_2} \Subset \mathcal{B}_1$, and $\chi = 2^*/2$. As usual, $2^* > 2$ is the Sobolev embedding exponent defined as in (2.1.5). Note that

$$\frac{\chi}{\chi - 1} = \frac{2^*}{2^* - 2} \qquad (2.11.8)$$

and that a crucial point here is that in (2.11.7) the dependence on L is made explicit. Via Lemma 2.4.6, and recalling (2.11.8), this yields

$$\|v\|_{L^\infty(\mathcal{B}_{\tau_1})} \lesssim_{n,p} \frac{L^{\frac{4}{p}\frac{2^*}{2^*-2}}}{(\tau_2 - \tau_1)^{\frac{4}{p}\frac{2^*}{2^*-2}}} \|v\|_{L^{p/2}(\mathcal{B}_{\tau_2})} . \qquad (2.11.9)$$

We now show how to use the above estimate, that holds when $p = q$, to treat the case $p < q$ straightaway. For this, observe that, when handling integral quantities evaluated on the ball $\mathcal{B}_{\tau_2}$ we have

$$|z| \equiv |Du| \leq \|Du\|_{L^\infty(\mathcal{B}_{\tau_2})}$$

and therefore we can always replace $(2.11.2)_3$ by

$$\begin{cases} |\partial_{zz} F(x, z)|[H_1(z)]^{1/2} + |\partial_{zx} F(x, z)| \leq \mathfrak{L}[H_1(z)]^{(p-1)/2} \text{ on } \mathcal{B}_{\tau_2}, \\ \mathfrak{L} := L\|H_1(Du)\|_{L^\infty(\mathcal{B}_{\tau_2})}^{(q-p)/2} \equiv L\|v\|_{L^\infty(\mathcal{B}_{\tau_2})}^{(q-p)/2} . \end{cases} \qquad (2.11.10)$$

Indeed, all we need to prove (2.11.7) is the Euler-Lagrange equation

$$\int_{\mathcal{B}_1} \partial_z F(x, Du) \cdot D\varphi \, dx = 0$$

with sutable test functions φ supported $\mathcal{B}_{\tau_2}$; this means that we can use (2.11.10). Note that the only part of (2.11.2) which is used to prove (2.11.7) is actually $(2.11.2)_{2,3}$. Applying (2.11.9) with L, replaced by $\mathfrak{L}$, we now gain

$$\|v\|_{L^\infty(\mathcal{B}_{\tau_1})} \lesssim_{n,p,L} \frac{\|v\|_{L^\infty(\mathcal{B}_{\tau_2})}^{\frac{2(q-p)}{p}\frac{2^*}{2^*-2}}}{(\tau_2 - \tau_1)^{\frac{4}{p}\frac{2^*}{2^*-2}}} \|v\|_{L^{p/2}(\mathcal{B}_{\tau_2})} . \qquad (2.11.11)$$

When $n > 2$ we note that

$$\frac{2(q-p)2^*}{p(2^*-2)} = \frac{(q-p)n}{p} \stackrel{(2.11.5)}{<} 1 .$$

Young's inequality in (2.11.11) then yields

$$h(\tau_1) \leq \frac{h(\tau_2)}{2} + \frac{c\|v\|_{L^{p/2}(\mathcal{B}_{\tau_2})}^{\frac{p}{(n+1)p-nq}}}{(\tau_2 - \tau_1)^{\frac{2n}{(n+1)p-nq}}} \qquad (2.11.12)$$

where $h(t) := \|v\|_{L^{\infty}(\mathcal{B}_t)}$ and $c \equiv c(n, p,)$. Lemma 2.10.2 finally gives

$$\|v\|_{L^{\infty}(\mathcal{B}_{1/2})} \lesssim_{\text{data}} \|v\|_{L^{p/2}(\mathcal{B}_1)}^{\frac{p}{(n+1)p-nq}} \lesssim_{\text{data}} \left(\int_{\mathcal{B}_1} (|Du|^p + 1)\,dx \right)^{\frac{2}{(n+1)p-nq}}$$

from which (2.11.3) follows using (2.11.2)$_1$. This proves (2.11.3) in the case $n > 2$ and $\varrho = 1$; scaling back we get the general case. When $n = 2$ note that $q/p < 3/2$, that is the assumed bound (2.11.5), implies $2(q - p)/p < 1$, therefore we take 2^* large enough (recall (2.1.5)) in order to achieve

$$\frac{2(q-p)2^*}{p(2^* - 2)} \leq \frac{2\mathfrak{b}(q-p)}{p} \stackrel{(2.11.4)}{<} 1$$

so that (2.11.11) is replaced by

$$\|v\|_{L^{\infty}(\mathcal{B}_{\tau_1})} \lesssim_p \frac{\|v\|_{L^{\infty}(\mathcal{B}_{\tau_2})}^{2\mathfrak{b}(q-p)/p}}{(\tau_2 - \tau_1)^{4\mathfrak{b}/p}} \|v\|_{L^{p/2}(\mathcal{B}_{\tau_2})}$$

and we can proceed as in the case $n > 2$.

> **The Key Takeaway**
> Renormalized energy inequalities can be used also in the setting of Moser iteration technique. This, together with quantitative knowledge of the constants along the interation schemes, allows in several cases to reduce the treatment of nonuniformly elliptic problems with polynomial growth to that of uniformly elliptic ones.

References

1. Adams, R.A., Fournier, J.J.F.: Sobolev Spaces. Pure Mathematics vs. Applied Mathematics, 2nd edn., vol. 140. Elsevier/Academic Press, Amsterdam (2003)
2. Adams, D.R., Hedberg, L.I.: Function Spaces and Potential Theory. Grundlehren der Mathematischen Wissenschaften, vol. 314. Springer, Berlin (1996)

3. Avelin, B., Kuusi, T., Mingione, G.: Nonlinear Calderón-Zygmund theory in the limiting case. Arch. Ration. Mech. Anal. **227**, 663–714 (2018)
4. Baasandorj, S., Byun, S.: Regularity for Orlicz phase problems. Memoirs Am. Math. Soc. (to appear)
5. Baasandorj, S., Byun, S., Oh, J.: Gradient estimates for multiphase problems. Calc. Var. PDE **60**, 104 (2021)
6. Balci, A., Surnachev, M.: Lavrentiev gap for some classes of generalized Orlicz functions. Nonlinear Anal. **207**, 112329 (2021)
7. Balci, A., Diening, L., Surnachev, M.: New examples on Lavrentiev gap using fractals. Calc. Var. PDE **59**, 180 (2020)
8. Balci, A., Diening, L., Giova, R., Passarelli di Napoli, A.: Elliptic equations with degenerate weights. Siam J. Math. Anal. **54**, 5017–5078 (2022)
9. Balci, A., Diening, L., Surnachev, M.: Scalar minimizers with maximal singular sets and lack of Meyers property (2023). https://arxiv.org/abs/2312.15772
10. Balci, A., Ortner, C., Storn, J.: Crouzeix-Raviart finite element method for nonautonomous variational problems with Lavrentiev gap. Numer. Math. **151**, 779–805 (2022)
11. Banerjee, A., Munive, H.I.: Gradient continuity estimates for the normalized p-Poisson equation. Commun. Contemp. Math. **22**, 1950069, 24pp. (2020)
12. Baroni, P.: Riesz potential estimates for a general class of quasilinear equations. Calc. Var. PDE **53**, 803–846 (2015)
13. Baroni, P., Colombo, M., Mingione, G.: Regularity for general functionals with double phase. Calc. Var. PDE **57**, 62 (2018)
14. Beck, L., Mingione, G.: Lipschitz bounds and nonuniform ellipticity. Commun. Pure Appl. Math. **73**, 944–1034 (2020)
15. Beck, L., Schmidt, T.: On the Dirichlet problem for variational integrals in BV. J. Reine Angew. Math. (Crelle J.) **674**, 113–194 (2013)
16. Beck, L., Schmidt, T.: Interior gradient regularity for BV minimizers of singular variational problems. Nonlinear Anal. **120**, 86–106 (2015)
17. Bella, P., Schäffner, M.: On the regularity of minimizers for scalar integral functionals with (p,q)-growth. Anal. PDE **13**, 2241–2257 (2020)
18. Bella, P., Schäffner, M.: Local boundedness and Harnack inequality for solutions of linear nonuniformly elliptic equations. Commun. Pure Appl. Math. **74**, 453–477 (2021)
19. Bella, P., Schäffner, M.: Lipschitz bounds for integral functionals with (p,q)-growth conditions. Adv. Calc. Var. https://doi.org/10.1515/acv-2022-0016
20. Benilan, P., Boccardo, L., Gallouët, T., Gariepy, R., Pierre, M., Vázquez, J.L.: An L^1-theory of existence and uniqueness of solutions of nonlinear elliptic equations. Ann. Scuola Norm. Sup. Pisa Cl. Sci. (IV) **22**, 241–273 (1995)
21. Bezerra, E.C., Jr., da Silva, J.V., Ricarte, G.C.: Fully nonlinear singularly perturbed models with non-homogeneous degeneracy. Rev. Mat. Iberoam. **39**, 123–164 (2022)
22. Biagi, S., Esposito, F., Vecchi, E.: Symmetry and monotonicity of singular solutions of double phase problems. J. Differ. Equ. **280**, 435–463 (2021)
23. Biagi, S., Esposito, F., Vecchi, E.: Symmetry of intrinsically singular solutions of double phase problems. Differ. Int. Equ. **36**, 229–246 (2023)
24. Bildhauer, M.: Convex Variational Problems. Linear, Nearly Linear and Anisotropic Growth Condition. Lecture Notes in Mathematics, vol. 1818, x+217pp. Springer, Berlin (2003)
25. Bildhauer, M., Fuchs, M.: $C^{1,\alpha}$-solutions to nonautonomous anisotropic variational problems. Calc. Var. PDE **24**, 309–340 (2005)
26. Bildhauer, M., Fuchs, M.: Splitting-type variational problems with mixed linear-superlinear growth conditions. J. Math. Anal. Appl. **501**, 124452 (2021)
27. Boccardo, L., Gallouët, T.: Nonlinear elliptic and parabolic equations involving measure data. J. Funct. Anal. **87**, 149–169 (1989)
28. Boccardo, L., Gallouët, T.: Nonlinear elliptic equations with right-hand side measures. Commun. PDE **17**, 641–655 (1992)

29. Bouchitté, G., Fonseca, I., Malý, J.: The effective bulk energy of the relaxed energy of multiple integrals below the growth exponent. Proc. R. Soc. Edinburgh Sect. A Math. **128**, 463–479 (1998)
30. Bousquet, P., Brasco, L.: Lipschitz regularity for orthotropic functionals with nonstandard growth conditions. Rev. Mat. Iber. **36**, 19892032 (2020)
31. Brasco, L., Lindgren, E.: Higher Sobolev regularity for the fractional p-Laplace equation in the superquadratic case. Adv. Math. **304**, 300–354 (2017)
32. Brasco, L., Lindgren, E., Schikorra, A.: Higher Hölder regularity for the fractional p-Laplacian in the superquadratic case. Adv. Math. **338**, 782–846 (2018)
33. Breit, D., Cianchi, A., Diening, L., Kuusi, T., Schwarzacher, S.: The p-Laplace system with right-hand side in divergence form: inner and up to the boundary pointwise estimates. Nonlinear Anal. **153**, 200–212 (2017)
34. Breit, D., Cianchi, A., Diening, L., Kuusi, T., Schwarzacher, S.: Pointwise Caldern-Zygmund gradient estimates for the p-system. J. Math. Pures Appl. (IX) **114**, 146–190 (2018)
35. Bulek, M., Gwiazda, P., Skrzeczkowski, J.: On a range of exponents for absence of Lavrentiev phenomenon for double phase functionals. Arch. Ration. Mech. Anal. **246**, 209–240 (2022)
36. Byun, S., Oh, J.: Regularity results for generalized double phase functionals. Anal. PDE **13**, 1269–1300 (2020)
37. Byun, S., Oh, J.: Caldern-Zygmund estimates for generalized double phase problems. J. Funct. Anal. **279**, 108670 (2020)
38. Byun, S., Youn, Y.: Potential estimates for elliptic systems with subquadratic growth. J. Math. Pures Appl. (IX) **131**, 193–224 (2019)
39. Byun, S., Kim, K., Ok, J.: Local Hlder continuity for fractional nonlocal equations with general growth. Math. Ann. **387**, 807–846 (2022)
40. Byun, S., Ok, J., Song, K.: Hlder regularity for weak solutions to nonlocal double phase problems. J. Math. Pures Appl. (IX) **168**, 110–142 (2022)
41. Byun, S., Kim, K., Kumar, D.: Regularity results for a class of nonlocal double phase equations with VMO coefficients. Publicacions Matematiques (to appear)
42. Caccioppoli, R.: Sulle equazioni ellittiche a derivate parziali con n variabili indipendenti. Atti Acc. Naz. Lincei. Rendiconti Lincei Matematica e Appl. (VI) **19**, 83–89 (1934)
43. Caffarelli, L., Vasseur, A.: Drift diffusion equations with fractional diffusion and the quasi-geostrophic equation. Ann. Math. (II) **171**, 1903–1930 (2010)
44. Campanato, S.: Equazioni ellittiche del II ordine e spazi $\mathfrak{L}^{(2,\lambda)}$. Ann. Mat. Pura Appl. (IV) **69**, 321–381 (1965)
45. Carozza, M., Kristensen, J., Passarelli di Napoli, A.: Higher differentiability of minimizers of convex variational integrals. Ann. Inst. H. Poincar Anal. Non Linaire **28**, 395–411 (2011)
46. Carozza, M., Kristensen, J., Passarelli di Napoli, A.: Regularity of minimizers of autonomous convex variational integrals. Ann. Sc. Norm. Super. Pisa Cl. Sci. (V) **13**, 1065–1089 (2014)
47. Cellina, A.: A case of regularity of solutions to nonregular problems. SIAM J. Control Optim. **53**, 28352845 (2015)
48. Chaker, J., Kim, M.: Regularity estimates for fractional orthotropic p-Laplacians of mixed order. Adv. Nonlinear Anal. **11**, 1307–1331 (2022)
49. Chaker, J., Kim, M., Weidner, M.: Regularity for nonlocal problems with nonstandard growth. Calc. Var. PDE **61**, 227 (2022)
50. Chlebicka, I., De Filippis, C.: Removable sets in nonuniformly elliptic problems. Ann. Mat. Pura Appl. (IV) **199**, 619–649 (2020)
51. Chlebicka, I., De Filippis, C., Koch, L.: Boundary regularity for manifold constrained $p(x)$-harmonic maps. J. Lond. Math. Soc. **104**, 2335–2375 (2021)
52. Choe, H.J.: Interior behaviour of minimizers for certain functionals with nonstandard growth. Nonlinear Anal. **19**, 933–945 (1992)
53. Cianchi, A.: Maximizing the L^∞-norm of the gradient of solutions to the Poisson equation. J. Geom. Anal. **2**, 499–515 (1992)
54. Cianchi, A.: Nonlinear potentials, local solutions to elliptic equations and rearrangements. Ann. Scu. Norm. Sup. Cl. Sci. (V) **10**, 335–361 (2011)

55. Cianchi, A., Maz'ya, V.: Global Lipschitz regularity for a class of quasilinear equations. Commun. PDE **36**, 100–133 (2011)
56. Cianchi, A., Maz'ya, V.: Global boundedness of the gradient for a class of nonlinear elliptic systems. Arch. Ration. Mech. Anal. **212**, 129–177 (2014)
57. Cianchi, A., Maz'ya, V.: Gradient regularity via rearrangements for p-Laplacian type elliptic boundary value problems. J. Eur. Math. Soc. **16**, 571–595 (2014)
58. Cianchi, A., Maz'ya, V.: Global gradient estimates in elliptic problems under minimal data and domain regularity. Commun. Pure Appl. Anal. **14**, 285–311 (2015)
59. Cianchi, A., Maz'ya, V.: Optimal second order regularity for the p-Laplace system. J. Math. Pures Appl. (IX) **132**, 41–78 (2019)
60. Colombo, M., Mingione, G.: Regularity for double phase variational problems. Arch. Ration. Mech. Anal. **215**, 443–496 (2015)
61. Colombo, M., Mingione, G.: Bounded minimisers of double phase variational integrals. Arch. Ration. Mech. Anal. **218**, 219–273 (2015)
62. Colombo, M., Mingione, G.: Calderón-Zygmund estimates and nonuniformly elliptic operators. J. Funct. Anal. **270**, 1416–1478 (2016)
63. Colombo, M., Tione, R.: Non-classical solutions to the p-Laplace equation. J. Europ. Math. Soc. (JEMS). https://doi.org/10.4171/JEMS/1462
64. Cozzi, M.: Regularity results and Harnack inequalities for minimizers and solutions of nonlocal problems: a unified approach via fractional De Giorgi classes. J. Funct. Anal. **272**, 4762–4837 (2017)
65. Dal Maso, G., Murat, F., Orsina, L., Prignet, A.: Renormalization solutions of elliptic equations with general measure data. Ann. Sc. Norm. Super. Pisa, Cl. Sci. (IV) **28**, 741–808 (1999)
66. da Silva, J.V., Ricarte, G.C.: Geometric gradient estimates for fully nonlinear models with non-homogeneous degeneracy and applications. Calc. Var. PDE **59**, 161 (2020)
67. da Silva, J.V., Rampasso, G.C., Ricarte, G.C., Vivas, H.A.: Free boundary regularity for a class of one-phase problems with non-homogeneous degeneracy. Isr. J. Math. **254**, 155–200 (2023)
68. Daskalopoulos, P., Kuusi, T., Mingione, G.: Borderline estimates for fully nonlinear elliptic equations. Commun. Partial Differ. Equ. **39**, 574–590 (2014)
69. De Filippis, C.: Partial regularity for manifold constrained $p(x)$-harmonic maps. Calc. Var. PDE **58**, 47 (2019)
70. De Filippis, C.: Regularity for solutions of fully nonlinear elliptic equations with nonhomogeneous degeneracy. Proc. Roy. Soc. Edinburgh Sect. A **151**, 110–132 (2021)
71. De Filippis, C.: Optimal gradient estimates for multiphase integrals. Math. Eng. **4**, 1–36 (2021)
72. De Filippis, C.: Fully nonlinear free transmission problems with nonhomogeneous degeneracies. Inter. Free Bound. **24**, 197–233 (2022)
73. De Filippis, C.: Quasiconvexity and partial regularity via nonlinear potentials. J. Math. Pures Appl. **163**, 11–82 (2022)
74. De Filippis, C., Mingione, G.: A borderline case of Caldern-Zygmund estimates for nonuniformly elliptic problems. St. Petersburg Math. J. **31**, 455–477 (2020)
75. De Filippis, C., Mingione, G.: Manifold constrained nonuniformly elliptic problems. J. Geom. Anal. **30**, 1661–1723 (2020)
76. De Filippis, C., Mingione, G.: On the regularity of minima of nonautonomous functionals. J. Geom. Anal. **30**, 1584–1626 (2020)
77. De Filippis, C., Mingione, G.: Interpolative gap bounds for nonautonomous integrals. Anal. Math. Phys. **11**, 117 (2021)
78. De Filippis, C., Mingione, G.: Lipschitz bounds and nonautonomous integrals. Arch. Ration. Mech. Anal. **242**, 973–1057 (2021)
79. De Filippis, C., Mingione, G.: Nonuniformly elliptic Schauder theory. Invent. Math. **234**, 1109–1196 (2023)

80. De Filippis, C., Mingione, G.: Regularity for double phase problems at nearly linear growth. Arch. Ration. Mech. Anal. **247**, 85 (2023)
81. De Filippis, C., Mingione, G.: Gradient regularity in mixed local and nonlocal problems. Math. Ann. **388**, 261–328 (2024)
82. De Filippis, C., Mingione, G.: The sharp growth rate in nonuniformly elliptic Schauder theory (2024). https://arxiv.org/abs/2401.07160
83. De Filippis, C., Oh, J.: Regularity for multiphase variational problems. J. Differ. Equ. **267**, 1631–1670 (2019)
84. De Filippis, C., Palatucci, G.: Hlder regularity for nonlocal double phase equations. J. Differ. Equ. **267**, 547–586 (2019)
85. De Filippis, C., Piccinini, M.: Borderline global regularity for nonuniformly elliptic systems. Int. Math. Res. Not. **20**, 17324–17376 (2023)
86. De Filippis, C., Stroffolini, B.: Singular multiple integrals and nonlinear potentials. J. Funct. Anal. **285**, 109952 (2023)
87. De Filippis, C., Koch, L., Kristensen, J.: Quantified Legendreness and the regularity of minima. Arch. Ration. Mech. Anal. **248**, 69 (2024)
88. De Filippis, F., Piccinini, M.: Regularity for multi-phase problems at nearly linear growth. J. Diff. Equ. **410**, 832–868 (2024)
89. De Giorgi, E.: Sulla differenziabilità e l'analiticità delle estremali degli integrali multipli regolari. Mem. Accad. Sci. Torino Cl. Sci. Fis. Mat. Nat. (III) **125**, 25–43 (1957)
90. Di Benedetto, E.: $C^{1+\alpha}$ local regularity of weak solutions of degenerate elliptic equations. Nonlinear Anal. **7**, 827–850 (1983)
91. Diening, L., Stroffolini, B., Verde, A.: Everywhere regularity of functionals with ϕ-growthm. Manuscript. Math. **129**, 449–481 (2009)
92. Di Marco, T., Marcellini, P.: A-priori gradient bound for elliptic systems under either slow or fast growth conditions. Calc. Var. PDE **59**, 120 (2020)
93. Di Nezza, E., Palatucci, G., Valdinoci, E.: Hitchhiker's guide to the fractional Sobolev spaces. Bull. Sci. Math. **136**, 521–573 (2012)
94. Domokos, A.: Differentiability of solutions for the non-degenerate p-Laplacian in the Heisenberg group. J. Differ. Equ. **204**, 439–470 (2004)
95. Dong, H., Zhu, H.: Gradient estimates for singular parabolic p-Laplace type equations with measure data. Calc. Var. PDE **61**, 86 (2022)
96. Duc, D.M., Eells, J.: Regularity of exponentially harmonic functions. Int. J. Math. **2**, 395–408 (1991)
97. Duzaar, F., Mingione, G.: Local Lipschitz regularity for degenerate elliptic systems. Ann. Inst. Henri Poincar Anal. Non Linaire **27**, 1361–1396 (2010)
98. Duzaar, F., Mingione, G.: Gradient estimates via linear and nonlinear potentials. J. Funct. Anal. **259**, 2961–2998 (2010)
99. Duzaar, F., Mingione, G.: Gradient estimates via non-linear potentials. Am. J. Math. **133**, 1093–1149 (2011)
100. Eleuteri, M., Marcellini, P., Mascolo, E.; Lipschitz estimates for systems with ellipticity conditions at infinity. Ann. Mat. Pura Appl. (IV) **195**, 1575–1603 (2016)
101. Eleuteri, M., Marcellini, P., Mascolo, E., Perrotta, S.: Local Lipschitz continuity for energy integrals with slow growth. Ann. Mat. Pura Appl. (IV) **201**, 1005–1032 (2022)
102. Esposito, L., Leonetti, F., Mingione, G.: Sharp regularity for functionals with (p, q)-growth. J. Differ. Equ. **204**, 5–55 (2004)
103. Evans, L.C.: Some new PDE methods for weak KAM theory. Calc. Var. PDE **17**, 159–177 (2003)
104. Fernndez Bonder, J., Salort, A., Vivas, H.: Interior and up to the boundary regularity for the fractional g-Laplacian: the convex case. Nonlinear Anal. **223**, 113060 (2022)
105. Fonseca, I., Malý, J.: Relaxation of multiple integrals below the growth exponent. Ann. Inst. H. Poincaré Anal. Non Linéaire **14**, 309–338 (1997)
106. Fonseca, I., Malý, J., Mingione, G.: Scalar minimizers with fractal singular sets. Arch. Ration. Mech. Anal. **172**, 295–307 (2004)

107. Fonseca, I., Marcellini, P.: Relaxation of multiple integral in subcritical Sobolev spaces. J. Geom. Anal. **7**, 57–81 (1997)
108. Frehse, J., Seregin, G.: Regularity of solutions to variational problems of the deformation theory of plasticity with logarithmic hardening. In: Proceedings of the St. Petersburg Mathematical Society. American Mathematical Society Translations: Series 2, vol. V, pp. 127–152. American Mathematical Society, Providence (1999)
109. Fuchs, M., Mingione, G.: Full $C^{1,\alpha}$-regularity for free and constrained local minimizers of elliptic variational integrals with nearly linear growth. Manuscript. Math. **102**, 227–250 (2000)
110. Fuchs, M., Seregin, G.A.: Some remarks on non-Newtonian fluids including nonconvex perturbations of the Bingham and Powell-Eyring model for viscoplastic fluids. Math. Models Methods Appl. Sci. **7**, 405–433 (1997)
111. Fuchs, M., Seregin, G.A.: A regularity theory for variational integrals with $L \log L$-growth. Calc. Var. PDE **6**, 171–187 (1998)
112. Fuchs, M., Seregin, G.A.: Variational Methods for Problems from Plasticity Theory and for Generalized Newtonian Fluids. Lecture Notes in Mathematics, vol. 1749, vi+269pp. Springer, Berlin (2000)
113. Garain, P., Lindgren, E., Higher Hölder regularity for mixed local and nonlocal degenerate elliptic equations. Calc. Var. PDE **62**, 67 (2023)
114. Giaquinta, M.: A counter-example to the boundary regularity of solutions to elliptic quasilinear systems. Manuscript. Math. **24**, 217–220 (1978)
115. Giaquinta, M., Giusti, E.: On the regularity of the minima of variational integrals. Acta Math. **148**, 31–46 (1982)
116. Giaquinta, M., Giusti, E.: Differentiability of minima of nondifferentiable functionals. Invent. Math. **72**, 285–298 (1983)
117. Giaquinta, M., Giusti, E.: Global $C^{1,\alpha}$-regularity for second order quasilinear elliptic equations in divergence form. J. Reine Angew. Math. (Crelle J.) **351**, 55–65 (1984)
118. Gilbarg, D., Trudinger, N.: Elliptic Partial Differential Equations of Second Order. Grundlehren der Mathematischen Wissenschaften, 2nd edn. vol. 224, xiii+513pp. Springer, Berlin (1983)
119. Giraud, G.: Sur le problme de Dirichlet gnralis; quations non linaires 'a m variables. Ann. Sci. ENS **43**, 1–128 (1926)
120. Giraud, G.: Sur le problme de Dirichlet gnralis (Deuxime mmoire). Ann. Sci. ENS **46**, 131–245 (1929)
121. Giusti, E.: Direct Methods in the Calculus of Variations. World Scientific Publishing, River Edge (2003)
122. Gmeineder, F.: The regularity of minima for the Dirichlet problem on BD. Arch. Ration. Mech. Anal. **237**, 1099–1171 (2020)
123. Gmeineder, F.: Partial regularity for symmetric quasiconvex functionals on BD. J. Math. Pures Appl. (IX) **145**, 83–129 (2021)
124. Gmeineder, F., Kristensen, J.: Partial regularity for BV minimizers. Arch. Ration. Mech. Anal. **232**, 1429–1473 (2019)
125. Gmeineder, F., Kristensen, J.: Sobolev regularity for convex functionals on BD. Calc. Var. PDE **58**, 56 (2019)
126. Grafakos, L.: Classical Fourier Analysis. Graduate Texts in Mathematics, 2nd edn., vol. 249, xvi+489pp. Springer, New York (2008)
127. Harjulehto, P., Hästö, P.: Orlicz Spaces and Generalized Orlicz Spaces. Lecture Notes in Mathematics, vol. 2236. Springer, Berlin (2019)
128. Harjulehto, P., Hästö, P.: Double phase image restoration. J. Math. Anal. Appl. **501**, 123832 (2021)
129. Harjulehto, P., Hästö, P., Toivanen, O.: Hlder regularity of quasiminimizers under generalized growth conditions. Calc. Var. PDE **56**, 22 (2017)
130. Hartman, P., Stampacchia, G.: On some non-linear elliptic differential-functional equations. Acta Math. **115**, 271–310 (1966)

131. Hästö, P.: The maximal operator on generalized Orlicz spaces. J. Funct. Anal. **269**, 4038–4048 (2015). Corrections in 271, 240–243 (2016)
132. Havin, M., Mazya, V.G.: Nonlinear potential theory. Russ. Math. Surv. **27**, 71–148 (1972)
133. Hedberg, L., Wolff, T.H.: Thin sets in nonlinear potential theory. Ann. Inst. Fourier (Grenoble) **33**, 161–187 (1983)
134. Heinonen, J., Kilpeläinen, T., Martio, O.: Nonlinear Potential Theory of Degenerate Elliptic Equations. Oxford Mathematical Monographs. Oxford, New York (1993)
135. Hirsch, J., Schäffner, M.: Growth conditions and regularity, an optimal local boundedness result. Commun. Cont. Math. **23**, 2050029 (2020)
136. Hopf, E.: Bemerkungen zu einem satze von S. Bernstein aus del' theorie del' elliptischen differentialgleichungen. Math. Z. **29**, 744–745 (1928)
137. Hästö, P., Ok, J.: Maximal regularity for local minimizers of nonautonomous functionals. J. Eur. Math. Soc. **24**, 1285–1334 (2022)
138. Hästö, P., Ok, J.: Regularity theory for nonautonomous partial differential equations without Uhlenbeck structure. Arch. Ration. Mech. Anal. **245**, 1401–1436 (2022)
139. Hästö, P., Ok, J.: Regularity theory for nonautonomous problems with a priori assumptions. Calc. Var. PDE **62**, 251 (2023)
140. Ivanov, A.V.: Local estimates of the maximum modulus of the first derivatives of the solutions of quasilinear nonuniformly elliptic and nonuniformly parabolic equations and of systems of general form. Proc. Steklov Inst. Math. **110**, 48–71 (1970)
141. Ivanov, A.V.: The Dirichlet problem for second order quasilinear nonuniformly elliptic equations. Trudy Mat. Inst. Steklov. **116**, 34–54 (1971)
142. Ivanov, A.V.: Quasilinear Degenerate and Nonuniformly Elliptic and Parabolic Equations of Second Order. Proceedings of Steklov Institute of Mathematics, vol. 160, xi+287pp. (1984)
143. Ivokina, N.M., Oskolkov, A.P.: Nonlocal estimates of the first derivatives of solutions of the Dirichlet problem for nonuniformly elliptic quasilinear equations. Zap. Naun. Sem. Leningrad. Otdel. Mat. Inst. Steklov. (LOMI) **5**, 37–109 (1967)
144. Iwaniec, T., Sbordone, C.: Weak minima of variational integrals. J. Reine Angew. Math. (Crelle J.) **454**, 143–161 (1994)
145. Kilpeläinen, T., Malý, J.: Degenerate elliptic equations with measure data and nonlinear potentials. Ann. Scuola Norm. Sup. Pisa Cl. Sci. (IV) **19**, 591–613 (1992)
146. Kilpeläinen, T., Malý, J.: The Wiener test and potential estimates for quasilinear elliptic equations. Acta Math. **172**, 137–161 (1994)
147. Koch, L.: Global higher integrability for minimisers of convex functionals with (p, q)-growth. Calc. Var. PDE **60**, 63 (2021)
148. Koch, L.: On global absence of Lavrentiev gap for functionals with (p, q)-growth (2022). https://arxiv.org/abs/2210.15454
149. Kristensen, J., Mingione, G.: The singular set of ω-minima. Arch. Ration. Mech. Anal. **177**, 93–114 (2005)
150. Kuusi, T., Mingione, G.: Gradient regularity for nonlinear parabolic equations. Ann. Sc. Norm. Super. Pisa, Cl. Sci. (V) **12**, 755–822 (2013)
151. Kuusi, T., Mingione, G.: Linear potentials in nonlinear potential theory. Arch. Ration. Mech. Anal. **207**, 215–246 (2013)
152. Kuusi, T., Mingione, G.: The Wolff gradient bound for degenerate parabolic equations. J. Eur. Math. Soc. **16**, 835–892 (2014)
153. Kuusi, T., Mingione, G.: Riesz potentials and nonlinear parabolic equations. Arch. Ration. Mech. Anal. **212**, 727–780 (2014)
154. Kuusi, T., Mingione, G.: Guide to nonlinear potential estimates. Bull. Math. Sci. **4**, 1–82 (2014)
155. Kuusi, T., Mingione, G.: A nonlinear Stein theorem. Calc. Var. PDE **51**, 45–86 (2014)
156. Kuusi, T., Mingione, G.: Partial regularity and potentials. J. Ecol. Polytech. Math. **3**, 309–363 (2016)
157. Kuusi, T., Mingione, G.: Vectorial nonlinear potential theory. J. Eur. Math. Soc. **20**, 929–1004 (2018)

158. Ladyzhenskaya, O.A., Ural'tseva, N.N.: Linear and Quasilinear Elliptic Equations. Academic Press, New York (1968)
159. Ladyzhenskaya, O.A., Ural'tseva, N.N.: Local estimates for gradients of solutions of nonuniformly elliptic and parabolic equations. Commun. Pure Appl. Math. **23**, 677–703 (1970)
160. Lieberman, G.M.: Interior gradient bounds for nonuniformly parabolic equations. Ind. Univ. Math. J. **32**, 579–601 (1983)
161. Lieberman, G.M.: The natural generalization of the natural conditions of Ladyzhenskaya and Ural'tseva for elliptic equations. Commun. PDE **16**, 311–361 (1991)
162. Lieberman, G.M.: On the regularity of the minimizer of a functional with exponential growth. Comment. Math. Univ. Carolinae **33**, 45–49 (1992)
163. Littman, W., Stampacchia, G., Weinberger, H.F.: Regular points for elliptic equations with discontinuous coefficients. Ann. Scu. Norm. Sup. Pisa (III) **17**, 43–77 (1963)
164. Malý, J., Ziemer, W.P.: Fine Regularity of Solutions of Elliptic Partial Differential Equations. Mathematical Surveys and Monographs, vol. 51. American Mathematical Society, Providence (1997)
165. Manfredi, J.J.: Regularity of the gradient for a class of nonlinear possibly degenerate elliptic equations. Ph.D. Thesis, University of Washington, St. Louis (1986)
166. Manfredi, J.J.: Regularity for minima of functionals with p-growth. J. Differ. Equ. **76**, 203–212 (1988)
167. Marcellini, P.: On the definition and the lower semicontinuity of certain quasiconvex integrals. Ann. Inst. H. Poincaré Anal. Non Linéaire **3**, 391409 (1986)
168. Marcellini, P.: The stored-energy for some discontinuous deformations in nonlinear elasticity. In: Partial Differential Equations and the Calculus of Variations, vol. II. Birkhäuser, Boston (1989)
169. Marcellini, P.: Regularity of minimizers of integrals of the calculus of variations with non standard growth conditions. Arch. Ration. Mech. Anal. **105**, 267–284 (1989)
170. Marcellini, P.: Regularity and existence of solutions of elliptic equations with p, q-growth conditions. J. Differ. Equ. **90**, 1–30 (1991)
171. Marcellini, P.: Regularity for elliptic equations with general growth conditions. J. Differ. Equ. **105**, 296–333 (1993)
172. Marcellini, P.: Everywhere regularity for a class of elliptic systems without. Ann. Sc. Norm. Super. Pisa Cl. Sci. (V) **23**, 1–25 (1996)
173. Marcellini, P.: Regularity for some scalar variational problems under general growth conditions. J. Optim. Theory Appl. **90**, 161–181 (1996)
174. Marcellini, P.: A variational approach to parabolic equations under general and p, q-growth conditions. Nonlinear Anal. **194**, 111456 (2020)
175. Marcellini, P.: Growth conditions and regularity for weak solutions to nonlinear elliptic pdes. J. Math. Anal. Appl. **501**, 124408 (2021)
176. Marcellini, P.: Local Lipschitz continuity for p, q-PDEs with explicit u-dependence. Nonlinear Anal. **226**, 113066 (2022)
177. Marcellini, P., Papi, G.: Nonlinear elliptic systems with general growth. J. Differ. Equ. **221**, 412–443 (2006)
178. Maz'ya, V.: The continuity at a boundary point of the solutions of quasi-linear elliptic equations. Vestnik Leningrad. Univ. **25**, 42–55 (1970)
179. Mingione, G.: Bounds for the singular set of solutions to non linear elliptic systems. Calc. Var. PDE **18**, 373–400 (2003)
180. Mingione, G.: Regularity of minima: an invitation to the dark side of the calculus of variations. Appl. Math. **51**, 355–425 (2006)
181. Mingione, G.: Gradient potential estimates. J. Eur. Math. Soc. **13**, 459–486 (2011)
182. Mingione, G.: Short Tales from Nonlinear Calderón-Zygmund Theory. Lecture Notes in Mathematics, Springer, vol. 2186, pp. 159–204 (2017)
183. Mingione, G., Radulescu, V.: Recent developments in problems with nonstandard growth and nonuniform ellipticity. J. Math. Anal. Appl. **501**, 125197 (2021)

184. Nakamura, K.: Local properties of fractional parabolic De Giorgi classes of order s, p. J. Funct. Anal. **285**, 110049 (2023)
185. Nguyen, Q.H., Phuc, N.C.: Existence and regularity estimates for quasilinear equations with measure data: the case $1 < p <= (3n-2)/(2n-1)$. Analysis PDE **15**, 1879–1895 (2022)
186. Nguyen Q.H., Phuc, N.C.: A comparison estimate for singular p-Laplace equations and its consequences. Arch. Rat. Mech. Anal. **247**, 49 (2023)
187. Nguyen, Q.H., Phuc, N.C.: Universal potential estimates for $1 < p \leq 2 - \frac{1}{n}$. Math. Eng. **5**, 1–24 (2023). https://doi.org/10.3934/mine.2023057
188. Ok, J.: Local Hlder regularity for nonlocal equations with variable powers. Calc. Var. PDE **62**, 32 (2023)
189. Ok, J., Scilla, G., Stroffolini, B.: Regularity theory for parabolic systems with Uhlenbeck structure. J. Math. Pures Appl. (IX) **182**, 116–163 (2024)
190. O'Neil, R.: Integral transforms and tensor products on Orlicz spaces and $L(p, q)$ spaces. J. Anal. Math. **21**, 1–276 (1968)
191. Oskolkov, A.P.: A priori estimates for the first derivatives-of solutions of the Dirichlet problem for nonuniformly elliptic quasilinear equations. Trudy Mat. Inst. Steklov **102**, 105–127 (1967)
192. Phillips, D.: A minimization problem and the regularity of solutions in the presence of a free boundary. Ind. Univ. Math. J. **32**, 1–17 (1983)
193. Phuc, N.C., Verbitsky, I.E.: Quasilinear and Hessian equations of Lane-Emden type. Ann. Math. (II) **168**, 859–914 (2008)
194. Phuc, N.C., Verbitsky, I.E.: Singular quasilinear and Hessian equations and inequalities. J. Funct. Anal. **256**, 1875–1906 (2009)
195. Pimentel, E.A., Walker, M.: Potential estimates for fully nonlinear elliptic equations with bounded ingredients. Math. Eng. **5**, 1–16 (2023). https://doi.org/10.3934/mine.2023063
196. Schäffner, M.: Higher integrability for variational integrals with nonstandard growth. Calc. Var. PDE **60**, 77 (2021)
197. Schauder, J.: ber lineare elliptische differentialgleichungen zweiter ordnung. Math. Z. **38**, 257–282 (1934)
198. Schauder, J.: Numerische abshätzunger in ellitischen linearen differentialgleichungen equations. Stud. Math. **5**, 34–42 (1934)
199. Schäffner, M.: Lipschitz bounds for nonuniformly elliptic integral functionals in the plane. https://arxiv.org/abs/2402.06252
200. Schmidt, T.: Regularity of relaxed minimizers of quasiconvex variational integrals with (p, q)-growth. Arch. Ration. Mech. Anal. **193**, 311–337 (2009)
201. Scott, J.M., Mengesha, T.: Self-improving inequalities for bounded weak solutions to nonlocal double phase equations. Commun. Pure Appl. Anal. **21**, 183–212 (2022)
202. Serrin, J.: The problem of Dirichlet for quasilinear elliptic differential equations with many independent variables. Philos. Trans. R. Soc. Lond. Ser. A **264**, 413–496 (1969)
203. Sil, S.: Nonlinear Stein theorem for differential forms. Calc. Var. PDE **58**, 154 (2019)
204. Simon, L.: Interior gradient bounds for nonuniformly elliptic equations of divergence form. Ph.D. Thesis. University of Adelaide (1971)
205. Simon, L.: Interior gradient bounds for nonuniformly elliptic equations. Ind. Univ. Math. J. **25**, 821–855 (1976)
206. Simon, L.: Schauder estimates by scaling. Calc. Var. PDE **5**, 391–407 (1997)
207. Stampacchia, G.: On some regular multiple integral problems in the calculus of variations. Commun. Pure Appl. Math. **16**, 383–421 (1963)
208. Stein, E.M.: Editor's note: the differentiability of functions in $\mathbb{R}^n$. Ann. Math. (II) **113**, 383–385 (1981)
209. Stein, E.M., Weiss, G.: Introduction to Fourier Analysis on Euclidean Spaces. Princeton University Press, Princeton (1971)
210. Šverák, V., Yan, X.: Non-Lipschitz minimizers of smooth uniformly convex variational integrals. Proc. Natl. Acad. Sci. USA **99/24**, 15269–15276 (2002)
211. Triebel, H.: The Structure of Functions. Monographs in Mathematics, vol. 97, xii+425pp. Birkhuser, Basel (2001)

212. Trudinger, N.: The Dirichlet problem for nonuniformly elliptic equations. Bull. Am. Math. Soc. **73**, 410–413 (1967)
213. Trudinger, N.: Harnack inequalities for nonuniformly elliptic divergence structure equations. Invent. Math. **64**, 517–531 (1981)
214. Trudinger, N.: A new approach to the Schauder estimates for linear elliptic equations. Proc. Centre Math. Appl. **1986**, 52–59 (1986)
215. Trudinger, N., Wang, X.J.: On the weak continuity of elliptic operators and applications to potential theory. Am. J. Math. **124**, 369–410 (2002)
216. Uhlenbeck, K.: Regularity for a class of nonlinear elliptic systems. Acta Math. **138**, 219–240 (1977)
217. Ural'tseva, N.N.: Degenerate quasilinear elliptic systems. Zap. Na. Sem. Leningrad. Otdel. Mat. Inst. Steklov. (LOMI) **7**, 184–222 (1968)
218. Ural'tseva, N.N., Urdaletova, A.B.: The boundedness of the gradients of generalized solutions of degenerate quasilinear nonuniformly elliptic equations. Vestnik Leningrad Univ. Math. **16**, 263–270 (1984)
219. Zhikov, V.V.: Averaging of functionals of the calculus of variations and elasticity theory. Izv. Akad. Nauk SSSR Ser. Mat. **50**, 675–710 (1986)
220. Zhikov, V.V.: On Lavrentiev's Phenomenon. Russ. J. Math. Phys. **3**, 249–269 (1995)
221. Zhikov, V.V.: On some variational problems. Russ. J. Math. Phys. **5**, 105–116 (1997)
222. Zhikov, V.V., Kozlov, S.M., Olenik, O.A.: Homogenization of Differential Operators and Integral Functionals, xii+570pp. Springer, Berlin (1994). ISBN: 3-540-54809-2

Chapter 3
Reduction Principles

Luboš Pick

Abstract This text contains lecture notes to the series of lectures delivered at the C.I.M.E. course "Geometric and analytic aspects of functional variational principles" held in Cetraro (Cosenza), June 27–June 1, 2022. I am very indebted to Paolo Salani and also to the Scientific Directors Andrea Cianchi, Vladimir Maz'ya and Tobias Weth for the very kind invitation. Delivering a course on a legendary C.I.M.E. school was a great honor and immense pleasure for me.

The central topic of this text is the idea of the so-called reduction principle. In its original form it is a powerful technique which can be used in order to replace difficult inequalities involving functions of several variables and differential operators such as gradients applied to them, by equivalent inequalities involving integral operators and functions of a single variable. A prototype of such problem is a Sobolev embedding.

While classical material such as rearrangement-invariant spaces, Hardy-type inequalities, symmetrization techniques, the Pólya–Szegő inequality and so on, are crucial ingredients of the theory, most of the results that will be mentioned have been developed during the last 25 years.

One of the most interesting observations is that a reduction principle usually bears some kind of information about optimality of function spaces involved in inequalities, embeddings, and boundedness of operators. It is mainly this fact that makes reduction principles so important in applications. This way one can obtain a very valuable information about the location of a threshold in certain inequalities.

Optimality of spaces works on both sides of an inequality or an embedding (the domain side and the target side) and they affect each other. A considerable disadvantage of the initial formulas for the optimal domain and the optimal target spaces is that they are rather implicit. Therefore, some effort will be spent to obtain

---

L. Pick (✉)
Department of Mathematical Analysis, Faculty of Mathematics and Physics, Charles University, Prague, Czech Republic

more manageable characterization of optimal function spaces, or at least to provide good sets of examples.

It turns out that optimal spaces have an intimate connection to other features such as their location as interpolation spaces, or boundedness of the so-called supremum operators either on them or on their Köthe duals. These facts have some interesting consequences that will be pointed out, too.

The text is structured as follows. We begin with a rather substantial introductory section (Sect. 3.1), in which we explain the motivation for the questions that will be studied in detail later. We illustrate the problems on various examples of differential operators, including those acting on an Euclidean space endowed with the Gaussian measure, and also some customary integral operators. One of the most notable features pointed out here is the need for a substantial extension of the pool of function spaces.

In the next section, Sect. 3.2, thus, we develop a theory of the so-called rearrangement-invariant spaces. This class of function spaces will serve in the sequel as our general natural working environment. It is, on the one hand, strikingly versatile, and, on the other, it enjoys certain truly remarkable features such as, for instance, a very profound collection of duality relations, powerful principles such as Hardy's lemma, saturated Hölder's inequality, the Hardy–Littlewood inequality, the Hardy–Littlewood–Pólya principle, the Luxemburg representation theorem, and more.

In Sect. 3.3, we present, as a prototype, the reduction principle for a first-order Sobolev embedding involving functions vanishing on the boundary. The main purpose of this section is to illustrate, on this particular example, some of the key ideas that can be then extended and used in various disguises further in the theory. In particular, we explain why the environment of rearrangement-invariant spaces is so successful, and how the Pólya–Szegő principle comes to the picture. We will briefly describe how to overcome certain technical difficulties by extending the theory to quasinorms. We will also point out how the optimality of function spaces can be obtained from a reduction principle.

The principal aim of Sect. 3.4 is to describe key ideas of an interpolation argument which can be successfully applied to obtain optimal Sobolev embeddings of higher order for functions acting on regular domains, involving rearrangement-invariant norms. The main techniques used here involve, among others, K-functional inequalities, Holmstedt's formulae, DeVore–Scherer theorem and the supremum operators. Again, details of proofs are sketched here.

In the last section, Sect. 3.5, we survey reduction principles suitable for various other situations, some related results, and examples. In this section we omit proofs and technical details, and concentrate instead on principal ideas and relations.

3 Reduction Principles

3.1 Introduction, Motivation and Examples

We begin by explaining what we mean by a reduction principle and what lies behind the scenes.

3.1.1 Euclidean Differential Operators

Consider a partial differential equation. For instance,

$$\begin{cases} -\Delta u = f & \text{in } \Omega, \\ u = 0 & \text{on } \partial\Omega, \end{cases} \qquad (3.1.1)$$

in which Ω is a domain (a bounded and connected set) in $\mathbb{R}^n$, $n \in \mathbb{N}$. The function f is known (it comes with the equation). The function u is unknown (if and when we find it, we can call it a *solution* to the equation).

But even without finding u, we know, through the equation, something about Δu. The principal task we pursue here is the *transfer of regularity* from f to u, hence, from Δu to u and the ultimate task is that this transfer should be, in a suitable sense, *sharp*.

Suppose that we a-priori know that f has certain quality. Then this information has commercial potential. It can be traded for something else. Useful marketable qualities are, for example, the degree of integrability, boundedness, continuity, smoothness, control of oscillation, control of singularities, and more. An important question is, how can we measure a quality, or, say, how can we assess the degree of regularity, of a given function? The best way how to do this seems to be to express it in terms of the membership in some *function space*.

So the task can be now rephrased as follows: Suppose that we know that $f \in X$, and we want to obtain that $u \in Y$, where X and Y are some function spaces. Thus, the goal is to establish the implication

$$f \in X \quad \Rightarrow \quad u \in Y. \qquad (3.1.2)$$

On the top of the implication (3.1.2), we would like to obtain some *quantification* of the transfer. Namely, we would like to show that there exists a positive constant C such that

$$\|u\|_Y \leq C \|f\|_X \qquad (3.1.3)$$

for every f for which there exists a unique solution u. This makes sense whenever X and Y are structures that are furnished, at least, with quasi-norms.

Owing to the Eq. (3.1.1), the inequality (3.1.3) can be interpreted as follows: there exists a positive constant C such that

$$\|u\|_Y \leq C \|\Delta u\|_X \tag{3.1.4}$$

for every twice weakly differentiable u of compact support on Ω such that $\Delta u \in X$.

Similar questions are of interest also for other differential operators in place of Δu, for example the first-order gradient, ∇.

Question For which X, Y, one has

$$|\nabla u| \in X \quad \Rightarrow \quad u \in Y$$

and

$$\|u\|_Y \leq C \| |\nabla u| \|_X ?$$

For instance, let $X = L^2(\Omega)$, where Ω is a ball in $\mathbb{R}^3$. So we know that $|\nabla u| \in L^2(\Omega)$. For what information on u we can trade it? The classical Poincaré inequality then tells us that also $u \in L^2(\Omega)$. But it would be silly to sell it so cheap. We can gain more from this trade. If nothing else, we should aim for a *higher degree of integrability*. The classical theory tells us that the implication

$$|\nabla u| \in L^2(\Omega) \quad \Rightarrow \quad u \in L^6(\Omega)$$

is true. On the other hand, simple examples can be used to show that the implication

$$|\nabla u| \in L^2(\Omega) \quad \Rightarrow \quad u \in L^7(\Omega) \tag{3.1.5}$$

is false. So there has to be some kind of threshold. One of the most important and interesting questions both in the theory and in applications is to locate this threshold and to express it in a manageable way.

If we stick just to the degree of integrability represented by membership to Lebesgue spaces, then the threshold can be easily nailed down owing to the *classical Sobolev inequality*. One of its simplest versions can be formulated as follows.

Assume that $n \in \mathbb{N}$, $n \geq 2$, $p \in [1, n)$, and Ω is a bounded domain in $\mathbb{R}^n$ with Lipschitz boundary. Then,

$$|\nabla u| \in L^p(\Omega) \quad \Rightarrow \quad u \in L^q(\Omega) \tag{3.1.6}$$

for every weakly differentiable u such that $|\nabla u| \in L^p(\Omega)$, where

$$\frac{1}{q} = \frac{1}{p} - \frac{1}{n}.$$

3 Reduction Principles

In particular, the implication

$$|\nabla u| \in L^p(\Omega) \quad \Rightarrow \quad u \in L^{\frac{np}{n-p}}(\Omega) \tag{3.1.7}$$

holds for every $p \in [1, n)$ and there exists a constant C such that for every weakly-differentiable scalar function u of n variables vanishing in an appropriate sense on the boundary of Ω and having the modulus of its weak first-order gradient in the space $L^p(\Omega)$, one has

$$\|u\|_{L^{\frac{np}{n-p}}(\Omega)} \leq C \||\nabla u|\|_{L^p(\Omega)}. \tag{3.1.8}$$

Moreover, the degree of integrability on the right-hand side of (3.1.7) (and on the left-hand side of (3.1.8)) is the *largest possible* in the sense that $L^{\frac{np}{n-p}}(\Omega)$ cannot be effectively replaced with $L^q(\Omega)$ in either (3.1.7) or (3.1.8) for any $q > \frac{np}{n-p}$ without violating these statements.

If ∇ is replaced by Δ and n is at least 3, then we get analogous inequalities, at least for $p \in (1, \frac{n}{2})$, namely,

$$\Delta u \in L^p(\Omega) \quad \Rightarrow \quad u \in L^{\frac{np}{n-2p}}(\Omega) \tag{3.1.9}$$

for every $p \in (1, \frac{n}{2})$, and

$$\|u\|_{L^{\frac{np}{n-2p}}(\Omega)} \leq C\|\Delta u\|_{L^p(\Omega)}. \tag{3.1.10}$$

Once again, given $p \in (1, \frac{n}{2})$, then, for any $q > \frac{np}{n-2p}$, the implication

$$\Delta u \in L^p(\Omega) \quad \Rightarrow \quad u \in L^q(\Omega) \tag{3.1.11}$$

is false.

It should be noticed that there is a considerable *gain* in integrability in the transfer from $|\nabla u|$ to u in (3.1.7). By this we mean that u itself belongs to a Lebesgue space with a higher degree of integrability than $|\nabla u|$, since $\frac{np}{n-p} > p$ for $p \in [1, n)$. There is a corresponding gain also in the transfer of regularity from Δu to u in (3.1.9), as $\frac{np}{n-2p} > p$ for each $p \in (1, \frac{n}{2})$.

It is important to observe that the magnitude of the gain in integrability *depends on the dimension*.

Let us ask ourselves how to prove a Sobolev inequality. In order to establish (3.1.9), one can use, for example, the classical *representation formula* (if n is at least 3), namely

$$u(x) = \frac{1}{(2-n)n\omega_n} \int_{\mathbb{R}^n} \frac{\Delta u(y)}{|x-y|^{n-2}} \, dy,$$

which holds for a.e. $x \in \mathbb{R}^n$ and for every compactly supported function u on $\mathbb{R}^n$ whose distributional Laplacian Δu is an integrable function, and in which ω_n denotes the measure of the unit ball in $\mathbb{R}^n$, that is,

$$\omega_n = \frac{\pi^{\frac{n}{2}}}{\Gamma(\frac{n}{2}+1)}.$$

The case $n = 2$ requires specific treatment, but can be handled in a similar way. In the negative direction, that is, in order to disprove false claims such as (3.1.5) or (3.1.11) when q is too large, one can easily construct counterexamples using radially decreasing functions.

The representation formula leads us to the idea of using special integral operators, called *Riesz potentials*. It will allow us to treat differential operators of all orders at once. Given $\gamma \in (0, n)$, we define the *Riesz potential*, I_γ, by

$$I_\gamma f(x) = \int_{\mathbb{R}^n} \frac{f(y)}{|x-y|^{n-\gamma}} \, dy \quad \text{for } x \in \mathbb{R}^n,$$

for all functions f for which the integral makes sense. The representation formula then reads as

$$u(x) = \frac{1}{(2-n)n\omega_n} I_2(\Delta u)(x) \quad \text{for a.e. } x \in \mathbb{R}^n.$$

Inequalities involving the first-order gradient can be handled similarly with the help of the Riesz potential I_1.

It follows from what has been mentioned above that if we know certain boundedness properties of I_2, say,

$$I_2 \colon X \to Y \tag{3.1.12}$$

for some function spaces X, Y that are at least quasinormed, then we would obtain

$$\|u\|_Y = \frac{1}{(n-2)n\omega_n} \|I_2(\Delta u)\|_Y \leq \|I_2\|_{X \to Y} \|\Delta u\|_X, \tag{3.1.13}$$

where $\|I_2\|_{X \to Y}$ is the operator norm. In this way, our ultimate task to obtain a quantified transfer regularity from Δu to u has been delegated, owing to (3.1.13), from validity of an inequality involving differential operators such as (3.1.4), to the (presumably easier) question of boundedness properties of an integral operator such as (3.1.12).

3.1.2 The Ornstein–Uhlenbeck Operator

Let us now switch for a time being to a different world and exhibit one more example of an inequality involving a differential operator, albeit in completely different circumstances. The point is to demonstrate that one should not dwell on certain presuppositions.

In Gaussian harmonic analysis and in various its applications to infinite-dimensional analysis, the role of the Laplacian, treated in the preceding subsection, is taken over by the *Ornstein–Uhlenbeck operator*, $\mathcal{L}$, given by

$$\mathcal{L}u(x) = \Delta u(x) - x \cdot \nabla u(x) \quad \text{for } x \in \mathbb{R}^n.$$

Here $\mathbb{R}^n$ is still the ambient Euclidean space, but it is endowed with the *Gaussian measure*, $d\gamma_n$, defined by

$$d\gamma_n(x) = \pi^{-\frac{n}{2}} e^{-\frac{|x|^2}{2}} \, dx \quad \text{for } x \in \mathbb{R}^n,$$

instead of the Lebesgue one. Put very roughly, applications enforce one often to study inequalities involving differential operators in infinitely many variables, for which the Lebesgue measure is meaningless. So the principal idea is to replace the Lebesgue measure with the Gaussian one and to require that relevant inequalities hold in $(\mathbb{R}^n, \gamma_n)$ for every n, with quantifying parameters independent on n (this idea then enables one to send $n \to \infty$). Resulting relations between function structures are then called *dimension-free embeddings*.

More precisely, we seek sharp estimates of the form

$$\|u\|_{Y(\mathbb{R}^n, \gamma_n)} \leq C \|\mathcal{L}u\|_{X(\mathbb{R}^n, \gamma_n)} \tag{3.1.14}$$

with C independent of u, *but also of n.*

Compared to the Euclidean setting studied in the preceding subsection, there are important differences that should not be missed. First, there is no representation formula involving a potential operator. Next, here nothing should depend on n. This stands in a sharp contrast with the Euclidean setting, where everything depends on the dimension, in particular, for instance, the constants in inequalities $((n-2)n\omega_n)$, the camaraderie between function spaces in embeddings (note that in (3.1.7) and (3.1.9), each of the respective partner target spaces for $L^p(\Omega)$, that is, $L^{\frac{np}{n-p}}(\Omega)$ and $L^{\frac{np}{n-2p}}(\Omega)$, both depend on n), or the appropriate integral operator (see the definition of the Riesz potential I_2), if there is one at all. Finally, it is not a-priori clear whether there will be any gain in integrability at all in (3.1.14) in the transfer from $\mathcal{L}u$ to u.

So the big question is, how to handle inequalities involving the Ornstein–Uhlenbeck operator. Similarly as in the Euclidean setting, it will be useful to begin by studying dimension-free inequalities involving the first-order gradient in place of the second-order differential operator $\mathcal{L}$.

3.1.3 Limiting Cases of the Euclidean–Sobolev Inequalities

Let us switch back for a moment to the Euclidean setting and have a closer look at optimality of function spaces. Recall (3.1.9). What happens at the *endpoints* $p = 1$ and $p = \frac{n}{2}$ and why have they been left out? Extrapolating from what we know for $p \in (1, \frac{n}{2})$, can we hope for something like

$$\Delta u \in L^1(\Omega) \quad \Rightarrow \quad u \in L^{\frac{n}{n-2}}(\Omega)$$

or

$$\Delta u \in L^{\frac{n}{2}}(\Omega) \quad \Rightarrow \quad u \in L^\infty(\Omega)?$$

It is not difficult to see that the answer is actually *negative* in both cases, that is, none of these implications holds. What we do have instead, in the Lebesgue environment, is the following:

$$\Delta u \in L^1(\Omega) \quad \Rightarrow \quad u \in L^q(\Omega) \quad \text{for every } q < \frac{n}{n-2}$$

and

$$\Delta u \in L^{\frac{n}{2}}(\Omega) \quad \Rightarrow \quad u \in L^q(\Omega) \quad \text{for every } q < \infty.$$

In other words, there is no *optimal degree of Lebesgue integrability* of u provided that Δu is either in $L^1(\Omega)$ or in $L^{\frac{n}{2}}(\Omega)$.

This situation is clearly very unsatisfactory, and it brings new challenges. For instance, what should we do with Δu in order to get $u \in L^\infty(\Omega)$? Here is, once again, a possibility:

$$\Delta u \in L^p(\Omega) \quad \text{for some } p \in (\tfrac{n}{2}, \infty] \quad \Rightarrow \quad u \in L^\infty(\Omega),$$

and, once again, it is not of much use, since the rank of p's for which this is valid is a left-open interval. This shows that the scale of Lebesgue spaces, although being of primary interest in analysis, is not sufficiently rich.

We can summarize some of the particular goals that surfaced in the above analysis:

1. to find a sensible function space $Y(\Omega)$ such that

$$\Delta u \in L^1(\Omega) \quad \Rightarrow \quad u \in Y(\Omega),$$

2. to find a sensible function space $Y(\Omega)$ such that

$$\Delta u \in L^{\frac{n}{2}}(\Omega) \quad \Rightarrow \quad u \in Y(\Omega),$$

3. to find a sensible function space $X(\Omega)$ such that

$$\Delta u \in X(\Omega) \quad \Rightarrow \quad u \in L^{\infty}(\Omega).$$

What should be clear by now is that we need more function spaces.

3.1.4 Limiting Sobolev Inequalities for the First-Order Gradient

To get a better and more complete picture, let us now return to the question of how does all this work for the first-order gradient. The quantified version of (3.1.7) states that, given $p \in [1, n)$ and setting $q = \frac{np}{n-p}$, there is a constant $C > 0$ such that

$$\left(\int_{\Omega} |u(x)|^q \, dx \right)^{\frac{1}{q}} \leq C \left(\int_{\Omega} |(\nabla u)(x)|^p \, dx \right)^{\frac{1}{p}} \tag{3.1.15}$$

for every $u \in C_0^1(\Omega)$. Standard examples show that in the *limiting case*, that is, when $p = n$, one cannot take the L^{∞}-norm on the left-hand side of (3.1.15) without violating it. Instead, one has, for every $q < \infty$,

$$\left(\int_{\Omega} |u(x)|^q \, dx \right)^{\frac{1}{q}} \leq C(q) \left(\int_{\Omega} |(\nabla u)(x)|^n \, dx \right)^{\frac{1}{n}} \tag{3.1.16}$$

for every $u \in C_0^1(\Omega)$. However, independently of one another, Yudovich [129], Pohozhaev [111] and Trudinger [126] have shown that if A_n is an increasing convex function equal to $\exp t^{n'}$ for large t, then there corresponds to each bounded domain Ω in $\mathbb{R}^n$ a constant $C = C(|\Omega|) > 0$ such that

$$\inf \left\{ \lambda > 0 : \int_{\Omega} A_n \left(\frac{|u(x)|}{\lambda} \right) dx \leq 1 \right\} \leq C \left(\int_{\Omega} |(\nabla u)(x)|^n \, dx \right)^{\frac{1}{n}} \tag{3.1.17}$$

for every $u \in C_0^1(\Omega)$. A simple proof of a more general result was later given by Strichartz in [121].

The inequality (3.1.17) cannot be improved by replacing A_n with any essentially faster growing function (see [27, 72]). Improvement is however possible on the *domain* side in the sense that, given a convex increasing function A such that

$$\inf\left\{\lambda > 0 : \int_\Omega A_n\left(\frac{|u(x)|}{\lambda}\right) dx \leq 1\right\} \\ \leq C \inf\left\{\lambda > 0 : \int_\Omega A\left(\frac{|\nabla u(x)|}{\lambda}\right) dx \leq 1\right\}, \tag{3.1.18}$$

then there exists another one, A_1, satisfying

$$\lim_{t\to\infty} \frac{A_1(t)}{A(t)} = 0$$

and (3.1.18) is still valid with A replaced by A_1. This was first observed in [109] and later extended to a general principle in [100] and [101]. These ideas are based on working in Orlicz spaces, which will be introduced later.

However, there is yet another way how to improve (3.1.17), this time on the target side. Results mentioned in the preceding paragraph show that there is no room for improvement within Orlicz spaces, but if we settle for a yet wider class of function spaces, then we stand a chance. More precisely, one can show that, to each bounded domain Ω, there is a positive constant $C = C(\Omega)$ such that

$$\left(\int_0^{|\Omega|} \left(\frac{u^*(t)}{\log\frac{|\Omega|}{t}}\right)^n \frac{dt}{t}\right)^{\frac{1}{n}} \leq C \left(\int_\Omega |(\nabla u)(x)|^n \, dx\right)^{\frac{1}{n}} \tag{3.1.19}$$

for every $u \in C_0^1(\Omega)$, where u^* is the nonincreasing rearrangement of u, see [16, 18, 46, 71, 95, 106], and it is not difficult to verify that this is an essential improvement of (3.1.17). Various proofs are available of this result, for example an elegant interpolation argument from [46] applies a theorem of [9] to the pair of elementary estimates

$$\begin{cases} I_1 : L^1(\mathbb{R}^n) \to L^{\frac{n}{n-1},\infty}(\mathbb{R}^n), \\ I_1 : L^{n,1}(\mathbb{R}^n) \to L^\infty(\mathbb{R}^n), \end{cases}$$

combining it with an appropriate first-order analogue of the representation formula (3.1.13). This approach requires Lorentz–Zygmund spaces, which will be also introduced soon.

Interestingly, even (3.1.19) is still not the end of the story in the sense that the target norm can still be essentially improved. It was shown in [92], that, first, to each bounded domain Ω, there is a $C = C(\Omega)$ such that

$$\left(\int_0^{|\Omega|} \frac{(u^*(\frac{t}{2}) - u^*(t))^n}{t} \, dt\right)^{\frac{1}{n}} \leq C \left(\int_\Omega |(\nabla u)(x)|^n \, dx\right)^{\frac{1}{n}} \tag{3.1.20}$$

3 Reduction Principles   161

for every $u \in C_0^1(\Omega)$, and, second, that the expression on the left-hand side of (3.1.20) is essentially larger than that on the left-hand side of (3.1.19), so (3.1.20) provides a nontrivial enhancement of (3.1.19) on the target side.

In view of the examples mentioned so far, we would like to develop a universal approach that would cover all the mentioned instances in terms of function spaces.

3.1.5 Integral Operators and Function Spaces

To finish the introductory section, let us just briefly recall that the need for more function spaces than just Lebesgue ones is not exclusive to Sobolev embeddings. We will illustrate this with the help of some notoriously known integral operators.

First, recall the *Laplace transform* $\mathcal{L}$, defined as

$$\mathcal{L}f(t) = \int_0^\infty e^{-ts} f(s)\,ds, \quad t \in (0, \infty),$$

and consider the following (of course rather artificially chosen) tasks:

(i) find a space $Y(0, \infty)$ such that $\mathcal{L}: L^3(0, \infty) \to Y(0, \infty)$,
(ii) find a space $X(0, \infty)$ such that $\mathcal{L}: X(0, \infty) \to L^{3/2}(0, \infty)$.

Then it can be easily shown that although there do exist function spaces $X(0, \infty)$ and $Y(0, \infty)$ fulfilling the requirements, *none* of them is a Lebesgue space.

Next, consider the *Hardy–Littlewood maximal operator* M, defined as

$$Mf(x) = \sup_{Q \ni x} \frac{1}{|Q|} \int_Q |f(y)|\,dy, \quad x \in \mathbb{R}^n, \quad n \in \mathbb{N},$$

and the problems:

(i) find a space $Y(\mathbb{R}^n)$ such that $M: L^1(\mathbb{R}^n) \to Y(\mathbb{R}^n)$,
(ii) find a space $X(\Omega)$ such that $M: X(\Omega) \to L^1(\Omega)$,

in which $\Omega \subset \mathbb{R}^n$ is a domain of finite measure. Then there are function spaces $Y(\mathbb{R}^n)$ fulfilling the first task, but *none* of them is a Lebesgue space. The second task is slightly more subtle as it *can* be fulfilled by a Lebesgue space, but not in a satisfactory (*sharp*) way (similarly as in (3.1.16)).

3.2 Function Spaces

Our aim in this section is to develop a reasonable portfolio of function spaces suitable for our purposes. In search for an appropriate pool of competing function spaces in our inequalities, we will use as inspiration some of the quite useful properties of Lebesgue spaces.

In result, instead of grappling with specific difficulties concerning individual classes of function spaces, we shall begin with defining the general environment of the so-called *rearrangement-invariant spaces* (sometimes called in literature also *symmetric spaces*) and then single out some of their most important examples such as Lorentz, Orlicz, Lorentz–Zygmund spaces, and their various modifications, refinements and combinations.

Most of the theory of the rearrangement-invariant spaces has been developed from 1950s to 1970s, see e.g. [26, 47, 48, 68, 89–91, 113, 114]. Detailed accounts can be found in several monographs, see e.g. [10, 85, 88], For more details and also for proofs of all the statements, our standard general reference is [10].

3.2.1 Spaces of Measurable Functions

Let $(\mathcal{R}, \nu)$ be a finite positive measure space. We denote by $\mathcal{M}(\mathcal{R}, \nu)$ the set of all ν-measurable functions on $\mathcal{R}$ taking values in $[-\infty, \infty]$. We also define

$$\mathcal{M}_+(\mathcal{R}, \nu) = \{u \in \mathcal{M}(\mathcal{R}, \nu) : u \geq 0\}$$

and

$$\mathcal{M}_0(\mathcal{R}, \nu) = \{u \in \mathcal{M}(\mathcal{R}, \nu) : u \text{ is finite } \nu\text{-a.e. on } \mathcal{R}\}.$$

In cases when no confusion can arise we write $\mathcal{R}$ in place of $(\mathcal{R}, \nu)$.

Given any function $u \in \mathcal{M}(\mathcal{R}, \nu)$, we define its *distribution function*, $u_* : [0, \infty) \to [0, \infty]$, by

$$u_*(t) = \nu(\{x \in \mathcal{R} : |u(x)| > t\}) \quad \text{for } t \in [0, \infty),$$

and its *non-increasing rearrangement*,

$$u^* : [0, \infty) \to [0, \infty],$$

by

$$u^*(t) = \inf\{\lambda \geq 0 : u_*(\lambda) \leq t\} \quad \text{for } t \in [0, \infty).$$

We also define $u^{**} : (0, \infty) \to [0, \infty]$ as

$$u^{**}(t) = \frac{1}{t} \int_0^t u^*(s)\, ds \quad \text{for } t \in (0, \infty).$$

3 Reduction Principles

Note that u^{**} is also non-increasing, and $u^*(t) \leq u^{**}(t)$ for $t \in (0, \infty)$. Moreover,

$$\int_0^t (u+v)^*(s)\,ds \leq \int_0^t u^*(s)\,ds + \int_0^t v^*(s)\,ds \quad \text{for } t \in [0, \infty), \qquad (3.2.1)$$

for every $u, v \in \mathcal{M}_+(\mathcal{R}, \nu)$.

Two measurable functions u and v (defined possibly on two different measure spaces) are said to be *equimeasurable* (or *equidistributed*) if $u^* = v^*$.

A basic property of rearrangements is the *Hardy-Littlewood inequality*, which tells us that, if $u, v \in \mathcal{M}(\mathcal{R}, \nu)$, then

$$\int_{\mathcal{R}} |u(x)v(x)|\,d\nu(x) \leq \int_0^{\nu(\mathcal{R})} u^*(t)v^*(t)\,dt. \qquad (3.2.2)$$

We say that a functional $\|\cdot\|_{X(0,1)}\colon \mathcal{M}_+(0,1) \to [0, \infty]$ is a *function norm*, if for all f, g and $\{f_n\}_{n\in\mathbb{N}} \in \mathcal{M}_+(0,1)$ and every $a \in [0, \infty)$, the following properties hold:

(P1) $\|f\|_{X(0,1)} = 0$ if and only if $f = 0$; $\|af\|_{X(0,1)} = a\|f\|_{X(0,1)}$,
$\|f+g\|_{X(0,1)} \leq \|f\|_{X(0,1)} + \|g\|_{X(0,1)}$,
(P2) $f \leq g$ a.e. implies $\|f\|_{X(0,1)} \leq \|g\|_{X(0,1)}$,
(P3) $f_n \nearrow f$ a.e. implies $\|f_n\|_{X(0,1)} \nearrow \|f\|_{X(0,1)}$,
(P4) $\|1\|_{X(0,1)} < \infty$,
(P5) a constant C exists such that $\int_0^1 f(x)\,d\nu(x) \leq C\|f\|_{X(0,1)}$.

If, in addition,

(P6) $\|f\|_{X(0,1)} = \|g\|_{X(0,1)}$ whenever $f^* = g^*$,

we say that $\|\cdot\|_{X(0,1)}$ is a *rearrangement-invariant function norm*.

We say that a functional $\|\cdot\|_{X(0,1)}$ is a *function quasinorm* if it satisfies all axioms of function norm except (P5) and with possibly a constant C bigger than 1 in $\|f+g\|_{X(0,1)} \leq C(\|f\|_{X(0,1)} + \|g\|_{X(0,1)})$. The smallest such C is called the *modulus of concavity* of $\|\cdot\|_{X(0,1)}$. If, moreover, (P6) holds, then we say that $\|\cdot\|_{X(0,1)}$ is a *rearrangement-invariant quasinorm*.

Given a function norm (quasinorm) $\|\cdot\|_{X(0,1)}$, the *Banach function space, BFS* (*quasi-Banach function space, QBFS*) $X = X(\mathcal{R}, \mu)$ is the collection of all μ-measurable functions $f\colon \mathcal{R} \to \mathbb{R}$ such that

$$\|f\|_{X(0,1)} < \infty.$$

If the function norm $\|\cdot\|_{X(0,1)}$ is rearrangement invariant, then we call X the *rearrangement-invariant space*, or, for short, *r.i. space*. Similarly we use the notion *quasi-rearrangement-invariant space*, or *quasi-r.i. space*.

Given a function norm $\|\cdot\|_{X(0,1)}$, we introduce another functional on $\mathcal{M}_+(0, 1)$, denoted by $\|\cdot\|_{X'(0,1)}$ and defined as

$$\|f\|_{X'(0,1)} = \sup_{\substack{g \in \mathcal{M}_+(0,1) \\ \|g\|_{X(0,1)} \leq 1}} \int_0^1 f(t)g(t)\,dt.$$

Then $\|\cdot\|_{X'(0,1)}$ is also a function norm on $\mathcal{M}_+(0,1)$. We shall call it the *associate norm* of $\|\cdot\|_{X(0,1)}$. A simple application of the Hardy–Littlewood inequality shows that also

$$\|f\|_{X'(0,1)} = \sup_{\substack{g \in \mathcal{M}_+(0,1) \\ \|g\|_{X(0,1)} \leq 1}} \int_0^1 f^*(t)g^*(t)\,dt.$$

An important property of the associate norm is the symmetric relation

$$\|f\|_{X(0,1)} = \sup_{\substack{g \in \mathcal{M}_+(0,1) \\ \|g\|_{X'(0,1)} \leq 1}} \int_0^1 f^*(t)g^*(t)\,dt. \tag{3.2.3}$$

As a consequence, we get

$$\|\cdot\|_{(X')'(0,1)} = \|\cdot\|_{X(0,1)}. \tag{3.2.4}$$

Let $\|\cdot\|_{X(0,1)}$ be a rearrangement-invariant function norm. Then the space $X(\mathcal{R}, \nu)$ is defined as the collection of all functions $u \in \mathcal{M}(\mathcal{R}, \nu)$ such that the quantity

$$\|u\|_{X(\mathcal{R},\nu)} = \|u^*(\nu(\mathcal{R})s)\|_{X(0,1)} \tag{3.2.5}$$

is finite. The space $X(\mathcal{R}, \nu)$ is a Banach space, endowed with the norm given by (3.2.5). With a slight abuse of notation, if $\mathcal{R} = (0, 1)$ and ν is the Lebesgue measure, we denote $X(\mathcal{R}, \nu)$ simply by $X(0, 1)$. The space $X(0, 1)$ is called the *representation space* of $X(\mathcal{R}, \nu)$.

Given a rearrangement-invariant space $X(\mathcal{R}, \nu)$, the rearrangement-invariant space $X'(\mathcal{R}, \nu)$ built upon the function norm $\|\cdot\|_{X'(0,1)}$ is called the *associate space* of $X(\mathcal{R}, \nu)$. It turns out that $X''(\mathcal{R}, \nu) = X(\mathcal{R}, \nu)$; hence, any rearrangement-invariant space $X(\mathcal{R}, \nu)$ is always the associate space of another rearrangement-invariant space, namely $X'(\mathcal{R}, \nu)$. Furthermore, the *Hölder inequality*

$$\int_0^1 f(t)g(t)\,dt \leq \|f\|_{X(0,1)} \|g\|_{X'(0,1)}, \tag{3.2.6}$$

3 Reduction Principles

holds for every $f, g \in \mathcal{M}_+(0, 1)$, and hence

$$\frac{1}{\nu(\mathcal{R})} \int_{\mathcal{R}} |u(x)v(x)| \, d\nu(x) \leq \|u\|_{X(\mathcal{R},\nu)} \|v\|_{X'(\mathcal{R},\nu)}$$

for every u and v in $\mathcal{M}(\mathcal{R}, \nu)$.

Let $X(\mathcal{R}, \nu)$ and $Y(\mathcal{R}, \nu)$ be rearrangement-invariant spaces. We write $X(\mathcal{R}, \nu) \to Y(\mathcal{R}, \nu)$ to denote that $X(\mathcal{R}, \nu)$ is continuously embedded into $Y(\mathcal{R}, \nu)$. One has that

$$X(\mathcal{R}, \nu) \subset Y(\mathcal{R}, \nu) \quad \text{if and only if} \quad X(\mathcal{R}, \nu) \to Y(\mathcal{R}, \nu).$$

Note that the embedding $X(\mathcal{R}, \nu) \to Y(\mathcal{R}, \nu)$ holds if and only if there exists a constant C such that $\|g\|_{Y(0,1)} \leq C \|g\|_{X(0,1)}$ for every $g \in \mathcal{M}_+(0, 1)$. Moreover, for any rearrangement-invariant spaces $X(\mathcal{R}, \nu)$ and $Y(\mathcal{R}, \nu)$, one has

$$X(\mathcal{R}, \nu) \to Y(\mathcal{R}, \nu) \quad \text{if and only if} \quad Y'(\mathcal{R}, \nu) \to X'(\mathcal{R}, \nu), \tag{3.2.7}$$

with the same embedding constants.

In the more general case when T is an operator defined on $X(\mathcal{R}_1, \nu_1)$ and taking values in $Y(\mathcal{R}_2, \nu_2)$, by

$$T : X(\mathcal{R}_1, \nu_1) \to Y(\mathcal{R}_2, \nu_2)$$

we denote the fact that T is *bounded* from $X(\mathcal{R}_1, \nu_1)$ to $Y(\mathcal{R}_2, \nu_2)$, that is, there exists a positive constant C such that

$$\|Tu\|_{Y(\mathcal{R}_2,\nu_2)} \leq C \|u\|_{X(\mathcal{R}_1,\nu_1)} \quad \text{for every } u \in X(\mathcal{R}_1, \nu_1).$$

The infimum over such C will be called the *operator norm* of T.

Given any $s > 0$, the *dilation operator* E_s, defined at $f \in \mathcal{M}(0, 1)$ by

$$(E_s f)(t) = \begin{cases} f(t/s) & \text{if } 0 < t \leq s \\ 0 & \text{if } s < t < 1. \end{cases}$$

Then, for any rearrangement-invariant space $X(0, 1)$, one has

$$\|E_s f\|_{X(0,1)} \leq \max\{1, 1/s\} \|f\|_{X(0,1)} \quad \text{for every } f \in \mathcal{M}(0, 1). \tag{3.2.8}$$

Hardy's Lemma tells us that if $f, g \in \mathcal{M}_+(0, 1)$, and

$$\int_0^t f(s) \, ds \leq \int_0^t g(s) \, ds \quad \text{for } t \in (0, 1),$$

then

$$\int_0^1 f(t)h(t)\,dt \le \int_0^1 g(t)h(t)\,dt$$

for every non-increasing function $h\colon (0,1) \to [0,\infty]$. A consequence of this result is the *Hardy–Littlewood–Pólya principle*, which asserts that if the functions $u, v \in \mathcal{M}(\mathcal{R}, \nu)$ satisfy

$$\int_0^t u^*(s)\,ds \le \int_0^t v^*(s)\,ds \quad \text{for } s \in (0,1),$$

then

$$\|u\|_{X(\mathcal{R},\nu)} \le \|v\|_{X(\mathcal{R},\nu)}$$

for every rearrangement-invariant space $X(\mathcal{R}, \nu)$.

Since $\nu(\mathcal{R}) < \infty$, for every rearrangement-invariant space $X(\mathcal{R}, \nu)$ one has that

$$L^\infty(\mathcal{R}, \nu) \to X(\mathcal{R}, \nu) \to L^1(\mathcal{R}, \nu). \tag{3.2.9}$$

Throughout, we use the convention that $\frac{1}{\infty} = 0$.

A basic example of a function norm is the *Lebesgue norm* $\|\cdot\|_{L^p(0,1)}$, defined as usual for $p \in [1, \infty]$.

Assume that $0 < p, q \le \infty$. We define the functional $\|\cdot\|_{L^{p,q}(0,1)}$ by

$$\|f\|_{L^{p,q}(0,1)} = \left\| t^{\frac{1}{p}-\frac{1}{q}} f^*(t) \right\|_{L^q(0,1)}$$

for $f \in \mathcal{M}_+(0,1)$. If either $1 < p < \infty$ and $1 \le q \le \infty$, or $p = q = 1$, or $p = q = \infty$, then $\|\cdot\|_{L^{p,q}(0,1)}$ is equivalent to a rearrangement-invariant function norm, and

$$(L^{p,q})'(0,1) = L^{p',q'}(0,1). \tag{3.2.10}$$

Here and throughout, by the equality $X = Y$ for two quasi-normed function spaces X and Y we will always mean that X and Y are *equivalent*, that is, they coincide in the set-theoretical sense and there exists a positive constant C such that

$$C^{-1}\|u\|_X \le \|u\|_Y \le C\|u\|_X \quad \text{for every } u \in X.$$

We further define the functional $\|\cdot\|_{L^{(p,q)}(0,1)}$ as

$$\|f\|_{L^{(p,q)}(0,1)} = \left\| t^{\frac{1}{p}-\frac{1}{q}} f^{**}(t) \right\|_{L^q(0,1)}$$

3 Reduction Principles

for $f \in \mathcal{M}_+(0, 1)$. If either $0 < p < \infty$ and $1 \leq q \leq \infty$, or $p = q = \infty$, then $\|\cdot\|_{L^{(p,q)}(0,1)}$ is a rearrangement-invariant function norm (see e.g. [110, Theorem 9.7.5]). The norms $\|\cdot\|_{L^{p,q}(0,1)}$ and $\|\cdot\|_{L^{(p,q)}(0,1)}$ are called *Lorentz function norms*, and the corresponding spaces $L^{p,q}(\mathcal{R}, \nu)$ and $L^{(p,q)}(\mathcal{R}, \nu)$ are called *Lorentz spaces*.

Lorentz spaces $\{L^{p,q}\}_{p,q \in (0,\infty]}$ form a two-parameter family of function spaces. They generalize Lebesgue spaces in the sense that $L^{p,p} = L^p$ for every $p \in (0, \infty]$. Their most important instances occur when $q = 1$, $q = p$ and $q = \infty$. The space $L^{p,\infty}$ is of special interest. In the literature it is often called a *weak Lebesgue space* and it is also a particular instance of more general *Marcinkiewicz spaces*. The Lorentz spaces are *nested* in the sense that $L^{p,q} \subsetneq L^{p,r}$ whenever $q < r$. Lorentz functionals are quasi-norms and also α-norms, and the sharp α is known in some cases.

Suppose now that $0 < p, q \leq \infty$ and $\alpha \in \mathbb{R}$. We define the functional $\|\cdot\|_{L^{p,q;\alpha}(0,1)}$ by

$$\|f\|_{L^{p,q;\alpha}(0,1)} = \left\| t^{\frac{1}{p} - \frac{1}{q}} \log^\alpha \left(\frac{e}{t}\right) f^*(t) \right\|_{L^q(0,1)} \tag{3.2.11}$$

for $f \in \mathcal{M}_+(0, 1)$. For suitable choices of p, q, α, $\|\cdot\|_{L^{p,q;\alpha}(0,1)}$ is equivalent to a function norm. If this is the case, then $\|\cdot\|_{L^{p,q;\alpha}(0,1)}$ is called a *Lorentz–Zygmund function norm*, and the corresponding space $L^{p,q;\alpha}(\mathcal{R}, \nu)$ is called a *Lorentz–Zygmund space*. A detailed study of (generalized) Lorentz-Zygmund spaces can be found in [105] or [56], see also [110, Chapter 9]. A useful limiting duality relation states that

$$(L^{\infty,q;-1})'(0, 1) = L^{(1,q')}(0, 1) \quad \text{for } q > 1,$$

see e.g. [115]. Owing to a classical Hardy inequality, one also has

$$L^{(p,q)}(0, 1) = L^{p,q}(0, 1) \tag{3.2.12}$$

if either $p \in (1, \infty)$ and $1 \leq q \leq \infty$ or $p = q = \infty$. Let us note that if one of the following conditions

$$\begin{cases} 1 < p < \infty, \ 1 \leq q \leq \infty, \\ p = 1, \ q = 1, \ \alpha \geq 0, \\ p = \infty, \ q = \infty, \ \alpha \leq 0, \\ p = \infty, \ 1 \leq q < \infty, \ \alpha + \frac{1}{q} < 0, \end{cases} \tag{3.2.13}$$

is satisfied, then $L^{p,q;\alpha}$ is equivalent to a rearrangement-invariant Banach function space.

It might be useful to notice that, with the help of the distribution function, the classical Lebesgue quasinorms can be rewritten in the form

$$\|u\|_{L^p} = \begin{cases} \left(\int_0^\infty p t^{p-1} u_*(t) \, dt\right)^{\frac{1}{p}} & \text{if } p \in (0, \infty), \\ \inf\{t \geq 0 : u_*(t) = 0\} & \text{if } p = \infty. \end{cases}$$

Similarly, given $p, q \in (0, \infty]$, one has, for the Lorentz functionals,

$$\|u\|_{L^{p,q}} = \begin{cases} \left(p \int_0^\infty \left[t\, u_*(t)^{\frac{1}{p}}\right]^q \frac{dt}{t}\right)^{\frac{1}{q}} & \text{if } q \in (0, \infty), \\ \sup_{t \in (0,\infty)} t\, u_*(t)^{\frac{1}{p}} & \text{if } q = \infty \end{cases}$$

for every u or, for short,

$$\|u\|_{L^{p,q}} = p^{\frac{1}{q}} \left\| t^{1-\frac{1}{q}} u_*(t)^{\frac{1}{p}} \right\|_{L^q(0,\infty)}.$$

A function $A \colon [0, \infty) \to [0, \infty]$ is called a *Young function* if it is convex (non trivial), left-continuous and vanishes at 0. Thus, any such function takes the form

$$A(t) = \int_0^t a(\tau) \, d\tau \quad \text{for } t \geq 0, \tag{3.2.14}$$

for some non-decreasing, left-continuous function $a \colon [0, \infty) \to [0, \infty]$. The *Luxemburg function norm* $\|\cdot\|_{L^A(0,1)}$ is defined by

$$\|f\|_{L^A(0,1)} = \inf\left\{\lambda > 0 : \int_0^1 A\left(\frac{f(t)}{\lambda}\right) dt \leq 1\right\},$$

for $f \in \mathcal{M}_+(0, 1)$. The corresponding rearrangement-invariant space $L^A(\mathcal{R}, \nu)$ is called an *Orlicz space*. In particular, $L^A(0, 1) = L^p(0, 1)$ if $A(t) = t^p$ for some $p \in [1, \infty)$, and $L^A(0, 1) = L^\infty(0, 1)$ if $A(t) = 0$ for $t \in [0, 1]$ and $A(t) = \infty$ for $t > 1$.

We will use the following notation for special instances of Orlicz spaces. If $p \in [1, \infty)$ and $\alpha \in \mathbb{R}$, then we denote by $L^p(\log L)^\alpha$ the Orlicz space L^A generated by the Young function $A(t) = t^p (\log t)^\alpha$ near ∞. If $\beta > 0$, then we denote by $\exp L^\beta$ the Orlicz space L^B generated by the Young function $B(t) = e^{t^\beta}$ near ∞.

Given two Young functions A and B, the function norms $\|\cdot\|_{L^A(0,1)}$ and $\|\cdot\|_{L^B(0,1)}$ are equivalent if and only if A and B are equivalent near infinity, in the sense that there exist constants $c \geq 1$ and $t_0 \geq 0$ such that

$$A(t/c) \leq B(t) \leq A(ct) \quad \text{for } t \geq t_0.$$

3 Reduction Principles

A common extension of Orlicz and Lorentz spaces is provided by the family of Orlicz–Lorentz spaces. Given $p \in (1, \infty]$, $q \in [1, \infty)$ and a Young function D such that

$$\int^\infty \frac{D(t)}{t^{1+p}}\, dt < \infty,$$

we denote by $\|\cdot\|_{L(p,q,D)(0,1)}$ the Orlicz–Lorentz rearrangement-invariant function norm defined as

$$\|f\|_{L(p,q,D)(0,1)} = \left\| t^{-\frac{1}{p}} f^*(t^{\frac{1}{q}}) \right\|_{L^D(0,1)} \tag{3.2.15}$$

for $f \in \mathcal{M}_+(0, 1)$. The fact that (3.2.15) actually defines a function norm follows from simple variants in the proof of [29, Proposition 2.1]. Given a measure space $(\mathcal{R}, \nu)$, we denote by $L(p, q, D)(\mathcal{R}, \nu)$ the *Orlicz–Lorentz space* associated with the rearrangement-invariant function norm $\|\cdot\|_{L(p,q,D)(0,1)}$. Note that this class of Orlicz–Lorentz spaces includes (up to equivalent norms) the Orlicz spaces and various instances of Lorentz and Lorentz-Zygmund spaces.

Given a rearrangement-invariant space X, the *fundamental function*, φ_X, of X, is defined by

$$\varphi_X(t) = \|\chi_E\|_X \quad \text{whenever } \mu(E) = t.$$

Note that the fundamental function is well defined thanks to the rearrangement invariance of X.

A very important fact is that, given any rearrangement-invariant space S, there always exists a *unique* Orlicz space $L(X)$ (called the *fundamental Orlicz space* of X) such that

$$\varphi_X = \varphi_{L(X)}.$$

Given a rearrangement-invariant space X, then the family of rearrangement-invariant spaces enjoying the same fundamental function as X is called a *fundamental level*.

For example, all Lorentz spaces $L^{p,q}$ with fixed p and varying q belong to the same fundamental level and their fundamental Orlicz space is the Lebesgue space $L^{p,p} = L^p$.

Apart from the Orlicz space, two other spaces on each fundamental level are of crucial importance. The *Lorentz endpoint space* $\Lambda(X)$ is given through the functional

$$\|f\|_{\Lambda(X)} = \int_0^\infty f^*(t)\, d\varphi_X(t),$$

in which the integral is to be understood in the Lebsgue–Stieltjes sense. The *Marcinkiewicz endpoint space* $M(X)$ is given by

$$\|f\|_{M(X)} = \sup_{t \in (0,\infty)} \frac{\varphi_X(t)}{t} \int_0^t f^*(s)\,ds.$$

One always has the *fundamental sandwich*:

$$\Lambda(X) \hookrightarrow X \hookrightarrow M(X).$$

In particular, we have, for any X,

$$\Lambda(X) \hookrightarrow L(X) \hookrightarrow M(X),$$

in which it might happen that one of the embeddings (in exceptional cases both of them) turn to equivalences. Typically, this happens for the first embedding when the space X is close to L^1, and for the second embedding when the space X is close to L^∞. The relation $\Lambda(X) = M(X)$ holds if (and only if) X is either L^1 or L^∞ ([22]).

For example, one has the following:

- If $\varphi_X(t) = t^{\frac{1}{p}}$, then $\Lambda(X) = L^{p,1}$, $L(X) = L^p$, and $M(X) = L^{p,\infty}$.
- If $\alpha > 0$ and $\varphi_X(t) = (1 + \log \frac{1}{t})^{-1/\alpha}$, then $L(X) = M(X) = \exp L^\alpha$.
- If $\varphi_X(t) = t$, then $\Lambda(X) = L(X) = M(X) = L^1$.
- If $\varphi_X(t) = 1$, then $\Lambda(X) = L(X) = M(X) = L^\infty$.

3.2.2 Sobolev Spaces

An open set Ω in $\mathbb{R}^n$ is said to have *the cone property* if there exists a finite cone Λ such that each point in Ω is the vertex of a finite cone contained in Ω and congruent to Λ.

An open set Ω is called a *Lipschitz domain* if it is bounded and each point of $\partial\Omega$ has a neighborhood $\mathcal{U}$ such that $\Omega \cap \mathcal{U}$ is the subgraph of a Lipschitz continuous function of $n-1$ variables.

Let $m \in \mathbb{N}$ and let $X(\Omega)$ be a rearrangement-invariant space. We define the m-th order Sobolev type space $W^m X(\Omega)$ as

$$W^m X(\Omega) = \{u : u \text{ is } m\text{-times weakly differentiable in } \Omega,$$
$$\text{and } |\nabla^k u| \in X(\Omega) \text{ for } k = 0, \ldots, m\},$$

3 Reduction Principles

equipped with the norm

$$\|u\|_{W^m X(\Omega)} = \sum_{k=0}^{m} \|\nabla^k u\|_{X(\Omega)}.$$

Here, $\nabla^m u$ denotes the vector of all m-th order weak derivatives of u. In particular, $\nabla^0 u$ stands for u, and $\nabla^1 u$ will also be simply denoted by ∇u.

The subspace $W^m_\perp X(\Omega)$ of $W^m X(\Omega)$ is defined as

$$W^m_\perp X(\Omega) = \left\{ u \in W^m X(\Omega) : \int_\Omega \nabla^k u \, dx = 0 \quad \text{for } 0 \leq k \leq m-1 \right\}.$$

The notation $V^m X(\Omega)$ will be employed to denote the space

$$V^m X(\Omega) = \left\{ u : u \text{ is } m\text{-times weakly differentiable in } \Omega, \text{ and } |\nabla^m u| \in X(\Omega) \right\},$$

equipped with the norm

$$\|u\|_{V^m X(\Omega)} = \sum_{k=0}^{m-1} \|\nabla^k u\|_{L^1(\Omega)} + \|\nabla^m u\|_{X(\Omega)}.$$

Note that, if $u \in V^m X(\Omega)$, then $|\nabla^m u| \in X(\Omega) \subset L^1(\Omega)$, by property (P5) of rearrangement-invariant spaces. Hence, one actually has that $|\nabla^k u| \in L^1(\Omega)$ for every $k = 0, \ldots, m-1$, by a standard Sobolev embedding on open sets with the cone property. The subspace $V^m_\perp X(\Omega)$ of $V^m X(\Omega)$ is defined analogously to $W^m_\perp X(\Omega)$. The spaces $W^m X(\Omega)$ and $V^m X(\Omega)$ are easily verified to be Banach spaces.

3.3 Reduction Principle: A Prototype

3.3.1 Reduction Principle

In Sobolev inequalities one compares the size of a function to the size of its gradient (or some general differential operator), measuring both in (possibly different) rearrangement-invariant (quasi-)norms. We focus on the question when such inequalities are optimal.

Suppose that we have a pair of rearrangement-invariant norms, $\|\cdot\|_{X(0,1)}$ and $\|\cdot\|_{Y(0,1)}$, for which there corresponds to each bounded domain $\Omega \subset \mathbb{R}^n$ a constant $C = C(|\Omega|) > 0$ such that

$$\|u^*(|\Omega|t)\|_{Y(0,1)} \leq C \||\nabla u|^*(|\Omega|t)\|_{X(0,1)} \quad (3.3.1)$$

for every $u \in C_0^1(\Omega)$. We would like to know that in (3.3.1) $\|\cdot\|_{X(0,1)}$ cannot be effectively decreased nor $\|\cdot\|_{Y(0,1)}$ effectively increased. In case when quasinorms are allowed, we work also with an inequality slightly different from (3.3.1), namely

$$\|u^*(|\Omega|t)\|_{Y(0,1)} \leq C \left\| \frac{d}{dt} \int_{\{x \in \mathbb{R}^n : |u(x)| > u^*(|\Omega|t)\}} |\nabla u(x)| \, dx \right\|_{X(0,1)} \quad (3.3.2)$$

for every $u \in C_0^1(\Omega)$. Here $n \geq 2$, $u \colon \Omega \to \mathbb{R}$ is a scalar function of n variables, $\nabla u = \left(\frac{\partial u}{\partial x_1}, \ldots, \frac{\partial u}{\partial x_n} \right)$ is the first-order gradient of u, $|\nabla u|$ is its Euclidean length, $X(\Omega)$ and $Y(\Omega)$ are rearrangement-invariant spaces over Ω, and u^*, $|\nabla u|^*$ are non-increasing rearrangements of u and $|\nabla u|$, respectively, on $(0, |\Omega|)$. Once a function u is rearranged on $(0, |\Omega|)$, we normalize it to $(0, 1)$ by considering $u^*(|\Omega|t)$. This way we ensure that we can measure its size by a rearrangement-invariant norm $\|\cdot\|_{X(0,1)}$.

We first reduce inequalities of type (3.3.1) or (3.3.2) to the boundedness of certain integral operators. This process is called a *reduction principle*.

A prototype result is the first-order reduction principle, which was proved in [54] and which involves functions vanishing on the boundary. Hence the geometry of Ω plays no role here.

Theorem 3.3.1 (Reduction Principle—A Prototype) *Let $n \in \mathbb{N}$, $n \geq 2$, and let $\|\cdot\|_{X(0,1)}$ and $\|\cdot\|_{Y(0,1)}$ be rearrangement-invariant norms. Then the following two statements are equivalent.*

(i) *To each bounded domain $\Omega \subset \mathbb{R}^n$, there exists a positive constant $C = C(|\Omega|)$ such that*

$$\|u^*(|\Omega|t)\|_{Y(0,1)} \leq C \||\nabla u|^*(|\Omega|t)\|_{X(0,1)} \quad \text{for every } u \in C_0^1(\Omega).$$

(ii) *There exists a positive constant $K > 0$ such that*

$$\left\| \int_t^1 f(s) s^{\frac{1}{n}-1} \, ds \right\|_{Y(0,1)} \leq K \|f\|_{X(0,1)} \quad \text{for every } f \in \mathcal{M}_+(0,1).$$

Sketch of the Proof (i)$\Rightarrow$(ii) In this direction, we use radially symmetric functions. If $g \in \mathcal{M}_+(0, 1)$ and $B \subset \mathbb{R}^n$ is the ball about origin with radius $\omega_n^{-\frac{1}{n}}$ (hence having measure equal to one), then we define the function $h \colon B \to \mathbb{R}$ by $h(x) = g(\omega_n |x|^n)$ for $x \in B$. Then h is a nonnegative radial function with

$$|\{x \in B : h(x) > \lambda\}| = |\{t \in (0, 1) : g(t) > \lambda\}| \quad \text{for } \lambda > 0,$$

which implies that $h^* = g^*$ on $(0, 1)$. In other words, h and g are equimeasurable.

3 Reduction Principles

Now, suppose that $f \in \mathcal{M}_+(0, 1)$ is given. Then we define

$$u(x) = \begin{cases} \int_{\omega_n |x|^n}^{1} f(s) s^{\frac{1}{n}-1} ds & \text{on } B \\ 0 & \text{outside } B. \end{cases}$$

One can then show by calculation that the integral inequality (ii) applied to f follows from the Sobolev inequality (i) applied to u.

(ii)$\Rightarrow$(i) We depart from the *generalized Pólya–Szegő principle*, suitable for rearrangement-invariant environment, which was proved in [32, Lemma 4.1] using ideas from [124, 125] and which states that

$$n\omega_n^{\frac{1}{n}} \int_0^t \left[-\tau^{1-\frac{1}{n}} \frac{du^*}{d\tau}(\tau) \right]^* (s) \, ds \leq \int_0^t |\nabla u|^*(s) \, ds \quad \text{for } t > 0. \tag{3.3.3}$$

We note that the inequality (3.3.3) is a consequence of the coarea formula in its form for functions of bounded variation. Once (3.3.3) is established, then one gets from the Hardy–Littlewood–Pólya principle for any rearrangement-invariant norm $\|\cdot\|_{X(0,1)}$ the estimate

$$n\omega_n^{\frac{1}{n}} \left\| t^{1-\frac{1}{n}} \left(-\frac{du^*}{dt}(t) \right) \right\|_{X(0,1)} \leq \||\nabla u|\|_{X(0,1)}. \tag{3.3.4}$$

Then (see [27])

$$u^*(|\Omega|t) = |\Omega| \int_t^1 -\left(\frac{du^*}{ds} \right) (|\Omega|s) \, ds$$

$$= |\Omega| \int_t^1 \left(-s^{1-\frac{1}{n}} \left(\frac{du^*}{ds} \right) (|\Omega|s) \right) s^{\frac{1}{n}-1} \, ds.$$

Hence, by the property (P2) of $\|\cdot\|_{Y(0,1)}$ and (ii), one obtains

$$\left\| u^*(|\Omega|t) \right\|_{Y(0,1)} \leq |\Omega| \left\| \int_t^1 \left| -s^{1-\frac{1}{n}} \left(\frac{du^*}{ds} \right) (|\Omega|s) \right| s^{\frac{1}{n}-1} \, ds \right\|_{Y(0,1)}$$

$$\leq K |\Omega| \left\| t^{1-\frac{1}{n}} \left(\frac{du^*}{dt} \right) (|\Omega|t) \right\|_{X(0,1)}$$

$$\leq K n^{-1} \omega_n^{-\frac{1}{n}} |\Omega| \, \| |\nabla u|^*(|\Omega|t) \|_{X(0,1)},$$

and the claim follows. □

It is worth noticing that the generalized version of the Pólya–Szegő principle which was used in the proof of the 'only if' part of Theorem 3.3.1, namely the estimate (3.3.4), can be also written in the following form, in which the operation $\star$, applied to both functions and domains, denotes the classical Steiner symmetrization.

Lemma 3.3.2 (Generalized Pólya–Szegő Principle) *Let $n \geq 1$, let $\Omega \subset \mathbb{R}^n$ be a bounded open set, let X be a rearrangement-invariant space, and $u \in W_0^1 X(\Omega)$. Then the symmetric rearrangement $u^\star$ of u belongs to $W_0^1 X(\Omega^\star)$ and*

$$\|\nabla u^\star\|_{X(\Omega^\star)} \leq \|\nabla u\|_{X(\Omega)}.$$

Remark 3.3.3 It should be noted that Lemma 3.3.2 is intimately connected to the Rayleigh–Faber–Krahn inequality which provides a lower bound for the first eigenvalue of the Laplace operator with Dirichlet boundary condition, which was conjectured by Lord Rayleigh in 1894 in connection with the sound of a vibrating membrane (see [112]), then a weaker result was proved by Courant in [44], and then it was finally solved independently by G. Faber [57] and E. Krahn [83].

The proof of the Rayleigh–Faber–Krahn inequality rests upon the combination of the variational characterization of the lowest eigenvalue and the fine properties of rearrangements, in particular on the fact that the size of a gradient of a function does not increase with symmetrization, as we have seen in (3.3.3), (3.3.4) and Lemma 3.3.2. For further details, see also [3, 5, 17, 73, 123].

In [54], a version of the reduction principle was proved for the case when the functionals involved are not norms but merely quasinorms.

Theorem 3.3.4 (Reduction Principle for Quasinorms) *Let $n \in \mathbb{N}$, $n \geq 2$, and let $\|\cdot\|_{X(0,1)}$ and $\|\cdot\|_{Y(0,1)}$ be rearrangement-invariant quasinorms. Then the following two statements are equivalent.*

(i) To each bounded domain $\Omega \subset \mathbb{R}^n$, there exists a positive constant $C = C(|\Omega|)$ such that

$$\|u^*(|\Omega|t)\|_{Y(0,1)} \leq C \left\| \frac{d}{dt} \int_{\{x \in \mathbb{R}^n : |u(x)| > u^*(t)\}} |\nabla u(x)| \, dx \right\|_{X(0,1)}$$

for every function $u \in C_0^1(\Omega)$.

(ii) There exists a positive constant K such that

$$\left\| \int_t^1 f(s) s^{\frac{1}{n}-1} \, ds \right\|_{Y(0,1)} \leq K \|f\|_{X(0,1)} \quad \text{for every } f \in \mathcal{M}_+(0,1).$$

Sketch of the Proof We depart from the following inequality, due to Talenti (see [125, p. 203]):

$$n \omega_n^{\frac{1}{n}} \left(-\frac{du^*}{dt} \right) \leq \frac{d}{dt} \int_{\{x \in \mathbb{R}^n : |u(x)| > u^*(t)\}} |\nabla u(x)| \, dx \quad \text{for a.e. } t > 0.$$

Using this and (ii), we get

$$\|u^*(|\Omega|t)\|_{Y(0,1)} \leq Kn\omega_n^{\frac{1}{n}} \left\|\frac{d}{dt} \int_{\{x \in \mathbb{R}^n : |u(x)| > u^*(t)\}} |\nabla u(x)|\, dx\right\|_{X(0,1)},$$

and (i) follows. In the converse direction the proof is analogous to that of the corresponding implication in Theorem 3.3.1. □

Remark 3.3.5 The Sobolev inequality (3.3.1) always implies (3.3.2), and they are equivalent when $\|\cdot\|_{X(0,1)}$ and $\|\cdot\|_{Y(0,1)}$ are rearrangement-invariant norms.

3.3.2 Optimal Function Spaces

We next apply the integral operators obtained from the reduction principle to associate to a given range quasinorm $\|\cdot\|_{Y(0,1)}$ the domain quasinorm $\|\cdot\|_{X(0,1)}$ which is optimal for it in (3.3.2). Let us note that the idea of pairing rearrangement-invariant norms was used in [77] in connection with integral operators of convolution type.

Theorem 3.3.6 (Optimal Domain Quasinorm) *Let $n \in \mathbb{N}$, $n \geq 2$, and let $\|\cdot\|_{Y(0,1)}$ be a rearrangement-invariant quasinorm. Define the functional $\|\cdot\|_{X(0,1)}$ on $\mathcal{M}_+(0,1)$ by*

$$\|f\|_{X(0,1)} = \left\|\int_t^1 f(s) s^{\frac{1}{n}-1}\, ds\right\|_{Y(0,1)} \quad \text{for } f \in \mathcal{M}_+(0,1).$$

Then $\|\cdot\|_{X(0,1)}$ is a rearrangement-invariant quasinorm with the property that to each bounded domain Ω in $\mathbb{R}^n$ there corresponds $C = C(|\Omega|) > 0$ such that (3.3.2) holds. Moreover, it is the smallest such quasinorm in the sense that when $\|\cdot\|_{Z(0,1)}$ is another, then there exists $K > 0$ for which $\|f\|_{X(0,1)} \leq K\|f\|_{Z(0,1)}$ for every $f \in \mathcal{M}_+(0,1)$.

Sketch of the Proof One has just to verify that the axioms of the rearrangement-invariant quasinorms are satisfied, which is relatively straightforward. The optimality of $\|\cdot\|_{X(0,1)}$ then follows immediately from its construction. □

In the next step, duality principles are used to describe the optimal range norm when the domain quasinorm is given.

Theorem 3.3.7 (Optimal Target Norm) *Let $n \in \mathbb{N}$, $n \geq 2$, and let $\|\cdot\|_{X(0,1)}$ be a rearrangement-invariant quasinorm such that there exists $C > 0$ such that*

$$\int_0^1 f(t) t^{\frac{1}{n}}\, dt \leq C\|f\|_{X(0,1)} \quad \text{for every } f \in \mathcal{M}_+(0,1).$$

Then, the functional σ defined by

$$\sigma(g) = \left\| t^{\frac{1}{n}} g^{**}(t) \right\|_{X'(0,1)}$$

is a rearrangement-invariant norm on $\mathcal{M}_+(0,1)$*. Moreover, (3.3.2) holds for* $\|\cdot\|_{Y(0,1)} = \sigma'$*, and* $\|\cdot\|_{Y(0,1)}$ *is the largest such rearrangement-invariant norm.*

Sketch of the Proof Once again, the verification of the validity of the axioms of a rearrangement-invariant (quasi)norm is straightforward. The fact that (3.3.2) holds and the optimality of $\|\cdot\|_{Y(0,1)}$ are consequences of Theorems 3.3.1 and 3.3.4. □

3.3.3 Examples: Lorentz–Karamata Quasinorms

In order to obtain a reasonably rich pool of examples, we develop the scale of function spaces based on the involvement of slowly-varying functions.

Definition 3.3.8 A positive function b is said to be *slowly varying* on $(1, \infty)$, in the sense of Karamata, if for each $\varepsilon > 0$, $t^\varepsilon b(t)$ is eventually increasing and $t^{-\varepsilon} b(t)$ is eventually decreasing.

Definition 3.3.9 Let $1 \le p, q \le \infty$ and suppose b is slowly varying on $(1, \infty)$. Assume that

$$\left\| t^{-\frac{1}{q}} b(\tfrac{1}{t}) \right\|_{L^q(0,1)} < \infty \quad \text{when } p = \infty. \tag{3.3.5}$$

The *Lorentz–Karamata* quasinorm $\|\cdot\|_{L^{p,q;b}(0,1)}$ is given at $f \in \mathcal{M}_+(0,1)$ by

$$\|f\|_{L^{p,q;b}(0,1)} = \left\| t^{\frac{1}{p}-\frac{1}{q}} b(\tfrac{1}{t}) f^*(t) \right\|_{L^q(0,1)}.$$

Lorentz–Karamata spaces were first introduced in [54], their title reflecting that the structure combines Lorentz-like fine tuning of functions of given degree of integrability with the concept of the slowly-varying functions that had been studied by Karamata [76]. They have been extensively investigated ever since [8, 20, 53, 59, 66, 69, 74, 107, 108], and their field of applications has been widened e.g. to Gaussian logarithmic inequalities [34], interpolation theory [4], inequalities involving operators on probability spaces [40], trace problems [36], Bessel potentials [64, 103] and more. The properties of slowly-varying functions are in some detail discussed in [130, Chapter 2, p. 184] and also [78]. Further details and relations can be found in monographs [11, 52].

Lorentz–Karamata spaces shelter other types of function spaces, namely Lebesgue, Lorentz, and Lorentz–Zygmund spaces. On the other hand they are special cases of the so-called *classical Lorentz spaces*.

3 Reduction Principles

Equipped with Lorentz–Karamata quasinorms, we can formulate a rather general result which provides us with various examples of optimal Sobolev embeddings.

Theorem 3.3.10 (Optimal Pairs of Lorentz–Karamata Functionals) *Let* $1 \leq p, q \leq \infty$, *suppose b is slowly varying on $(1, \infty)$, and assume that (3.3.5) is satisfied. Define the functionals $\| \cdot \|_{X(0,1)}$ and $\| \cdot \|_{Y(0,1)}$ on $\mathcal{M}_+(0, 1)$ by*

$$\|f\|_{Y(0,1)} = \begin{cases} \left\| t^{\frac{1}{p}-\frac{1}{q}} b(\tfrac{1}{t}) f^*(t) \right\|_{L^q(0,1)} & \text{when } p > q, \\ \left\| t^{\frac{1}{p}-\frac{1}{q}} b(\tfrac{1}{t}) f^{**}(t) \right\|_{L^q(0,1)} & \text{when } p \leq q, \end{cases}$$

and

$$\|f\|_{X(0,1)} = \left\| \int_t^1 f(s) s^{\frac{1}{n}-1} \, ds \right\|_{Y(0,1)}.$$

Then $\| \cdot \|_{X(0,1)}$ and $\| \cdot \|_{Y(0,1)}$ are optimal in (3.3.2) as a rearrangement-invariant norm and a Banach function norm, respectively.

Sketch of the Proof The proof is rather involved and technical, it has to be divided into several cases depending on the comparison of p and q, and it uses plenty of knowledge from the theory of rearrangement-invariant spaces, but the principal idea is simple. One just has to show that

$$\|g\|_{Y'(0,1)} \approx \left\| t^{\frac{1}{n}} g^{**}(t) \right\|_{X'(0,1)} \quad \text{for } g \in \mathcal{M}_+(0, 1),$$

and then invoke Theorem 3.3.7 with $\sigma(g) = \|g\|_{Y'(0,1)}$. □

A natural next concern would be to construct the optimal rearrangement-invariant domain norm $\| \cdot \|_{X(0,1)}$ corresponding to a fixed rearrangement-invariant range norm $\| \cdot \|_{Y(0,1)}$. In view of Theorem 3.3.1, a natural candidate for $\| \cdot \|_{X(0,1)}$ is the functional

$$f \mapsto \left\| \int_t^1 f^*(s) s^{\frac{1}{n}-1} \, ds \right\|_{Y(0,1)} \quad \text{for } f \in \mathcal{M}_+(0, 1). \tag{3.3.6}$$

A partial answer to the question when this functional is equivalent to a rearrangement-invariant norm is supplied by the following observation.

Theorem 3.3.11 (Simplification of Optimal Domain Norm) *Let $\| \cdot \|_{Y(0,1)}$ be a rearrangement-invariant norm on $\mathcal{M}_+(0, 1)$ for which there exists $C > 0$ such that for every $f \in \mathcal{M}_+(0, 1)$ one has*

$$\left\| \int_t^1 f^{**}(s) s^{\frac{1}{n}-1} \, ds \right\|_{Y(0,1)} \leq C \left\| \int_t^1 f^*(s) s^{\frac{1}{n}-1} \, ds \right\|_{Y(0,1)}. \tag{3.3.7}$$

Then, the functional $\|\cdot\|_{X(0,1)}$, defined by

$$\|f\|_{X(0,1)} = \left\|\int_t^1 f^*(s)s^{\frac{1}{n}-1}\,ds\right\|_{Y(0,1)} \quad \text{for } f \in \mathcal{M}_+(0,1), \tag{3.3.8}$$

is a rearrangement-invariant norm on $\mathcal{M}_+(0,1)$. Moreover, $\|\cdot\|_{X(0,1)}$ is the optimal rearrangement-invariant domain norm for $\|\cdot\|_{Y(0,1)}$ in both (3.3.1) and (3.3.2).

The proof of Theorem 3.3.11 is easy, but its assumption (3.3.7) is too strong in the sense that it is not necessary for the functional defined in (3.3.8) to be a rearrangement-invariant norm. This can be illustrated for instance with the Lorentz norm

$$\|f\|_{Y(0,1)} = \int_0^1 f^*(t) t^{-\frac{1}{n}}\,dt,$$

for which one has

$$\left\|\int_t^1 f^*(s)s^{\frac{1}{n}-1}\,ds\right\|_{Y(0,1)} = \int_0^1 f^*(t)\,dt,$$

hence $Y = L^1$, while, however

$$\left\|\int_t^1 f^{**}(s)s^{\frac{1}{n}-1}\,ds\right\|_{Y(0,1)} = \int_0^1 f^*(t)\log\left(\frac{1}{t}\right) dt,$$

in consequence of which (3.3.7) obviously does not hold. We shall address this discrepancy soon.

Using the results described in this paragraph, we can now exhibit several examples of optimal Sobolev inequalities.

Example 3.3.12 Let $p \in (1, n)$. Define

$$\|f\|_{X(0,1)} = \|f\|_{L^p(0,1)}$$

and

$$\|f\|_{Y(0,1)} = \|t^{-\frac{1}{n}} f^*(t)\|_{L^p(0,1)}.$$

Then $\|\cdot\|_{X(0,1)}$ and $\|\cdot\|_{Y(0,1)}$ form an optimal pair of rearrangement-invariant norms in (3.3.1). Note that $Y(0,1) = L^{\frac{np}{n-p},p}(0,1)$. This means that we not only recovered the Sobolev embedding

$$W^1 L^p(\Omega) \to L^{\frac{np}{n-p},p}(\Omega),$$

3 Reduction Principles

but we added to it the information that both the domain space and the target space are optimal (the largest, respectively the smallest possible) rearrangement-invariant spaces in the sense that, for any rearrangement-invariant norm $\|\cdot\|_{Z(0,1)}$ such that

$$W^1 L^p(\Omega) \to Z(\Omega),$$

we necessarily have

$$L^{\frac{np}{n-p},p}(\Omega) \to Z(\Omega),$$

and, at the same time, for any rearrangement-invariant norm $\|\cdot\|_{\Upsilon(0,1)}$ such that

$$W^1 \Upsilon(\Omega) \to L^{\frac{np}{n-p},p}(\Omega),$$

we necessarily have

$$\Upsilon(\Omega) \to L^p(\Omega).$$

Example 3.3.13 Define

$$\|f\|_{X(0,1)} = \|f\|_{L^n(0,1)}$$

and

$$\|f\|_{Y(0,1)} = \|t^{-\frac{1}{n}} \log^{-1}(\tfrac{1}{t}) f^*(t)\|_{L^n(0,1)}.$$

Note that $Y(0, 1) = L^{\infty,n;-1}(0, 1)$. Then $\|\cdot\|_{X(0,1)}$ and $\|\cdot\|_{Y(0,1)}$ form a pair of rearrangement-invariant norms, $\|\cdot\|_{Y(0,1)}$ is optimal in (3.3.1) though $\|\cdot\|_{X(0,1)}$ is not, as it can be replaced by effectively larger ones. One such improvement was established in [56], where it was shown that one can, in fact, take

$$\|f\|_{X(0,1)} = \|f\|_{(L^n + L^{n,1;\frac{1}{n}-1})(0,1)},$$

which is essentially larger than $L^n(0, 1)$, but even this norm is not optimal. The optimal one turns out to be

$$\|f\|_{X(0,1)} = \left\| t^{-\frac{1}{n}} \log^{-1}(\tfrac{1}{t}) \int_t^1 f^{**}(s) s^{\frac{1}{n}-1} ds \right\|_{L^n(0,1)},$$

as shown in [109].

Example 3.3.14 Let b be a slowly-varying function. Set

$$\|f\|_{X(0,1)} = \left\| b\left(\frac{1}{t}\right) \int_t^1 f^{**}(s) s^{\frac{1}{n}-1} \, ds \right\|_{L^\infty(0,1)}$$

and

$$\|f\|_{Y(0,1)} = \left\| b\left(\frac{1}{t}\right) f^*(t) \right\|_{L^\infty(0,1)}.$$

Then $\|\cdot\|_{X(0,1)}$ and $\|\cdot\|_{Y(0,1)}$ form an optimal pair of rearrangement-invariant norms in (3.3.1). In the special case when $b \equiv 1$, this recovers the embedding

$$W^1 L^{n,1}(\Omega) \to L^\infty(\Omega),$$

established in [120], together with its optimality, which was proved in [32].

Example 3.3.15 Let b be a slowly-varying function. Set

$$\|f\|_{X(0,1)} = \int_0^1 b\left(\frac{1}{t}\right) f(t) \, dt$$

and

$$\|f\|_{Y(0,1)} = \int_0^1 t^{-\frac{1}{n}} b\left(\frac{1}{t}\right) f^*(t) \, dt.$$

Then $\|\cdot\|_{X(0,1)}$ and $\|\cdot\|_{Y(0,1)}$ form an optimal pair of norms in (3.3.2), but only $\|\cdot\|_{Y(0,1)}$ is rearrangement invariant unless $b \approx 1$, in which case $\|\cdot\|_{X(0,1)} = \|\cdot\|_{L^1(0,1)}$ and the pair is optimal in (3.3.1).

3.4 Reduction Principle for Higher-Order Sobolev Inequalities

It is natural to consider Sobolev inequalities for m-times continuously differentiable functions when $m \in \mathbb{N}$, $m \geq 2$. This requires an m-th order gradient, and a corresponding supply of function spaces. It is also time to abandon the restriction to functions vanishing at the boundary. Therefore the geometry of Ω will come to the picture. We will begin with domains having Lipschitz boundary.

First question one has to ask is how (and where) should the reduced estimate

$$\left\| \int_t^{|\Omega|} g(s) s^{-1+\frac{1}{n}} \, ds \right\|_{\overline{Y}(0,|\Omega|)} \leq C_2 \|g\|_{\overline{X}(0,|\Omega|)}$$

be modified for inequalities involving higher-order gradients? A reasonable guess seems to be, at least for $m \leq n - 1$, the inequality

$$\left\| \int_t^{|\Omega|} g(s) s^{-1+\frac{m}{n}} ds \right\|_{\overline{Y}(0,|\Omega|)} \leq C_2 \|g\|_{\overline{X}(0,|\Omega|)}.$$

It should be noticed, however, that the technique of Lemma 3.3.2, which was successfully applied in the case when $m = 1$, cannot be directly extended to the higher-order case, mainly because the Pólya–Szegő principle is not appropriate for extensions to higher-order derivatives. Second- and higher-order derivatives of u^* cannot be estimated in terms of derivatives of u because u^* does not have to be twice weakly differentiable, not even for very smooth u.

3.4.1 The Reduction Principle for Second-Order Embeddings

A remarkable exception from this rule was found by A. Cianchi, who established a positive one-dimensional result for functions whose second-order derivative is a bounded measure [28]. A simple example of a polynomial function shows that the first-order derivative of the decreasing rearrangement of a very smooth function does not necessarily possess a weak derivative. This discrepancy is circumvented in [28] by considering a larger class than a Sobolev space, namely, the space of functions whose second-order distributional derivative is a measure with finite total variation. Then it is shown that such space is closed under the operation of non-increasing rearrangement and the nonnegative functions satisfy the second-order version of the Pólya–Szegő estimate.

The following theorem is due to A. Cianchi [30, Theorem 1.1].

Theorem 3.4.1 (Second-Order Version of the Pólya–Szegő Principle) *Let* $n \in \mathbb{N}$, $n \geq 3$, *let* $\Omega \subset \mathbb{R}^n$ *be an open bounded set, let* $X(\Omega)$ *be an r.i. space over* Ω. *Then there is a positive constant* $C(n)$ *such that*

$$C(n) \left\| t^{1-\frac{2}{n}} \left(-\frac{du^*}{dt}(t) \right) \right\|_{\overline{X}(0,|\Omega|)} \leq \||\nabla^2 u|\|_{X(\Omega)} \tag{3.4.1}$$

for every $u \in W_0^2 X(\Omega)$.

It is of interest to compare the inequalities (3.3.4) and (3.4.1). As observed in [30], unlike the former one, the latter inequality does not have an isoperimetric nature in the sense that there does not exist a constant $C(n)$ for which equality holds in (3.4.1) for all spherically symmetric u. So it cannot be applied to obtaining best constants in Sobolev inequalities. However, it still can be efficiently used for proving the corresponding (second-order) reduction principle, which then reads as follows

[30, Theorem 1.3]. Note that it confirms, at least in part, our educated guess made above.

Theorem 3.4.2 (Second-Order Reduction Principle) *Let $n \in \mathbb{N}$, $n \geq 3$, let $\Omega \subset \mathbb{R}^n$ be open and bounded. Let $X(\Omega)$, $Y(\Omega)$ be r.i. spaces over Ω. Then the following are equivalent.*

(i) There is C_1 such that the Sobolev inequality

$$\|u\|_{Y(\Omega)} \leq C_1 \| |\nabla^2 u| \|_{X(\Omega)}$$

holds for every $u \in W_0^2 X(\Omega)$.
(ii) There is C_2 such that

$$\left\| \int_t^{|\Omega|} g(s) s^{-1+\frac{2}{n}} \, ds \right\|_{\overline{Y}(0,|\Omega|)} \leq C_2 \|g\|_{\overline{X}(0,|\Omega|)}$$

holds for every $g \in \mathcal{M}_+(0, |\Omega|)$.

3.4.2 The Reduction Principle for Higher-Order Embeddings via Interpolation

Our next aim is to extend the reduction principle to all orders m up to $n - 1$. To achieve that, we will use the methods of interpolation theory. The result is surveyed in the following assertion from [79, Theorem A].

Theorem 3.4.3 (Higher-Order Reduction Principle) *Let $m, n \in \mathbb{N}$, $n \geq 2$, $1 \leq m \leq n - 1$. Let $\|\cdot\|_{X(0,1)}$ and $\|\cdot\|_{Y(0,1)}$ be rearrangement-invariant norms on $\mathcal{M}_+(0, 1)$. Then, to each bounded domain $\Omega \subset \mathbb{R}^n$, with $\partial \Omega \in \text{Lip}_1$, there corresponds a constant $C > 0$, depending on Ω, n and m, such that*

$$\|u^*(|\Omega|t)\|_{Y(0,1)} \leq C \| |\nabla^m u|^*(|\Omega|t) \|_{X(0,1)} \quad \text{for all } u \in W^m X(\Omega) \quad (3.4.2)$$

if and only if there is a constant $K > 0$ such that

$$\left\| \int_t^1 f(s) s^{\frac{m}{n}-1} \, ds \right\|_{Y(0,1)} \leq K \|f\|_{X(0,1)} \quad \text{for every } f \in \mathcal{M}_+(0, 1). \quad (3.4.3)$$

Sketch of the Proof Assume first that (3.4.3) holds. We begin with the endpoint embeddings

$$\|u^*(|\Omega|t)\|_{L^{\frac{n}{n-m},1}(0,1)} \lesssim \| |\nabla^m u|^*(|\Omega|t) \|_{L^1(0,1)} \quad (3.4.4)$$

and

$$\|u^*(|\Omega|t)\|_{L^\infty(0,1)} \lesssim \||\nabla^m u|^*(|\Omega|t)\|_{L^{\frac{n}{m},1}(0,1)}, \tag{3.4.5}$$

which hold for every $u \in C_0^m(\Omega)$ and in which the constants depend on the Lebesgue measure of the bounded domain $\Omega \subset \mathbb{R}^n$, as well as n and m. For $m = 1$, inequalities (3.4.4) and (3.4.5) are just special cases of Examples 3.3.15 and 3.3.14, respectively, both applied with $b \equiv 1$ (see also [82]). Induction can be used to prove them for all m.

As the next step, we apply an appropriate version of Holmstedt's formula [75, 84] and the DeVore-Scherer theorem [50] to the endpoint estimates. We obtain the inequality

$$\int_0^t s^{-\frac{m}{n}} u^*(|\Omega|s)\,ds \lesssim \int_0^t s^{-\frac{m}{n}} \int_{\frac{s}{2}}^1 |\nabla^m u|^*(y) y^{\frac{m}{n}-1}\,dy\,ds, \tag{3.4.6}$$

in which the constant depends only on Ω, n and m.

Now we denote by $H_{\frac{n}{m}}$ the operator given at every $g \in \mathcal{M}_+(0,1)$ by

$$H_{\frac{n}{m}} g(t) = \int_t^1 g(s) s^{\frac{m}{n}-1}\,ds \quad \text{for } t \in (0,1),$$

and by $H'_{\frac{n}{m}}$ the dual operator to $H_{\frac{n}{m}}$ with respect to the L^1-pairing, namely

$$H'_{\frac{n}{m}} g(t) = t^{\frac{m}{n}-1} \int_0^t g(s)\,ds \quad \text{for } t \in (0,1).$$

Then, (3.4.3) can be rewritten as

$$H_{\frac{n}{m}}: X(0,1) \to Y(0,1) \tag{3.4.7}$$

with the embedding constant K. By duality,

$$H'_{\frac{n}{m}}: Y'(0,1) \to X'(0,1),$$

with the identical embedding constant. Thus,

$$\left\|H'_{\frac{n}{m}} g\right\|_{X'(0,1)} \leq K \|g\|_{Y'(0,1)} \quad \text{for every } g \in Y'(0,1). \tag{3.4.8}$$

One of the key ideas in this proof is to realize that, in fact, (3.4.8) is equivalent to

$$\left\|H'_{\frac{n}{m}} g^*\right\|_{X'(0,1)} \leq K \|g\|_{Y'(0,1)} \quad \text{for every } g \in Y'(0,1). \tag{3.4.9}$$

The implication (3.4.8)⇒(3.4.9) is just a restriction of an inequality to a smaller pool of functions combined with the fact that $\|g\|_{Y'(0,1)}$ equals $\|g^*\|_{Y'(0,1)}$ owing to the rearrangement invariance of the norm $\|\cdot\|_{Y'(0,1)}$. The nontrivial part of the observation is the converse implication, (3.4.9)⇒(3.4.8), which is true thanks to the special case of the Hardy–Littlewood inequality which states that $\int_0^t g \leq \int_0^t g^*$ for every $g \in \mathcal{M}_+$. It should be noted that an analogous argument would not work for the operator $H_{\frac{n}{m}}$ itself, which is the principal pitfall here and the main reason why one has to switch to duality.

Define the functional $\|\cdot\|_{Y'_X(0,1)}$ by

$$\|g\|_{Y'_X(0,1)} = \left\|H'_{\frac{n}{m}}g^*\right\|_{X'(0,1)} = \left\|t^{\frac{m}{n}}g^{**}(t)\right\|_{X'} \quad \text{for } g \in \mathcal{M}(0,1). \tag{3.4.10}$$

Then one can show that $\|\cdot\|_{Y'_X(0,1)}$ is a rearrangement-invariant norm, the key ingredient here being the subadditivity of the operation $g \mapsto g^{**}$. Moreover, (3.4.10) implies, in particular, that $H'_{\frac{n}{m}} : Y'_X(0,1) \to X'(0,1)$ (with constant one). Thus, using duality once again, we get

$$H_{\frac{n}{m}} : X(0,1) \to Y_X(0,1). \tag{3.4.11}$$

It is important to notice that $Y_X(0,1)$ is in fact the smallest such rearrangement-invariant space. Indeed, assume that

$$H_{\frac{n}{m}} : X(0,1) \to Z(0,1)$$

for some rearrangement-invariant space $Z(0,1)$. Then

$$H'_{\frac{n}{m}} : Z'(0,1) \to X'(0,1),$$

that is,

$$\left\|H'_{\frac{n}{m}}g\right\|_{X'(0,1)} \lesssim \|g\|_{Z'(0,1)}$$

for every $g \in \mathcal{M}(0,1)$, in which the constant depends on X and Z. Restricting this to decreasing functions and using the rearrangement invariance of $Z'(0,1)$, we get

$$\left\|H'_{\frac{n}{m}}g^*\right\|_{X'(0,1)} \lesssim \|g\|_{Z'(0,1)},$$

which is, by the definition of $\|\cdot\|_{Y'_X(0,1)}$,

$$Z'(0,1) \to Y'_X(0,1).$$

3 Reduction Principles

Another application of duality now yields

$$Y_X(0, 1) \to Z(0, 1),$$

establishing the optimality of Y_X in (3.4.11). Therefore, (3.4.7) necessarily entails

$$Y_X(0, 1) \to Y(0, 1). \tag{3.4.12}$$

Consequently, in particular,

$$\|u^*(|\Omega|t)\|_{Y(0,1)} \lesssim \|u^*(|\Omega|t)\|_{Y_X(0,1)} \quad \text{for every } u \in W^m X(\Omega). \tag{3.4.13}$$

We will need one more auxiliary operator, this time the one involving the operation of taking the supremum over an interval with a variable bound, instead of integration. Denote by $T_{\frac{n}{m}}$ the operator given at every $g \in \mathcal{M}(0, 1)$ by

$$T_{\frac{n}{m}} g(t) = t^{-\frac{m}{n}} \sup_{s \in (t,1)} s^{\frac{m}{n}} g^*(s) \quad \text{for } t \in (0, 1).$$

The key property of the operator $T_{\frac{n}{m}}$ is the fact that, for every $g \in \mathcal{M}(0, 1)$, one has

$$g^* \leq T_{\frac{n}{m}} g, \tag{3.4.14}$$

and at the same time both functions $t^{\frac{m}{n}}(T_{\frac{n}{m}} g)^*(t)$ and $t^{\frac{m}{n}}(T_{\frac{n}{m}} g)^{**}(t)$ are nonincreasing in t on $(0, 1)$. This way $T_{\frac{n}{m}} g$ constitutes a majorant of g^* which, together with its maximal nonincreasing rearrangement, stays nonincreasing even after being multiplied by the increasing power function $t^{\frac{m}{n}}$. Moreover, it can be easily proved [37, 65] that the operator $T_{\frac{n}{m}}$ satisfies the ('endpoint') boundedness estimates

$$\begin{aligned} T_{\frac{n}{m}} &: L^1(0, 1) \to L^1(0, 1), \\ T_{\frac{n}{m}} &: L^{\frac{n}{m},\infty}(0, 1) \to L^{\frac{n}{m},\infty}(0, 1). \end{aligned} \tag{3.4.15}$$

Interpolation argument involving the K-functional inequality arising from (3.4.15) can be then applied to prove that, for every rearrangement-invariant norm $\|\cdot\|_{Z(0,1)}$, one has

$$\left\| \sup_{s \in (t,1)} s^{\frac{m}{n}} f^{**}(s) \right\|_{Z(0,1)} \lesssim \left\| t^{\frac{m}{n}} f^{**}(t) \right\|_{Z(0,1)} \tag{3.4.16}$$

with constant independent of Z. An optimal decomposition technique (for details see [79, Theorem 3.8]) can be used to show that

$$(T_{\frac{n}{m}} g)^{**}(t) \lesssim T_{\frac{n}{m}} (g^{**})(t) \quad \text{for every } g \in \mathcal{M}(0, 1) \text{ and } t \in (0, 1). \tag{3.4.17}$$

Using (3.4.17) and (3.4.16), we establish the estimate

$$\left\| t^{\frac{m}{n}} \left(T_{\frac{n}{m}} g \right)^{**}(t) \right\|_{Z(0,1)} \lesssim \left\| t^{\frac{m}{n}} T_{\frac{n}{m}} \left(g^{**} \right)(t) \right\|_{Z(0,1)}$$
$$= \left\| \sup_{s \in (t,1)} s^{\frac{m}{n}} g^{**}(s) \right\|_{Z(0,1)} \lesssim \left\| t^{\frac{m}{n}} g^{**}(t) \right\|_{Z(0,1)}$$
(3.4.18)

for every rearrangement-invariant norm $\|\cdot\|_{Z(0,1)}$ and every $g \in \mathcal{M}(0,1)$. Applying (3.4.18) to the choice $Z(0,1) = X'(0,1)$, we obtain

$$T_{\frac{n}{m}} : Y'_X(0,1) \to Y'_X(0,1). \tag{3.4.19}$$

Owing to (3.4.6) and Hardy's lemma, we get, for every $h \in \mathcal{M}(0,1)$ and $u \in \mathcal{M}(\Omega)$,

$$\int_0^1 t^{-\frac{m}{n}} h^*(t) u^*(|\Omega|t) \, dt$$
$$\lesssim \int_0^1 t^{-\frac{m}{n}} h^*(t) \int_{\frac{t}{2}}^1 |\nabla^m u|^*(|\Omega|s) s^{\frac{m}{n}-1} \, ds \, dt.$$
(3.4.20)

Fix $g \in \mathcal{M}(0,1)$. We define h by

$$h(t) = \sup_{s \in (t,1)} s^{\frac{m}{n}} g^*(s) \quad \text{for } t \in (0,1).$$

Then $h = h^*$, and (3.4.20) turns into

$$\int_0^1 T_{\frac{n}{m}} g(t) u^*(|\Omega|t) \, dt \lesssim \int_0^1 T_{\frac{n}{m}} g(t) \int_{\frac{t}{2}}^1 |\nabla^m u|^*(|\Omega|s) s^{\frac{m}{n}-1} \, ds \, dt. \tag{3.4.21}$$

Altogether,

$$\|u^*(|\Omega|t)\|_{Y(0,1)} \lesssim \|u^*(|\Omega|t)\|_{Y_X(0,1)} \quad \text{(by (3.4.13))}$$

$$= \sup_{\|g\|_{Y'_X(0,1)} \le 1} \int_0^1 g^*(t) u^*(|\Omega|t) \, dt \quad \text{(by (3.2.3))}$$

$$\le \sup_{\|g\|_{Y'_X(0,1)} \le 1} \int_0^1 T_{\frac{n}{m}} g(t) u^*(|\Omega|t) \, dt \quad \text{(by (3.4.14))}$$

$$\lesssim \sup_{\|g\|_{Y'_X(0,1)} \le 1} \int_0^1 T_{\frac{n}{m}} g(t) \int_{\frac{t}{2}}^1 |\nabla^m u|^*(|\Omega|s) s^{\frac{m}{n}-1} \, ds \, dt \quad \text{(by (3.4.21))}$$

3 Reduction Principles

$$\leq \sup_{\|g\|_{Y'_X(0,1)} \leq 1} \|T_{\frac{n}{m}}g\|_{Y'_X(0,1)} \left\| \int_{\frac{t}{2}}^{1} |\nabla^m u|^*(|\Omega|s) s^{\frac{m}{n}-1} ds \right\|_{Y_X(0,1)} \quad \text{(by (3.2.6))}$$

$$\lesssim \sup_{\|g\|_{Y'_X(0,1)} \leq 1} \|g\|_{Y'_X(0,1)} \left\| \int_{\frac{t}{2}}^{1} |\nabla^m u|^*(|\Omega|s) s^{\frac{m}{n}-1} ds \right\|_{Y_X(0,1)} \quad \text{(by (3.4.19))}$$

$$= \left\| \int_{\frac{t}{2}}^{1} |\nabla^m u|^*(|\Omega|s) s^{\frac{m}{n}-1} ds \right\|_{Y_X(0,1)} \quad \text{(trivial)}$$

$$\lesssim \||\nabla^m u|^*(|\Omega|t)\|_{X(0,1)} \quad \text{(by (3.4.11) and (3.2.8))}.$$

This yields (3.4.2). The converse implication can be obtained by a higher-order analogue of the corresponding part of the proof of Theorem 3.3.1. □

3.4.3 Optimal Partner Spaces

Similarly as in the case of the first-order embeddings, the reduction principle yields almost straightforward the formulas for optimal partner spaces, on both range and domain sides.

Theorem 3.4.4 (Optimal Spaces) *Let $m, n \in \mathbb{N}$, $n \geq 2$, $1 \leq m \leq n - 1$. Assume that Ω is a bounded domain in $\mathbb{R}^n$ with $\partial\Omega \in \mathrm{Lip}_1$.*

(i) Let $\|\cdot\|_{X(0,1)}$ be a rearrangement-invariant norm. Then, the functional $\|\cdot\|_{Y_X(0,1)}$, defined by

$$\|g\|_{Y'_X(0,1)} = \left\| t^{\frac{m}{n}} g^{**}(t) \right\|_{X'(0,1)} \quad \text{for } g \in \mathcal{M}(0,1), \tag{3.4.22}$$

is a rearrangement-invariant norm on $\mathcal{M}_+(0, 1)$. Moreover, the Sobolev embedding

$$W^m X(\Omega) \to Y_X(\Omega) \tag{3.4.23}$$

holds, and $\|\cdot\|_{Y_X(0,1)}$ is essentially the largest such rearrangement-invariant norm (hence $Y_X(\Omega)$ is essentially the smallest rearrangement-invariant space which renders (3.4.23) true).

(ii) Let $\|\cdot\|_{Y(0,1)}$ be a rearrangement-invariant norm such that

$$Y(0, 1) \to L^{\frac{n}{n-m}, 1}(0, 1).$$

Define the functional $\|\cdot\|_{X_Y(0,1)}$ *on* $\mathcal{M}(0,1)$ *by*

$$\|f\|_{X_Y(0,1)} = \sup_{f \sim h} \left\| \int_t^1 h(s) s^{\frac{m}{n}-1} \, ds \right\|_{Y(0,1)} \quad \text{for } f \in \mathcal{M}_+(0,1). \quad (3.4.24)$$

Then $\|\cdot\|_{X_Y(0,1)}$ *is a rearrangement-invariant norm. Moreover,*

$$W^m X_Y(\Omega) \to Y(\Omega) \quad (3.4.25)$$

holds, and $\|\cdot\|_{X_Y(0,1)}$ *is essentially the smallest such rearrangement-invariant norm (hence* $X_Y(\Omega)$ *is essentially the largest rearrangement-invariant space which renders (3.4.25) true).*

Sketch of the Proof We begin with proving (i). The fact that $\|\cdot\|_{Y_X(0,1)}$ is a rearrangement-invariant norm has already been observed in the course of the proof of Theorem 3.4.3. The validity of the embedding (3.4.23) follows straightforward from (3.4.11) and Theorem 3.4.3, implication (3.4.3)$\Rightarrow$(3.4.2). Hence the only part of the assertion that needs proof is the optimality of $Y_X(\Omega)$ in (3.4.23). Suppose, thus, that

$$W^m X(\Omega) \to Y(\Omega)$$

for some rearrangement-invariant norm $\|\cdot\|_{Y(0,1)}$. Then, by Theorem 3.4.3, implication (3.4.2)$\Rightarrow$(3.4.3), one has also (3.4.7). As shown in the proof of Theorem 3.4.3, this enforces (3.4.12). In turn, we get

$$Y_X(\Omega) \to Y(\Omega),$$

establishing the desired optimality and finishing the proof of (i).

As for (ii), the validity of the embedding (3.4.25) and the optimality of $X_Y(\Omega)$ in it are clear. To verify that $\|\cdot\|_{X_Y(0,1)}$ is a rearrangement-invariant norm requires checking of the axioms separately, using properties of measure-preserving transformations [10, Chapter 2, Corollary 7.6]. □

Theorem 3.4.4 can be applied to verify whether a known Sobolev embedding is sharp or whether it can be essentially improved. Moreover, both (3.4.22) and (3.4.24) give constructions for optimal rearrangement-invariant partner range or domain spaces in a Sobolev embedding in case the domain or range space, respectively, is fixed. A principal disadvantage of these constructions is that they are quite implicit. It is therefore worth to spend some effort in order to make them more explicit.

3 Reduction Principles

In view of Theorem 3.3.11, a natural question is, under what (reasonable) conditions, one has

$$\|f\|_{X_Y(0,1)} \approx \left\| \int_t^1 f^*(s) s^{\frac{m}{n}-1} ds \right\|_{Y(0,1)} \quad \text{for } f \in \mathcal{M}_+(0,1). \tag{3.4.26}$$

First, note that the estimate

$$\|f\|_{X_Y(0,1)} \gtrsim \left\| \int_t^1 f^*(s) s^{\frac{m}{n}-1} ds \right\|_{Y(0,1)} \quad \text{for } f \in \mathcal{M}_+(0,1),$$

is trivial, since $f^* \sim f$, and, furthermore, $h \sim f$ means $h^* = f^*$, so the question can be reformulated as for which Y the inequality

$$\left\| \int_t^1 f(s) s^{\frac{m}{n}-1} ds \right\|_{Y(0,1)} \lesssim \left\| \int_t^1 f^*(s) s^{\frac{m}{n}-1} ds \right\|_{Y(0,1)} \tag{3.4.27}$$

holds for every $f \in \mathcal{M}_+(0,1)$. Similarly as in Theorem 3.3.11, it can be proved that a simple sufficient condition for (3.4.27) is

$$\left\| \int_t^1 f^{**}(s) s^{\frac{m}{n}-1} ds \right\|_{Y(0,1)} \lesssim \left\| \int_t^1 f^*(s) s^{\frac{m}{n}-1} ds \right\|_{Y(0,1)},$$

but, once again, it is too restrictive. It turns out that a reasonable condition can be given in terms of the operator $T_{\frac{n}{m}}$.

Theorem 3.4.5 *Let* $m, n \in \mathbb{N}$, $1 \leq m \leq n - 1$. *Assume that* $Y(0, 1)$ *is a rearrangement-invariant norm such that*

$$T_{\frac{n}{m}} : Y'(0,1) \to Y'(0,1). \tag{3.4.28}$$

Then (3.4.27) *holds.*

Sketch of the Proof Fix $f \in \mathcal{M}_+(0,1)$. Since $t \mapsto \int_t^1 f(s) s^{\frac{m}{n}-1} ds$ is decreasing in t, we have

$$\left\| \int_t^1 f(s) s^{\frac{m}{n}-1} ds \right\|_{Y(0,1)} = \sup_{g \geq 0} \frac{\int_0^1 g^*(t) \int_t^1 f(s) s^{\frac{m}{n}-1} ds\, dt}{\|g\|_{Y'(0,1)}}.$$

By Fubini theorem, this reads as

$$\left\| \int_t^1 f(s) s^{\frac{m}{n}-1} ds \right\|_{Y(0,1)} = \sup_{g \geq 0} \frac{\int_0^1 f(s) s^{\frac{m}{n}} g^{**}(s) ds}{\|g\|_{Y'(0,1)}}.$$

Using the pointwise estimate (3.4.14), we get

$$\left\|\int_t^1 f(s)s^{\frac{m}{n}-1}\,ds\right\|_{Y(0,1)} \leq \sup_{g\geq 0}\frac{\int_0^1 f(s)s^{\frac{m}{n}}(T_{\frac{n}{m}}g)^{**}(s)\,ds}{\|g\|_{Y'(0,1)}}.$$

Now here comes the key part of the proof. It is called 'adding a star', and it is based on the idea that, unlikely the function $s \mapsto s^{\frac{m}{n}}g^{**}(s)$, the function $s \mapsto s^{\frac{m}{n}}T_{\frac{n}{m}}g^{**}(s)$ is *always* decreasing, regardless of g. Therefore, by the Hardy–Littlewood inequality,

$$\left\|\int_t^1 f(s)s^{\frac{m}{n}-1}\,ds\right\|_{Y(0,1)} \leq \sup_{g\geq 0}\frac{\int_0^1 f^*(s)s^{\frac{m}{n}}(T_{\frac{n}{m}}g)^{**}(s)\,ds}{\|g\|_{Y'(0,1)}}.$$

Using (3.4.28), we get

$$\left\|\int_t^1 f(s)s^{\frac{m}{n}-1}\,ds\right\|_{Y(0,1)} \lesssim \sup_{g\geq 0}\frac{\int_0^1 f^*(s)s^{\frac{m}{n}}(T_{\frac{n}{m}}g)^{**}(s)\,ds}{\|T_{\frac{n}{m}}g\|_{Y'(0,1)}}.$$

The more so, by extending the supremum, one has

$$\left\|\int_t^1 f(s)s^{\frac{m}{n}-1}\,ds\right\|_{Y(0,1)} \lesssim \sup_{h\geq 0}\frac{\int_0^1 f^*(s)s^{\frac{m}{n}}h^{**}(s)\,ds}{\|h\|_{Y'(0,1)}}.$$

One more use of the Fubini theorem yields

$$\left\|\int_t^1 f(s)s^{\frac{m}{n}-1}\,ds\right\|_{Y(0,1)} \lesssim \sup_{h\geq 0}\frac{\int_0^1 h^*(t)\int_t^1 f^*(s)s^{\frac{m}{n}-1}\,ds\,dt}{\|h\|_{Y'(0,1)}},$$

and this can be evaluated, owing to (3.2.3), to obtain

$$\left\|\int_t^1 f(s)s^{\frac{m}{n}-1}\,ds\right\|_{Y(0,1)} \lesssim \left\|\int_t^1 f^*(s)s^{\frac{m}{n}-1}\,ds\right\|_{Y(0,1)},$$

establishing the claim. □

Interestingly, the condition (3.4.28) can be, perhaps rather unexpectedly, characterized in terms of at least two other statements of quite a different nature. This equivalence reveals an interesting link between optimality properties and interpolation properties of function spaces, and it also reflects certain extremal property of the supremum operator.

3 Reduction Principles

Theorem 3.4.6 *Let $m, n \in \mathbb{N}$, $1 \leq m \leq n - 1$. Assume that $Y(0, 1)$ is a rearrangement-invariant space. Then, the following three statements are equivalent.*

(i) The condition (3.4.28) holds.
(ii) The space $Y(0, 1)$ is an interpolation space with respect to the pair $(L^{\frac{n}{n-m}, 1}(0, 1), L^{\infty}(0, 1))$.
(iii) The space $Y(0, 1)$ is optimal in $H_{\frac{n}{m}} : X(0, 1) \to Y(0, 1)$ for some rearrangement-invariant space $X(0, 1)$.

Sketch of the Proof (iii)$\Rightarrow$(i) If $Y(0, 1)$ is optimal in $H_{\frac{n}{m}} : X(0, 1) \to Y(0, 1)$ for some rearrangement-invariant space, say $X(0, 1)$, then $\tilde{Y}(0, 1) = Y_X(0, 1)$, and therefore (i) follows immediately from (3.4.18) applied to $Z = X'$.

(i)$\Rightarrow$(ii) If T is a linear operator satisfying

$$\begin{cases} T : L^{\frac{n}{n-m}, 1}(0, 1) \to L^{\frac{n}{n-m}, 1}(0, 1) \\ T : L^{\infty}(0, 1) \to L^{\infty}(0, 1), \end{cases}$$

then the standard optimal decomposition technique can be used to obtain the pointwise estimate

$$\int_0^t s^{-\frac{m}{n}} (Tg)^*(s) \, ds \lesssim \int_0^t s^{-\frac{m}{n}} g^*(s) \, ds \qquad (3.4.29)$$

for every $g \in \mathcal{M}(0, 1)$ and $t \in (0, 1)$. Therefore, by Hardy's lemma, one has

$$\int_0^1 t^{-\frac{m}{n}} (Tg)^*(t) h^*(t) \, dt \lesssim \int_0^1 t^{-\frac{m}{n}} g^*(t) h^*(t) \, dt \qquad (3.4.30)$$

for every $g, h \in \mathcal{M}(0, 1)$. Now we fix $f \in Y'(0, 1)$ and set

$$h(t) = h^*(t) = \sup_{s \in (t, 1)} s^{\frac{m}{n}} f^*(t) \quad \text{for } t \in (0, 1).$$

Recalling (3.4.14) and inserting the function h into (3.4.30), we arrive at

$$\int_0^1 (Tg)^*(t) f^*(t) \, dt \leq \int_0^1 (Tg)^*(t) (T_{\frac{n}{m}} f)(t) \, dt \lesssim \int_0^1 g^*(t) (T_{\frac{n}{m}} f)(t) \, dt.$$

By Hölder's inequality followed by an application of (i), this yields

$$\int_0^1 (Tg)^*(t) f^*(t) \, dt \lesssim \|g\|_{Y(0,1)} \|T_{\frac{n}{m}} f\|_{Y'(0,1)} \lesssim \|g\|_{Y(0,1)} \|f\|_{Y'(0,1)}.$$

Finally, by (3.2.3), we obtain

$$\|Tg\|_{Y(0,1)} = \sup_{f \neq 0} \frac{\int_0^1 (Tg)^*(t) f^*(t)\, dt}{\|f\|_{Y'(0,1)}} \lesssim \|g\|_{Y(0,1)}.$$

Hence, T is bounded on $Y(0, 1)$, which establishes (ii).

(ii)$\Rightarrow$(iii) It can be shown by standard estimates that $Y(0, 1)$ is optimal, for instance, in $H_{\frac{n}{m}} \colon X_Y(0, 1) \to Y(0, 1)$, which is sufficient in order to verify the desired implication. A different and more elegant proof, based on an interpolation theorem by Cwikel and Nilsson [45], which works for a substantially more general situation, is due to Z. Mihula, and can be found in [98, Theorem 4.16]. $\square$

Certain simplification, of 'dual' nature, of the formula for $Y_X(0, 1)$ is available, too, but we need to introduce one more type of a supremum operator. For $m, n \in \mathbb{N}$, $1 \leq m \leq n - 1$, let the operator $S_{\frac{n}{m}}$ be defined as

$$S_{\frac{n}{m}} g(t) = t^{1-\frac{m}{n}} \sup_{s \in (0,t)} s^{\frac{m}{n}-1} g^*(s) \quad \text{for } g \in \mathcal{M}(0, 1) \text{ and } t \in (0, 1).$$

With the help of the operator $S_{\frac{n}{m}}$, we can show the next result.

Theorem 3.4.7 *Let $m, n \in \mathbb{N}$, $1 \leq m \leq n - 1$. Assume that $X(0, 1)$ is a rearrangement-invariant space. Then, the following three statements are equivalent.*

(i) The condition

$$S_{\frac{n}{m}} \colon X'(0, 1) \to X'(0, 1). \tag{3.4.31}$$

holds.

(ii) The space $X(0, 1)$ is an interpolation space with respect to the pair $(L^1(0, 1), L^{\frac{n}{m}, 1}(0, 1))$.

(iii) The space $X(0, 1)$ is optimal in $H_{\frac{n}{m}} \colon X(0, 1) \to Y(0, 1)$ for some rearrangement-invariant space $Y(0, 1)$.

Moreover, if any of these conditions is satisfied, then

$$\|f\|_{Y_X(0,1)} \approx \left\| t^{-\frac{m}{n}} \left(f^{**}(t) - f^*(t) \right) \right\|_{X(0,1)} + \int_0^1 f^*(t)\, dt \quad \text{for } f \in \mathcal{M}(0, 1).$$

It follows that the formula for the optimal partner $X_Y(0, 1)$ can be dramatically improved as long as we know that the space $Y(0, 1)$ is an optimal range partner for *some* rearrangement-invariant domain space, and that a similar principle works when domain and target spaces are swapped.

In practice, one usually starts with a Sobolev space, $W^m X(\Omega)$, and seeks to find its optimal target partner space, $Y_X(\Omega)$, and in the next step one creates $X_{Y_X}(\Omega)$.

3 Reduction Principles

The above results then guarantee that the embedding

$$W^m X_{Y_X}(\Omega) \to Y_X(\Omega)$$

has both the domain and the target space optimal. Then one always has $X(\Omega) \to X_{Y_X}(\Omega)$, and it might, or does not have to, happen that $X(\Omega) \subsetneq X_{Y_X}(\Omega)$.

Example 3.4.8 Let $n, m \in \mathbb{N}$, $n \geq 2$, $1 \leq m \leq n-1$, and let $\Omega \subset \mathbb{R}^n$ be a domain with a Lipschitz boundary. Let $p \in [1, n)$. Then

$$W^m L^p(\Omega) \to L^{\frac{np}{n-mp}, p}(\Omega), \tag{3.4.32}$$

and the spaces $(L^p(\Omega), L^{\frac{np}{n-mp}, p}(\Omega))$ form an optimal pair in (3.4.32).

Example 3.4.9 Let $n, m \in \mathbb{N}$, $n \geq 2$, $1 \leq m \leq n-1$, and let $\Omega \subset \mathbb{R}^n$ be a domain with a Lipschitz boundary. Then

$$W^m L^{\frac{n}{m}}(\Omega) \to L^{\infty, \frac{n}{m}; -1}(\Omega), \tag{3.4.33}$$

and the target space $L^{\infty, \frac{n}{m}; -1}(\Omega)$ is optimal pair in (3.4.33), though the domain space $L^{\frac{n}{m}}(\Omega)$ is not. It thus follows from the theory that $L^{\frac{n}{m}}(\Omega)$ is not an optimal domain to *any* rearrangement-invariant target space. In particular, in (3.4.33), the domain space can be enlarged as far as to the space normed by

$$\left\| H_{\frac{n}{m}} f^* \right\|_{L^{\infty, \frac{n}{m}; -1}(\Omega)},$$

which can be shown to be essentially bigger than $L^{\frac{n}{m}}(\Omega)$ (it even has a better fundamental function). This shows, in particular, that (3.1.19) has the optimal rearrangement-invariant partner target space for the limiting Lebesgue space $L^n(\Omega)$. The inequality (3.1.20) has a better target space, as already mentioned, but it is not a rearrangement-invariant space (it is not even a linear structure).

3.5 Survey of Miscellaneous Reduction Principles

3.5.1 The Laplacian

We can now return to the questions which we left open in Sect. 3.1.3. For this purpose, we first need to adapt the reduction principle to inequalities involving the Laplace operator rather than just the full gradient.

Theorem 3.5.1 (Reduction Principle for the Laplacian) *Let $n \in \mathbb{N}$, $n \geq 3$. Let $\|\cdot\|_{X(0,1)}$ and $\|\cdot\|_{Y(0,1)}$ be rearrangement-invariant norms on $\mathcal{M}_+(0, 1)$. Then,* to each bounded domain $\Omega \subset \mathbb{R}^n$, with $\partial\Omega \in \mathrm{Lip}_1$, there corresponds a constant $C > 0$, depending on Ω and n, such that

$$\|u^*(|\Omega|t)\|_{Y(0,1)} \leq C \||\Delta u|^*(|\Omega|t)\|_{X(0,1)} \quad \text{for all suitable } u \tag{3.5.1}$$

if and only if there is a constant $K > 0$ such that

$$\left\| \int_t^1 f^{**}(s) s^{\frac{2}{n}-1} \, ds \right\|_{Y(0,1)} \leq K \|f\|_{X(0,1)} \quad \text{for every } f \in \mathcal{M}_+(0,1). \tag{3.5.2}$$

It should be noted that, by Fubini's theorem, one has

$$\int_t^1 f^{**}(s) s^{\frac{2}{n}-1} \, ds \approx t^{\frac{2}{n}-1} \int_0^t f^*(s) \, ds + \int_t^1 f^*(s) s^{\frac{2}{n}-1} \, ds.$$

Therefore the condition (3.5.2) has one Hardy-type operator more compared to (3.4.3). This reflects the fact that the inequality (3.5.1) is stronger than the full Sobolev second-order embedding (3.4.2) (with $m = 2$) as it involves a smaller collection of derivatives on the right-hand side.

Optimal spaces can be characterized for the inequalities involving Laplacian in the same spirit as it was done for the first-order gradient.

Using Theorem 3.5.1, we can also return to our problem concerning the transfer of regularity. Recalling that

$$\Delta u \in L^1(\Omega) \quad \not\Rightarrow \quad u \in L^{\frac{n}{n-2}}(\Omega),$$

we now have instead

$$\Delta u \in L^1(\Omega) \quad \Rightarrow \quad u \in L^{\frac{n}{n-2},\infty}(\Omega).$$

This answers the question (i) from Sect. 3.1.3. Moreover, both the target space $L^{\frac{n}{n-2},\infty}(\Omega)$ and the domain space $L^1(\Omega)$ are optimal. Similarly,

$$\Delta u \in L^{\frac{n}{2}}(\Omega) \quad \not\Rightarrow \quad u \in L^\infty(\Omega),$$

but one has

$$\Delta u \in L^{\frac{n}{2},1}(\Omega) \quad \Rightarrow \quad u \in L^\infty(\Omega) \quad (u \in C(\Omega)),$$

where, once again, both the domain space $L^{\frac{n}{2},1}(\Omega)$ and the target space $L^\infty(\Omega)$ are optimal. This takes care of the question (ii). Note that these refinements make sense, since

$$L^{\frac{n}{n-2}}(\Omega) \subsetneq L^{\frac{n}{n-2},\infty}(\Omega) \quad \text{and} \quad L^{\frac{n}{2},1}(\Omega) \subsetneq L^{\frac{n}{2}}(\Omega).$$

3 Reduction Principles

As for the question (iii), we noted that

$$\Delta u \in L^{\frac{n}{2}}(\Omega) \not\Rightarrow u \in L^{\infty}(\Omega).$$

Now, equipped with other function spaces, we can show that [121]

$$\Delta u \in L^{\frac{n}{2}}(\Omega) \Rightarrow u \in \exp L^{\frac{n}{n-2}}(\Omega).$$

This result has optimal *Orlicz* target space [27], but within rearrangement-invariant spaces it can be enhanced still, in the spirit of (3.1.19), to

$$\Delta u \in L^{\frac{n}{2}}(\Omega) \Rightarrow u \in L^{\infty, \frac{n}{2}; -1}(\Omega),$$

and here the target norm is optimal (although the domain one is not).

3.5.2 Gaussian Sobolev Embeddings

In Sect. 3.1.2 we briefly mentioned certain specific problems related to the Gaussian harmonic analysis. Before we return to the (second-order) Ornstein–Uhlenbeck operator and present the corresponding reduction principle, we shall first deal with corresponding first-order embeddings.

In connection with the study of quantum fields and hypercontractivity semigroups, extensions of the classical Sobolev inequality in $\mathbb{R}^n$ to the setting when the underlying measure space is infinite-dimensional have been investigated. The main motivation for this research is that, in certain circumstances, the study of quantum fields can be reduced to operator or semigroup estimates which are in turn equivalent to inequalities of Sobolev type in infinitely many variables (see [102] and the references therein).

In attempting to generalize Euclidean Sobolev embeddings to the case where the underlying space is infinite-dimensional, one immediately meets two problems. First, $\frac{np}{n-p} \to p+$ as $n \to \infty$, so the gain in integrability is apparently being lost. Second, and more serious, the Lebesgue measure on an infinite-dimensional space is meaningless.

Both these problems were overcome at the same time in the fundamental paper by L. Gross [67], where the Lebesgue measure was replaced by the Gauss measure γ_n. Since $\gamma_n(\mathbb{R}^n) = 1$ for every $n \in \mathbb{N}$, the extension as $n \to \infty$ is meaningful. The idea was then to seek a version of the Sobolev inequality that would hold on the probability space $(\mathbb{R}^n, \gamma_n)$ with a constant independent of n. In [67] it is proved that

$$\|u - u_{\gamma_n}\|_{L^2 \log L(\mathbb{R}^n, \gamma_n)} \leq C \|\nabla u\|_{L^2(\mathbb{R}^n, \gamma_n)} \quad (3.5.3)$$

for some absolute constant C and for every weakly differentiable function u making the right-hand side finite. Here, $u_{\gamma_n} = \int_{\mathbb{R}^n} u(x) d\gamma_n(x)$. Observe that (3.5.3) still provides some slight gain in the integrability from $|\nabla u|$ to u, even though it is no longer a power-gain. Gross' result meant a breakthrough in the theory and as such it ignited an extensive research of Sobolev inequalities in Gaussian space. The paper was refined in many ways by many authors. In particular, new simplified proofs, various extensions and modifications, and plenty of further applications appeared.

In [34], a comprehensive treatment of optimal Sobolev embeddings in the Gauss space in the general form

$$\|u - u_{\gamma_n}\|_{Y(\mathbb{R}^n, \gamma_n)} \leq C \|\nabla u\|_{X(\mathbb{R}^n, \gamma_n)} \tag{3.5.4}$$

for some constant C and for every $u \in V^1 X(\mathbb{R}^n, \gamma_n)$, where $X(\mathbb{R}^n, \gamma_n)$ and $Y(\mathbb{R}^n, \gamma_n)$ are rearrangement-invariant spaces. The approach was of course based in a crucial way on a suitable version of a reduction principle.

Theorem 3.5.2 (Reduction Principle for Gaussian Embeddings) *Let $X(\mathbb{R}^n, \gamma_n)$ and $Y(\mathbb{R}^n, \gamma_n)$ be r.i. spaces. Then a constant C_1 exists such that*

$$\|u - u_{\gamma_n}\|_{Y(\mathbb{R}^n, \gamma_n)} \leq C_1 \|\nabla u\|_{X(\mathbb{R}^n, \gamma_n)} \tag{3.5.5}$$

for every $u \in V^1 X(\mathbb{R}^n, \gamma_n)$ if and only if a constant C_2 exists such that

$$\left\| \int_t^1 \frac{f(s)}{s\sqrt{1 + \log \frac{1}{s}}} ds \right\|_{Y(0,1)} \leq C_2 \|f\|_{X(0,1)} \tag{3.5.6}$$

for every $f \in X(0, 1)$. Moreover, C_1 and C_2 depend only on each other.

It should be noticed that the integral operator which appears in the reduction principle, namely

$$f \mapsto \int_t^1 \frac{f(s)}{s\sqrt{1 + \log \frac{1}{s}}} ds \quad \text{for } f \in \mathcal{M}_+(0, 1),$$

is dimension free, as expected. This is not the case in the reduction principles suitable for the Euclidean Sobolev embeddings.

The proof of Theorem 3.5.2 relies upon a specific version of the Pólya–Szegő principle tailored for the Gaussian symmetrization with arbitrary r.i. norms. Its statement involves the *Gaussian symmetral* $u^\bullet \colon \mathbb{R}^n \to \mathbb{R}$ of a measurable function u in $\mathbb{R}^n$ defined as

$$u^\bullet(x) = u^\circ(\Phi(x_1)) \quad \text{for } x = (x_1, \ldots, x_n) \in \mathbb{R}^n, \tag{3.5.7}$$

3 Reduction Principles

where $\Phi: \mathbb{R} \to (0, 1)$ is the function given by

$$\Phi(t) = \frac{1}{\sqrt{2\pi}} \int_t^\infty e^{-\frac{\tau^2}{2}} d\tau \quad \text{for } t \in \mathbb{R}. \tag{3.5.8}$$

Note that, actually, $u \sim u^\bullet$, since $u \sim u^\circ$, and

$$\Phi(t) = \gamma_n \left(\{x \in \mathbb{R}^n : x_1 \geq t\}\right) \quad \text{for } t \in \mathbb{R}. \tag{3.5.9}$$

An equivalent formulation of the Pólya–Szegő principle can be given in terms of u° and of the *isoperimetric function* of the Gauss space $I : (0, 1) \to (0, \infty)$ defined by

$$I(s) = \frac{1}{\sqrt{2\pi}} e^{-\frac{\Phi^{-1}(s)^2}{2}} \quad \text{for } s \in (0, 1), \tag{3.5.10}$$

and $I(0) = I(1) = 0$. The function I owes its name to the fact that the isoperimetric inequality in the Gauss space reads

$$P_{\gamma_n}(E) \geq I(\gamma_n(E)) \tag{3.5.11}$$

for every measurable set $E \subset \mathbb{R}^n$ [14], with equality if and only if E is (equivalent to) a half-space [21] (see also [39]). Here,

$$P_{\gamma_n}(E) = \frac{1}{(2\pi)^{\frac{n}{2}}} \int_{\partial^M E} e^{-\frac{|x|^2}{2}} d\mathcal{H}^{n-1}(x),$$

the Gaussian perimeter of E, where $\partial^M E$ stands for the essential boundary of E (in the sense of geometric measure theory), and $\mathcal{H}^{n-1}$ denotes the $(n-1)$-dimensional Hausdorff measure. Note that the function I is increasing in $[0, \frac{1}{2}]$, and fulfils

$$I(s) = I(1-s) \quad \text{for } s \in [0, 1]. \tag{3.5.12}$$

Moreover,

$$I(s) \approx s\sqrt{1 + \log \frac{1}{s}}, \quad \text{for } s \in (0, 1/2], \tag{3.5.13}$$

with absolute equivalence constants.

The appropriate version of the Pólya–Szegő principle reads as follows.

Theorem 3.5.3 *Let $u \in V^1 L^1(\mathbb{R}^n, \gamma_n)$. Then u° is locally absolutely continuous in $(0, 1)$. Moreover, if $X(\mathbb{R}^n, \gamma_n)$ is any r.i. space and $u \in V^1 X(\mathbb{R}^n, \gamma_n)$, then $u^\bullet \in V^1 X(\mathbb{R}^n, \gamma_n)$ and*

$$\|\nabla u\|_{X(\mathbb{R}^n, \gamma_n)} \geq \|\nabla u^\bullet\|_{X(\mathbb{R}^n, \gamma_n)} = \left\|I(s)(-u^{\circ\prime}(s))\right\|_{X(0,1)}. \tag{3.5.14}$$

The optimal range in the Gaussian Sobolev inequality (3.5.4) for a given domain is characterized in the following theorem.

Theorem 3.5.4 *Let $X(\mathbb{R}^n, \gamma_n)$ be an r.i. space, and let $Y(\mathbb{R}^n, \gamma_n)$ be the r.i. space whose associate norm is given by*

$$\|u\|_{Y'(\mathbb{R}^n, \gamma_n)} = \left\| \frac{u^{**}(s)}{\sqrt{1 + \log \frac{1}{s}}} \right\|_{X'(0,1)} \quad (3.5.15)$$

for any $u \in \mathcal{M}(\mathbb{R}^n)$. Then $Y(\mathbb{R}^n, \gamma_n)$ is the optimal range for $X(\mathbb{R}^n, \gamma_n)$ in the Gaussian Sobolev inequality (3.5.4).

The question of the optimal domain space for a fixed range has similar features as those that we have seen in connection with Euclidean Sobolev inequalities. In particular, the general formula for the optimal domain comes in an implicit way. In order to make it more explicit, one has to use an appropriate version of a supremum operator, namely

$$Tf(s) = \sqrt{1 + \log \frac{1}{s}} \sup_{s \le r \le 1} \frac{f^*(r)}{\sqrt{1 + \log \frac{1}{r}}} \quad \text{for } s \in (0, 1). \quad (3.5.16)$$

Then we have the following characterization.

Theorem 3.5.5 *Let $Y(\mathbb{R}^n, \gamma_n)$ be an r.i. space such that*

$$\exp L^2(\mathbb{R}^n, \gamma_n) \to Y(\mathbb{R}^n, \gamma_n), \quad (3.5.17)$$

and

$$T \text{ is bounded on } Y'(0, 1). \quad (3.5.18)$$

Then the optimal domain $X(\mathbb{R}^n, \gamma_n)$ for $Y(\mathbb{R}^n, \gamma_n)$ in the Gaussian Sobolev inequality (3.5.4) satisfies

$$\|u\|_{X(\mathbb{R}^n, \gamma_n)} \approx \left\| \int_s^1 \frac{u^*(r)}{r\sqrt{1 + \log \frac{1}{r}}} \, dr \right\|_{Y(0,1)} \quad (3.5.19)$$

for $u \in \mathcal{M}(\mathbb{R}^n)$, with absolute equivalence constants.

With the help of these results we can present some examples.

Example 3.5.6 *Let $p \in [1, \infty)$. Then a constant $C = C(p)$ exists such that*

$$\|u - u_{\gamma_n}\|_{L^p(\log L)^{\frac{p}{2}}(\mathbb{R}^n, \gamma_n)} \le C \|\nabla u\|_{L^p(\mathbb{R}^n, \gamma_n)} \quad (3.5.20)$$

3 Reduction Principles

for every $u \in V^1 L^p(\mathbb{R}^n, \gamma_n)$. Moreover, $(L^p(\mathbb{R}^n, \gamma_n), L^p(\log L)^{\frac{p}{2}}(\mathbb{R}^n, \gamma_n))$ is an optimal pair in (3.5.20).

Example 3.5.7 An absolute constant C exists such that

$$\|u - u_{\gamma_n}\|_{\exp L^2(\mathbb{R}^n, \gamma_n)} \leq C \|\nabla u\|_{L^\infty(\mathbb{R}^n, \gamma_n)} \qquad (3.5.21)$$

for every $u \in V^1 L^\infty(\mathbb{R}^n, \gamma_n)$. Moreover, $(L^\infty(\mathbb{R}^n, \gamma_n), \exp L^2(\mathbb{R}^n, \gamma_n))$ is an optimal pair in (3.5.21).

Example 3.5.8 Let $\beta \in (0, \infty)$. Then, a constant $C = C(\beta)$ exists such that

$$\|u - u_{\gamma_n}\|_{\exp L^{\frac{2\beta}{2+\beta}}(\mathbb{R}^n, \gamma_n)} \leq C \|\nabla u\|_{\exp L^\beta(\mathbb{R}^n, \gamma_n)} \qquad (3.5.22)$$

for every $u \in V^1 \exp L^\beta(\mathbb{R}^n, \gamma_n)$. Moreover, $(\exp L^\beta(\mathbb{R}^n, \gamma_n), \exp L^{\frac{2\beta}{2+\beta}}(\mathbb{R}^n, \gamma_n))$ is an optimal pair in (3.5.22).

These examples demonstrate the interesting phenomenon that while there is a *gain* in integrability when the domain is a Lebesgue space, there is actually a *loss* in integrability when the domain is close (or even equal) to $L^\infty(\mathbb{R}^n, \gamma_n)$ (observe that $\frac{2\beta}{2+\beta} < \beta$ when $\beta > 0$). Furthermore, we added to the Gross' original result some of its later generalizations the information that both the range and the domain were already sharp in the broad context of r.i. spaces.

3.5.3 Higher-Order Sobolev Embeddings and Isoperimetric Inequalities

We have seen in Sect. 3.4 that a reduction principle suitable for higher-order Euclidean Sobolev inequalities can be obtained via interpolation. To some extent this approach works also, for instance, in connection with the problem of traces of Sobolev functions that will be mentioned in Sect. 3.5.5 below. However, the approach is not universal enough to provide all answers. Typical examples of problems for which it does not work, are higher-order Gaussian Sobolev embeddings, Sobolev embeddings on domains with lower regularity, and Sobolev embeddings on general probability spaces. These questions have been attacked in [40] using a new approach based on certain sharp iteration scheme.

The intimate connection of Sobolev spaces to isoperimetric inequalities was first observed by Maz'ya [93, 94], who proved that quite general Sobolev inequalities are equivalent to either isoperimetric or isocapacitary inequalities. Before the two problems had been investigated separately, by Sobolev [118, 119], Gagliardo [62] and Nirenberg [104] on the one hand, and by De Giorgi [49] on the other. Independently, Federer and Fleming [58] also exploited De Giorgi's isoperimetric inequality to exhibit the best constant in the special case of the Sobolev inequality for functions whose gradient is integrable with power 1 in $\mathbb{R}^n$. These results led to an extensive research of the interplay between isoperimetric and Sobolev inequalities and also to many applications.

The approach to Sobolev embeddings via isoperimetric inequalities has some considerable advantages. First, it covers a fairly broad range of situations, including Lipschitz domains, John domains, Maz'ya domains, probability measure spaces, and more. Second, it typically leads to sharp results. The drawback is that the existing literature is often restricted to the first-order derivatives, except perhaps for a very few quite specific instances while many important problems require higher order of the derivatives involved. This is mainly caused by the techniques used such as truncation, symmetrization, Pólya–Szegő principles etc., which do not allow a direct generalization to the higher-order case. Alternative methods which can be employed to handle higher-order Sobolev inequalities such as representation formulas, potential estimates, Fourier transforms or atomic decomposition are not flexible enough to provide us with optimal results in sufficient generality.

The approach developed in [40] shows that in fact isoperimetric inequalities do imply optimal higher-order Sobolev embeddings in a very general framework. The central ingredient is the combination of the first-order reduction principle with an iteration method, which is sharp enough to produce optimal higher-order reduction principles from the first-order ones. As a consequence, optimal results within the class of rearrangement-invariant spaces are obtained for Sobolev embeddings of any order. More precisely, the best possible targets for arbitrary-order Sobolev embeddings can be characterized. A key step is the development of a sharp iteration method involving subsequent applications of optimal Sobolev embeddings.

The isoperimetric function is precisely known only in some rather rare situations, e.g. when Ω is an Euclidean ball [95] or when it is the entire $\mathbb{R}^n$ equipped with the Gauss measure [14]. For our purpose, however, only the asymptotic behavior of $I_{\Omega,\nu}$ near 0 is needed, and that can be evaluated for various classes of domains, for example for Lipschitz domains, John domains or for the space $\mathbb{R}^n$ equipped with product probability measures.

A natural worry here is whether optimality is preserved under iteration. Examples show that when optimality is considered within standard families of function spaces, this is not necessarily so, even in the basic setting of Euclidean domains with Lipschitz boundaries. For instance, for a regular domain Ω in $\mathbb{R}^2$, one has

$$V^2 L^1(\Omega) \to L^\infty(\Omega). \qquad (3.5.23)$$

On the other hand, iteration of two consequent first-order Sobolev embeddings, optimal within Lebesgue spaces, only gives

$$V^2 L^1(\Omega) \to V^1 L^2(\Omega) \to L^q(\Omega) \qquad (3.5.24)$$

for every $q < \infty$. The space $V^1 L^2(\Omega)$ contains unbounded functions, hence there is a loss of information in the iteration process.

3 Reduction Principles

One might relate the loss of optimality in (3.5.24) to the lack of an optimal Lebesgue target space for the first-order Sobolev embedding of $V^1L^2(\Omega)$ when $n = 2$. However, similar loss can happen in situations where the optimal first-order target spaces do exist. Consider, for example, Euclidean Sobolev embeddings involving Orlicz spaces. In this setting the optimal target space always exists, and it can be explicitly determined [27, 31]. Indeed, if Ω is a regular domain in $\mathbb{R}^n$ and $1 \leq m < n$, then

$$V^m L^{\frac{n}{m}}(\Omega) \to \exp L^{\frac{n}{n-m}}(\Omega). \tag{3.5.25}$$

The target space in (3.5.25) is known to be optimal in the class of all Orlicz spaces [27, 29]. Assume, for example, that $n \geq 3$ and $m = 2$. Then (3.5.25) reduces to

$$V^2 L^{\frac{n}{2}}(\Omega) \to \exp L^{\frac{n}{n-2}}(\Omega). \tag{3.5.26}$$

However, iterating first-order embeddings, optimal in Orlicz spaces, one gets only

$$V^2 L^{\frac{n}{2}}(\Omega) \to V^1 L^n(\Omega) \to \exp L^{\frac{n}{n-1}}(\Omega) \supsetneq \exp L^{\frac{n}{n-2}}(\Omega). \tag{3.5.27}$$

Thus, subsequent applications of first-order optimal Sobolev embeddings even in the class of Orlicz spaces, where optimal target space always exists, need not result in optimal higher-order Sobolev embeddings.

The underlying idea behind the iteration is that such a loss of optimality of the target space under iteration will not occur when first-order (in fact, any-order) Sobolev embeddings whose targets are *optimal among all rearrangement-invariant spaces are iterated*. We thus proceed via a two-step argument, which can be outlined as follows. Firstly, given any rearrangement-invariant space $X(\Omega)$ and the isoperimetric function $I_{\Omega,\nu}$ of (Ω, ν), the optimal target for the first-order Sobolev space $V^1X(\Omega, \nu)$ is characterized; secondly, first-order Sobolev embeddings with optimal targets are iterated.

Returning back to the examples, recall that the optimal rearrangement-invariant range partner for $V^1L^1(\Omega)$ is not $L^2(\Omega)$, but the essentially smaller space $L^{2,1}(\Omega)$. Thus,

$$V^2 L^1(\Omega) \to V^1 L^{2,1}(\Omega) \to L^\infty(\Omega), \tag{3.5.28}$$

and now there is no loss of information. Similarly, iterating the (optimal) embeddings

$$V^1 L^{\frac{n}{2}}(\Omega) \to V^1 L^{n,\frac{n}{2}}(\Omega) \quad \text{and} \quad V^1 L^{n,\frac{n}{2}}(\Omega) \to L^{\infty,\frac{n}{2};-1}(\Omega),$$

we get

$$V^2 L^1(\Omega) \to L^{\infty,\frac{n}{2};-1}(\Omega),$$

which, thanks to the (strict) embedding $L^{\infty,\frac{n}{2};-1}(\Omega) \to \exp L^{\frac{n}{n-2}}(\Omega)$ is even better result than (3.5.26). Again, no information is lost.

It turns out that the Lipschitz property of the underlying domain Ω is not necessary for obtaining sharp Sobolev embeddings. Both the reduction principle and the optimal range construction work just fine with no changes when Ω is only required that its isoperimetric function $I_\Omega(s)$ satisfies

$$I_\Omega(s) \approx s^{\frac{1}{n'}} \tag{3.5.29}$$

near 0, where $n' = \frac{n}{n-1}$. This condition, in fact, defines the class of the so-called *John domains*, which are more general than the Lipschitz ones.

The isoperimetric inequality relative to (Ω, ν) tells us that

$$P_\nu(E, \Omega) \geq I_{\Omega,\nu}(\nu(E)), \tag{3.5.30}$$

where E is any measurable set $E \subset \Omega$, and $P_\nu(E, \Omega)$ stands for its perimeter in Ω with respect to ν. Moreover, $I_{\Omega,\nu}$ denotes the largest non-decreasing function in $[0, \frac{1}{2}]$ for which (3.5.30) holds, called the isoperimetric function (or isoperimetric profile) of (Ω, ν), which was introduced in [93].

The results concerning optimality of function spaces in Sobolev embeddings depend only on a lower bound for the isoperimetric function $I_{\Omega,\nu}$ of (Ω, ν) in terms of some other non-decreasing function $I: [0, 1] \to [0, \infty)$; precisely, on the existence of a positive constant c such that

$$I_{\Omega,\nu}(s) \geq cI(cs) \quad \text{for } s \in [0, \tfrac{1}{2}]. \tag{3.5.31}$$

First, it can be observed that if $I_{\Omega,\nu}(s)$ does not decay at 0 faster than linearly, namely if there exists a positive constant C such that

$$I_{\Omega,\nu}(s) \geq Cs \quad \text{for } s \in [0, \tfrac{1}{2}], \tag{3.5.32}$$

then any function $u \in V^m X(\Omega, \nu)$ does at least belong to $L^1(\Omega, \nu)$, together with all its derivatives up to the order $m - 1$. In the light of (3.5.32), we can safely assume that

$$\inf_{t \in (0,1)} \frac{I(t)}{t} > 0. \tag{3.5.33}$$

We shall now state the main general reduction principle.

Theorem 3.5.9 *Assume that (Ω, ν) fulfils (3.5.31) for some non-decreasing function I satisfying (3.5.33). Let $m \in \mathbb{N}$, and let $\|\cdot\|_{X(0,1)}$ and $\|\cdot\|_{Y(0,1)}$ be*

3 Reduction Principles

rearrangement-invariant function norms. If there exists a constant C_1 such that

$$\left\| \int_t^1 \frac{f(s)}{I(s)} \left(\int_t^s \frac{dr}{I(r)} \right)^{m-1} ds \right\|_{Y(0,1)} \leq C_1 \|f\|_{X(0,1)} \tag{3.5.34}$$

for every nonnegative $f \in X(0, 1)$, then

$$V^m X(\Omega, \nu) \to Y(\Omega, \nu), \tag{3.5.35}$$

and there exists a constant C_2 such that

$$\|u\|_{Y(\Omega,\nu)} \leq C_2 \|\nabla^m u\|_{X(\Omega,\nu)} \tag{3.5.36}$$

for every $u \in V_\perp^m X(\Omega, \nu)$.

It turns out that inequality (3.5.34) holds for every nonnegative $f \in X(0, 1)$ if and only if it just holds for every nonnegative and non-increasing $f \in X(0, 1)$.

A major feature of Theorem 3.5.9 is the difference occurring in (3.5.34) between the first-order case ($m = 1$) and the higher-order case ($m > 1$). Indeed, the integral operator appearing in (3.5.34) when $m = 1$ is just a weighted Copson-type operator, namely a negative primitive function of f times a weight, whereas, in the higher-order case, one has to consider a genuine kernel with a considerably more complicated structure. This seems to be the first known instance where such a kernel operator is needed in a reduction principle for Sobolev-type embeddings.

The Sobolev embedding (3.5.35) (or the Poincaré inequality (3.5.36)) and inequality (3.5.34) are actually equivalent in customary families of measure spaces (Ω, ν). This fact allows us to determine the optimal rearrangement-invariant target spaces in Sobolev embeddings for these measure spaces. Incidentally, let us mention that when $m = 1$, this is the case whenever the geometry of (Ω, ν) allows the construction of a family of trial functions u in (3.5.35) or (3.5.36) characterized by the following properties: the level sets of u are isoperimetric (or almost isoperimetric) in (Ω, ν); $|\nabla u|$ is constant (or almost constant) on the boundary of the level sets of u. If $m > 1$, then the latter requirement has to be complemented by requiring that the derivatives of u up to the order m restricted to the boundary of the level sets satisfy certain conditions depending on I.

Such conditions have, however, a technical nature, and it is not worth to state them explicitly. In fact, heuristically speaking, properties (3.5.34), (3.5.36) and (3.5.35) turn out to be equivalent for every $m \geq 1$ on the same measure spaces (Ω, ν) as for $m = 1$. Such an equivalence certainly holds in any customary, non-pathological situation, including the three frameworks to which our results will be applied, namely John domains, Euclidean domains from Maz'ya classes and product probability spaces in $\mathbb{R}^n$ extending the Gauss space. In all these cases, we can characterize optimal arbitrary-order rearrangement-invariant target spaces.

Now we are in a position to characterize the space which is the optimal rearrangement-invariant target space in the Sobolev embedding (3.5.35). Such an optimal space is the one governed by the rearrangement-invariant function norm $\|\cdot\|_{X_{m,I}(0,1)}$, whose associate norm is defined as

$$\|f\|_{X'_{m,I}(0,1)} = \left\| \frac{1}{I(s)} \int_0^s \left(\int_t^s \frac{dr}{I(r)} \right)^{m-1} f^*(t) dt \right\|_{X'(0,1)} \tag{3.5.37}$$

for $f \in \mathfrak{M}_+(0,1)$.

Theorem 3.5.10 *Assume that (Ω, ν), m, I and $\|\cdot\|_{X(0,1)}$ are as in Theorem 3.5.9. Then the functional $\|\cdot\|_{X'_{m,I}(0,1)}$, given by (3.5.37), is a rearrangement-invariant function norm, whose associate norm $\|\cdot\|_{X_{m,I}(0,1)}$ satisfies*

$$V^m X(\Omega, \nu) \to X_{m,I}(\Omega, \nu), \tag{3.5.38}$$

and there exists a constant C such that

$$\|u\|_{X_{m,I}(\Omega,\nu)} \leq C \|\nabla^m u\|_{X(\Omega,\nu)} \tag{3.5.39}$$

for every $u \in V_\perp^m X(\Omega, \nu)$.

Moreover, if (Ω, ν) is such that (3.5.35), or equivalently (3.5.36), implies (3.5.34), and hence (3.5.34), (3.5.35) and (3.5.36) are equivalent, then the function norm $\|\cdot\|_{X_{m,I}(0,1)}$ is optimal in (3.5.38) and (3.5.39) among all rearrangement-invariant norms.

An important special case of Theorems 3.5.9 and 3.5.10 is enucleated in the following corollary.

Corollary 3.5.11 *Assume that (Ω, ν), m, I and $\|\cdot\|_{X(0,1)}$ are as in Theorem 3.5.9. If*

$$\left\| \frac{1}{I(s)} \left(\int_0^s \frac{dr}{I(r)} \right)^{m-1} \right\|_{X'(0,1)} < \infty, \tag{3.5.40}$$

then

$$V^m X(\Omega, \nu) \to L^\infty(\Omega, \nu), \tag{3.5.41}$$

and there exists a constant C such that

$$\|u\|_{L^\infty(\Omega,\nu)} \leq C \|\nabla^m u\|_{X(\Omega,\nu)} \tag{3.5.42}$$

for every $u \in V_\perp^m X(\Omega, \nu)$.

3 Reduction Principles

Moreover, if (Ω, ν) is such that (3.5.35), or equivalently (3.5.36), implies (3.5.34), and hence (3.5.34), (3.5.35) and (3.5.36) are equivalent, then (3.5.40) is necessary for (3.5.41) or (3.5.42) to hold.

If (Ω, ν) is such that (3.5.35), or equivalently (3.5.36), implies (3.5.34), and hence (3.5.34), (3.5.35) and (3.5.36) are equivalent, then (3.5.41) cannot hold, whatever $\|\cdot\|_{X(0,1)}$ is, if I decays so fast at 0 that

$$\int_0^{\cdot} \frac{dr}{I(r)} = \infty.$$

We shall now point out the preservation of optimality in targets among all rearrangement-invariant spaces under iteration of Sobolev embeddings of arbitrary order.

Theorem 3.5.12 *Assume that (Ω, ν), I and $\|\cdot\|_{X(0,1)}$ are as in Theorem 3.5.9. Let $k, h \in \mathbb{N}$. Then*

$$(X_{k,I})_{h,I}(\Omega, \nu) = X_{k+h,I}(\Omega, \nu), \qquad (3.5.43)$$

up to equivalent norms.

In many instances in practice, the function I satisfies the estimate

$$\int_0^s \frac{dr}{I(r)} \approx \frac{s}{I(s)} \quad \text{for } s \in (0, 1). \qquad (3.5.44)$$

We note that (3.5.44) is not true for every relevant case. It holds for instance for John domains and for domains from Maz'ya classes $\mathcal{J}_\alpha$ with $\alpha < 1$, but it does not hold for domains in $\mathcal{J}_1$ or for the Gaussian space.

It is useful to treat the cases for which (3.5.44) holds separately because then the results of Theorems 3.5.9, 3.5.10 and 3.5.12 can be considerably simplified. For example, the reduction theorem then reads as follows.

Corollary 3.5.13 *Let (Ω, ν), m, I, $\|\cdot\|_{X(0,1)}$ and $\|\cdot\|_{Y(0,1)}$ be as in Theorem 3.5.9. Assume, in addition, that I fulfills (3.5.44). If there exists a constant C_1 such that*

$$\left\| \int_t^1 f(s) \frac{s^{m-1}}{I(s)^m} ds \right\|_{Y(0,1)} \leq C_1 \|f\|_{X(0,1)} \qquad (3.5.45)$$

for every nonnegative $f \in X(0,1)$, then

$$V^m X(\Omega, \nu) \to Y(\Omega, \nu), \qquad (3.5.46)$$

and there exists a constant C_2 such that

$$\|u\|_{Y(\Omega,\nu)} \leq C_2 \|\nabla^m u\|_{X(\Omega,\nu)} \tag{3.5.47}$$

for every $u \in V^m_\perp X(\Omega, \nu)$.

Given a rearrangement-invariant space $X(\Omega)$, the corresponding optimal range partner in the Sobolev embedding on a John domain is then the space $X^\sharp_{m,I}(\Omega)$, whose associate space has norm

$$\|f\|_{(X^\sharp_{m,I})'(\Omega)} = \left\| \frac{t^{m-1}}{I(t)^m} \int_0^t f^*(s)\,ds \right\|_{X'(0,1)} \tag{3.5.48}$$

for every $f \in \mathcal{M}_+(0, 1)$.

Let us now deal with specific situations separately.

We say that a bounded open set Ω in $\mathbb{R}^n$ is called a *John domain* if there exist a constant $c \in (0, 1)$ and a point $x_0 \in \Omega$ such that for every $x \in \Omega$ there exists a rectifiable curve $\varpi : [0, l] \to \Omega$, parameterized by the arc length and such that $\varpi(0) = x$, $\varpi(l) = x_0$, and

$$\operatorname{dist}(\varpi(r), \partial\Omega) \geq cr \quad \text{for } r \in [0, l].$$

The class of John domains includes other more classical families of domains, such as Lipschitz domains, and domains with the cone property. The John domains arise in connection with the study of holomorphic dynamical systems and quasiconformal mappings. John domains are known to support a first-order Sobolev inequality with the same exponents as in the standard Sobolev inequality [13, 70, 81]. In fact, being a John domain is a necessary condition for such a Sobolev inequality to hold in the class of two-dimensional simply connected open sets, and in quite general classes of higher dimensional domains [19]. We note that, as a consequence of (3.5.29), the corresponding function I now satisfies (3.5.44). The reduction principle thus takes the following form.

Theorem 3.5.14 *Let* $n \in \mathbb{N}$, $n \geq 2$, *and let* $m \in \mathbb{N}$. *Assume that* Ω *is a John domain in* $\mathbb{R}^n$. *Let* $\|\cdot\|_{X(0,1)}$ *and* $\|\cdot\|_{Y(0,1)}$ *be rearrangement-invariant function norms. Then the following assertions are equivalent.*

(i) *The Hardy type inequality*

$$\left\| \int_t^1 f(s) s^{-1+\frac{m}{n}}\,ds \right\|_{Y(0,1)} \leq C_1 \|f\|_{X(0,1)} \tag{3.5.49}$$

holds for some constant C_1, *and for every nonnegative* $f \in X(0, 1)$.

(ii) *The Sobolev embedding*
$$V^m X(\Omega) \to Y(\Omega) \tag{3.5.50}$$
holds.

(iii) *The Poincaré inequality*
$$\|u\|_{Y(\Omega)} \leq C_2 \|\nabla^m u\|_{X(\Omega)} \tag{3.5.51}$$
holds for some constant C_2 and every $u \in V_\perp^m X(\Omega)$.

Given a rearrangement-invariant space $X(\Omega)$ where Ω is a John domain, the corresponding optimal range partner in the Sobolev embedding on a John domain is then the space $X_{m,John}(\Omega)$, whose associate space has norm

$$\|f\|_{X'_{m,John}(\Omega)} = \left\| s^{-1+\frac{m}{n}} \int_0^s f^*(r)dr \right\|_{X'(0,1)} \tag{3.5.52}$$

Note that the results in this subsection concerning John domains recover most of the results from Sect. 3.4, using a completely different method of proof. The approach using iteration seems to be more versatile than that based on interpolation. The main reason is that the interpolation technique can only work effectively when a suitable pair of endpoint estimates involving Lorentz and Marcinkiewicz spaces is available, and that is not always the case (see [34–36, 40–42]).

Our next set of instances will be *Maz'ya classes* of Euclidean domains. Given $\alpha \in [\frac{1}{n'}, 1]$, we denote by $\mathcal{J}_\alpha$ the *Maz'ya class* of all Euclidean domains Ω satisfying (3.5.31), with $I(s) = s^\alpha$ for $s \in [0, \frac{1}{2}]$, namely domains Ω in $\mathbb{R}^n$ such that

$$I_\Omega(s) \geq C s^\alpha \quad \text{for } s \in [0, \frac{1}{2}], \tag{3.5.53}$$

for some positive constant C. Thanks to (3.5.29), any John domain belongs to the class $\mathcal{J}_{\frac{1}{n'}}$.

The reduction theorem in the class $\mathcal{J}_\alpha$ takes the following form.

Theorem 3.5.15 *Let $n \in \mathbb{N}$, $n \geq 2$, $m \in \mathbb{N}$, and $\alpha \in [\frac{1}{n'}, 1]$. Let $\|\cdot\|_{X(0,1)}$ and $\|\cdot\|_{Y(0,1)}$ be rearrangement-invariant function norms. Assume that either $\alpha \in [\frac{1}{n'}, 1)$ and there exists a constant C_1 such that*

$$\left\| \int_t^1 f(s) s^{-1+m(1-\alpha)} ds \right\|_{Y(0,1)} \leq C_1 \|f\|_{X(0,1)} \tag{3.5.54}$$

for every nonnegative $f \in X(0, 1)$, or $\alpha = 1$ and there exists a constant C_1 such that

$$\left\| \int_t^1 f(s) \frac{1}{s} \left(\log \frac{s}{t} \right)^{m-1} ds \right\|_{Y(0,1)} \leq C_1 \|f\|_{X(0,1)} \tag{3.5.55}$$

for every nonnegative $f \in X(0, 1)$. Then the Sobolev embedding

$$V^m X(\Omega) \to Y(\Omega) \tag{3.5.56}$$

holds for every $\Omega \in \mathcal{J}_\alpha$ and, equivalently, the Poincaré inequality

$$\|u\|_{Y(\Omega)} \leq C_2 \left\| \nabla^m u \right\|_{X(\Omega)} \tag{3.5.57}$$

holds for every $\Omega \in \mathcal{J}_\alpha$, for some constant C_2 and every $u \in V_\perp^m X(\Omega)$.

Conversely, if the Sobolev embedding (3.5.56), or, equivalently, the Poincaré inequality (3.5.57), holds for every $\Omega \in \mathcal{J}_\alpha$, then either inequality (3.5.54), or (3.5.55) holds, according to whether $\alpha \in [\frac{1}{n'}, 1)$ or $\alpha = 1$.

Given a rearrangement-invariant space $X(\Omega)$ where $\Omega \in \mathcal{J}_\alpha$, the corresponding optimal range partner in the Sobolev embedding on a John domain is then the space $X_{m,\alpha}(\Omega)$, whose associate space has norm

$$\|f\|_{X'_{m,\alpha}(\Omega)} = \begin{cases} \left\| s^{-1+m(1-\alpha)} \int_0^s f^*(r) dr \right\|_{X'(0,1)} & \text{if } \alpha \in [\frac{1}{n'}, 1), \\ \left\| \frac{1}{s} \int_0^s \left(\log \frac{s}{r} \right)^{m-1} f^*(r) dr \right\|_{X'(0,1)} & \text{if } \alpha = 1. \end{cases} \tag{3.5.58}$$

We shall now apply the general results to some concrete function spaces. We shall mainly focus on Lebesgue, Lorentz and Orlicz spaces.

Theorem 3.5.16 *Let $n \in \mathbb{N}$, $n \geq 2$, and let $\Omega \in \mathcal{J}_\alpha$ for some $\alpha \in [\frac{1}{n'}, 1)$. Let $m \in \mathbb{N}$ and $p \in [1, \infty]$. Then*

$$V^m L^p(\Omega) \to \begin{cases} L^{\frac{p}{1-mp(1-\alpha)}}(\Omega) & \text{if } m(1-\alpha) < 1 \text{ and } 1 \leq p < \frac{1}{m(1-\alpha)}, \\ L^r(\Omega) & \text{for any } r \in [1, \infty), \text{ if } m(1-\alpha) < 1 \\ & \text{and } p = \frac{1}{m(1-\alpha)}, \\ L^\infty(\Omega) & \text{otherwise.} \end{cases} \tag{3.5.59}$$

Moreover, in the first and the third cases, the target spaces in (3.5.59) are optimal among all Lebesgue spaces, as Ω ranges in $\mathcal{J}_\alpha$.

Although the target spaces in (3.5.59) cannot be improved in the class of Lebesgue spaces, the conclusions of (3.5.59) can be strengthened if more general rearrangement-invariant spaces are employed. Such a strengthening can be obtained as a special case of a Sobolev embedding for Lorentz spaces which reads as follows.

Theorem 3.5.17 *Let $n \in \mathbb{N}$, $n \geq 2$, and let $\Omega \in \mathcal{J}_\alpha$ for some $\alpha \in [\frac{1}{n'}, 1)$. Let $m \in \mathbb{N}$ and $p, q \in [1, \infty]$. Assume that one of the conditions in (3.2.13) holds. Then*

$$V^m L^{p,q}(\Omega) \to \begin{cases} L^{\frac{p}{1-mp(1-\alpha)},q}(\Omega) & \text{if } m(1-\alpha) < 1 \text{ and } 1 \leq p < \frac{1}{m(1-\alpha)}, \\ L^{\infty,q;-1}(\Omega) & \text{if } m(1-\alpha) < 1, \; p = \frac{1}{m(1-\alpha)} \text{ and } q > 1, \\ L^\infty(\Omega) & \text{otherwise.} \end{cases}$$
(3.5.60)

Moreover, the target spaces in (3.5.60) are optimal among all rearrangement-invariant spaces, as Ω ranges in $\mathcal{J}_\alpha$.

The particular choice of parameters $p = q$, $1 \leq p < \frac{1}{m(1-\alpha)}$ in Theorem 3.5.17 shows that

$$V^m L^p(\Omega) \to L^{\frac{p}{1-mp(1-\alpha)},p}(\Omega).$$

This is a non-trivial strengthening of the first embedding in (3.5.59), since $L^{\frac{p}{1-mp(1-\alpha)},p}(\Omega) \subsetneq L^{\frac{p}{1-mp(1-\alpha)}}$.

Likewise, the choice $m(1-\alpha) < 1$ and $p = q = \frac{1}{m(1-\alpha)}$ shows that also the second embedding in (3.5.59) can be in fact essentially improved to

$$V^m L^p(\Omega) \to L^{\infty,p;-1}(\Omega).$$

Assume now that $\alpha = 1$. The embedding theorem in Lebesgue spaces takes the following form.

Theorem 3.5.18 *Let $n \in \mathbb{N}$, $n \geq 2$, and let $\Omega \in \mathcal{J}_1$. Let $m \in \mathbb{N}$ and $p \in [1, \infty]$. Then*

$$V^m L^p(\Omega) \to \begin{cases} L^p(\Omega) & \text{if } 1 \leq p < \infty, \\ L^r(\Omega) & \text{for any } r \in [1, \infty), \text{ if } p = \infty. \end{cases}$$
(3.5.61)

In the former case, the target space is optimal in (3.5.61) among all Lebesgue spaces, as Ω ranges in $\mathcal{J}_1$.

Optimal embeddings for Lorentz-Sobolev spaces are provided in the next theorem.

Theorem 3.5.19 *Let $n \in \mathbb{N}$, $n \geq 2$, and let $\Omega \in \mathcal{J}_1$. Let $m \in \mathbb{N}$ and $p, q \in [1, \infty]$. Assume that one of the conditions in (3.2.13) holds. Then*

$$V^m L^{p,q}(\Omega) \to \begin{cases} L^{p,q}(\Omega) & \text{if } 1 \leq p < \infty, \\ \exp L^{\frac{1}{m}}(\Omega) & \text{if } p = q = \infty. \end{cases} \quad (3.5.62)$$

The target spaces are optimal in (3.5.62) among all rearrangement-invariant spaces, as Ω ranges in $\mathcal{J}_1$.

Our last application in this section concerns Orlicz–Sobolev spaces. Let $n \in \mathbb{N}$, $n \geq 2$, $m \in \mathbb{N}$, $\alpha \in [\frac{1}{n'}, 1)$, and let A be a Young function. We may assume, without loss of generality, that

$$\int_0^{\cdot} \left(\frac{t}{A(t)}\right)^{\frac{m(1-\alpha)}{1-m(1-\alpha)}} dt < \infty. \quad (3.5.63)$$

Indeed, the function A can be modified near 0, if necessary, in such a way that (3.5.63) is fulfilled, on leaving the space $V^m L^A(\Omega)$ unchanged (up to equivalent norms).

If $m < \frac{1}{1-\alpha}$ and

$$\int^{\infty} \left(\frac{t}{A(t)}\right)^{\frac{m(1-\alpha)}{1-m(1-\alpha)}} dt = \infty, \quad (3.5.64)$$

we define the function $H_{m,\alpha} : [0, \infty) \to [0, \infty)$ as

$$H_{m,\alpha}(s) = \left(\int_0^s \left(\frac{t}{A(t)}\right)^{\frac{m(1-\alpha)}{1-m(1-\alpha)}} dt\right)^{1-m(1-\alpha)} \quad \text{for } s \geq 0, \quad (3.5.65)$$

and the Young function $A_{m,\alpha}$ as

$$A_{m,\alpha}(t) = A(H_{m,\alpha}^{-1}(t)) \quad \text{for } t \geq 0. \quad (3.5.66)$$

Theorem 3.5.20 *Assume that $n \in \mathbb{N}$, $n \geq 2$, $m \in \mathbb{N}$, $\alpha \in [\frac{1}{n'}, 1)$ and $\Omega \in \mathcal{J}_\alpha$. Let A be a Young function fulfilling (3.5.63). Then*

$$V^m L^A(\Omega) \to \begin{cases} L^{A_{m,\alpha}}(\Omega) & \text{if } m < \frac{1}{1-\alpha} \text{ and } (3.5.64) \text{ holds,} \\ L^{\infty}(\Omega) & \text{if either } m \geq \frac{1}{1-\alpha} \text{ or } m < \frac{1}{1-\alpha} \text{ and } (3.5.64) \text{ fails.} \end{cases} \quad (3.5.67)$$

Moreover, the target spaces in (3.5.67) are optimal among all Orlicz spaces, as Ω ranges in $\mathcal{J}_\alpha$.

3 Reduction Principles

Example 3.5.21 Consider the case when

$$A(t) \approx t^p (\log t)^\beta \text{ near infinity, where either } p > 1$$

$$\text{and } \beta \in \mathbb{R}, \text{ or } p = 1 \text{ and } \beta \geq 0.$$

Hence, $L^A(\Omega) = L^p \log^\beta(\Omega)$. An application of Theorem 3.5.20 tells us that

$$V^m L^p \log^\beta(\Omega)$$

$$\to \begin{cases} L^{\frac{p}{1-pm(1-\alpha)}} \log^{\frac{\beta}{1-pm(1-\alpha)}}(\Omega) & \text{if } mp(1-\alpha) < 1, \\ \exp L^{\frac{1}{1-(1+\beta)m(1-\alpha)}}(\Omega) & \text{if } mp(1-\alpha) = 1 \text{ and } \beta < \frac{1-m(1-\alpha)}{m(1-\alpha)}, \\ \exp\exp L^{\frac{1}{1-m(1-\alpha)}}(\Omega) & \text{if } mp(1-\alpha) = 1 \text{ and } \beta = \frac{1-m(1-\alpha)}{m(1-\alpha)}, \\ L^\infty(\Omega) & \text{if either } mp(1-\alpha) > 1, \\ & \text{or } mp(1-\alpha) = 1 \text{ and } \beta > \frac{1-m(1-\alpha)}{m(1-\alpha)}. \end{cases}$$

(3.5.68)

Moreover, the target spaces in (3.5.68) are optimal among all Orlicz spaces, as Ω ranges in $\mathcal{J}_\alpha$.

The first three embeddings in (3.5.68) can be improved on allowing more general rearrangement-invariant target spaces. Indeed, we have that

$$V^m L^p \log^\beta(\Omega)$$

$$\to \begin{cases} L^{\frac{p}{1-pm(1-\alpha)}, p; \frac{\beta}{p}}(\Omega) & \text{if } mp(1-\alpha) < 1, \\ L^{\infty, \frac{1}{m(1-\alpha)}; m(1-\alpha)\beta - 1}(\Omega) & \text{if } mp(1-\alpha) = 1 \text{ and } \beta < \frac{1-m(1-\alpha)}{m(1-\alpha)}, \\ L^{\infty, \frac{1}{m(1-\alpha)}; -m(1-\alpha), -1}(\Omega) & \text{if } mp(1-\alpha) = 1 \text{ and } \beta = \frac{1-m(1-\alpha)}{m(1-\alpha)}, \end{cases}$$

(3.5.69)

the targets being optimal among all rearrangement-invariant spaces in (3.5.69) as Ω ranges among all domains in $\mathcal{J}_\alpha$.

Our final set of examples will concern the *product probability spaces*.

The class of product probability measures in $\mathbb{R}^n$, $n \geq 1$, arises in connection with the study of generalized hypercontractivity theory and integrability properties of the associated heat semigroups. The isoperimetric problem in the corresponding probability spaces was studied in [6], see also [7, 12, 86, 87].

Assume that $\Phi: [0, \infty) \to [0, \infty)$ is a strictly increasing, twice continuously differentiable convex function in $(0, \infty)$ such that $\sqrt{\Phi}$ is concave, and $\Phi(0) = 0$. Let μ_Φ be the probability measure on $\mathbb{R}$ given by

$$d\mu_\Phi(x) = c_\Phi e^{-\Phi(|x|)} dx, \qquad (3.5.70)$$

where c_Φ is a constant chosen in such a way that $\mu_\Phi(\mathbb{R}) = 1$. The product measure $\mu_{\Phi,n}$ on $\mathbb{R}^n$, $n \geq 1$, generated by μ_Φ, is then defined as

$$\mu_{\Phi,n} = \underbrace{\mu_\Phi \times \cdots \times \mu_\Phi}_{n\text{-times}}. \tag{3.5.71}$$

Clearly, $\mu_{\Phi,1} = \mu_\Phi$, and $(\mathbb{R}^n, \mu_{\Phi,n})$ is a probability space for every $n \in \mathbb{N}$.

The main example of a measure μ_Φ is obtained by taking

$$\Phi(t) = \tfrac{1}{2}t^2.$$

This choice yields $\mu_{\Phi,n} = \gamma_n$, the Gauss measure which obeys

$$d\gamma_n(x) = (2\pi)^{-\frac{n}{2}} e^{-\frac{|x|^2}{2}} dx. \tag{3.5.72}$$

More generally, the *Boltzmann measures* $\gamma_{n,\beta}$ are associated with

$$\Phi(t) = \tfrac{1}{\beta} t^\beta$$

for some $\beta \in [1, 2]$. Let $H: \mathbb{R} \to (0, 1)$ be defined as

$$H(t) = \int_t^\infty c_\Phi e^{-\Phi(|r|)} \, dr \quad \text{for } t \in \mathbb{R}, \tag{3.5.73}$$

and let $F_\Phi : [0, 1] \to [0, \infty)$ be given by

$$F_\Phi(s) = c_\Phi e^{-\Phi(|H^{-1}(s)|)} \quad \text{for } s \in (0, 1), \text{ and } \quad F_\Phi(0) = F_\Phi(1) = 0. \tag{3.5.74}$$

Since μ_Φ is a probability measure and $\mu_{\Phi,n}$ is defined by (3.5.71), it is easily seen that, for each $i = 1, \ldots, n$,

$$\mu_{\Phi,n}(\{(x_1, \ldots, x_n) : x_i > t\}) = H(t) \quad \text{for } t \in \mathbb{R}, \tag{3.5.75}$$

and

$$P_{\mu_{\Phi,n}}(\{(x_1, \ldots, x_n) : x_i > t\}, \mathbb{R}^n) = c_\Phi e^{-\Phi(|t|)} = -H'(t) \quad \text{for } t \in \mathbb{R}. \tag{3.5.76}$$

Hence, $F_\Phi(s)$ agrees with the perimeter of any half-space of the form $\{x_i > t\}$, whose measure is s.

Next, define $L^\Phi : [0, 1] \to [0, \infty)$ as

$$L^\Phi(s) = s\Phi'\big(\Phi^{-1}(\log(\tfrac{2}{s}))\big) \quad \text{for } s \in (0, 1], \text{ and } \quad L^\Phi(0) = 0. \tag{3.5.77}$$

3 Reduction Principles

Then the isoperimetric function of $(\mathbb{R}^n, \mu_{\Phi,n})$ satisfies

$$I_{(\mathbb{R}^n,\mu_{\Phi,n})}(s) \approx F_{\Phi}(s) \approx L^{\Phi}(s) \quad \text{for } s \in [0, \tfrac{1}{2}] \tag{3.5.78}$$

(see [7, Proposition 13 and Theorem 15]). Furthermore, half-spaces, whose boundary is orthogonal to a coordinate axis, are "approximate solutions" to the isoperimetric problem in $(\mathbb{R}^n, \mu_{\Phi,n})$ in the sense that there exist constants C_1 and C_2, depending on n, such that, for every $s \in (0, 1)$, any such half-space V with measure s satisfies

$$C_1 P_{\mu_{\Phi,n}}(V, \mathbb{R}^n) \leq I_{(\mathbb{R}^n,\mu_{\Phi,n})}(s) \leq C_2 P_{\mu_{\Phi,n}}(V, \mathbb{R}^n).$$

In the special case when $\mu_{\Phi,n} = \gamma_n$, the Gauss measure, Eq. (3.5.78) yields

$$I_{(\mathbb{R}^n,\gamma_n)}(s) \approx s \left(\log \tfrac{2}{s}\right)^{\tfrac{1}{2}} \quad \text{for } s \in (0, \tfrac{1}{2}]. \tag{3.5.79}$$

Note that now (3.5.44) is not satisfied. Moreover, any half-space is, in fact, an exact minimizer in the isoperimetric inequality [13, 122].

The reduction theorem for Sobolev embeddings in product probability spaces reads as follows.

Theorem 3.5.22 *Let $n \in \mathbb{N}$, $m \in \mathbb{N}$, let $\mu_{\Phi,n}$ be the probability measure defined by (3.5.71), and let $\|\cdot\|_{X(0,1)}$ and $\|\cdot\|_{Y(0,1)}$ be rearrangement-invariant function norms. Then the following facts are equivalent.*

(i) *The inequality*

$$\left\| \int_s^1 f(r) \frac{\left(\Phi^{-1}\left(\log \tfrac{2}{s}\right) - \Phi^{-1}\left(\log \tfrac{2}{r}\right)\right)^{m-1}}{r \Phi'\left(\Phi^{-1}\left(\log \tfrac{2}{r}\right)\right)} \, dr \right\|_{Y(0,1)} \leq C_1 \|f\|_{X(0,1)} \tag{3.5.80}$$

holds for some constant C_1, and for every nonnegative $f \in X(0, 1)$.
(ii) *The embedding*

$$V^m X(\mathbb{R}^n, \mu_{\Phi,n}) \to Y(\mathbb{R}^n, \mu_{\Phi,n}) \tag{3.5.81}$$

holds.
(iii) *The Poincaré inequality*

$$\|u\|_{Y(\mathbb{R}^n,\mu_{\Phi,n})} \leq C_2 \left\|\nabla^m u\right\|_{X(\mathbb{R}^n,\mu_{\Phi,n})} \tag{3.5.82}$$

holds for some constant C_2, and for every $u \in V_\perp^m X(\mathbb{R}^n, \mu_{\Phi,n})$.

When $m = 1$ and the measure $\mu_{\Phi,n}$ agrees with the Gauss measure γ_n, the result of Theorem 3.5.22 is by now standard (see e.g. [34]).

The rearrangement-invariant function norm $\|\cdot\|_{X_{m,\Phi}(0,1)}$ which yields the optimal rearrangement-invariant target space $Y(\mathbb{R}^n, \mu_{\Phi,n})$ in embedding (3.5.81) is defined as follows. Let $\|\cdot\|_{X(0,1)}$ be a rearrangement-invariant function norm, and let $n, m \in \mathbb{N}$. Then $\|\cdot\|_{X_{m,\Phi}(0,1)}$ is the rearrangement-invariant function norm whose associate function norm is given by

$$\|f\|_{X'_{m,\Phi}(0,1)} = \left\| \int_0^r f^*(s) \frac{\left(\Phi^{-1}\left(\log \frac{2}{s}\right) - \Phi^{-1}\left(\log \frac{2}{r}\right)\right)^{m-1}}{r\Phi'\left(\Phi^{-1}\left(\log \frac{2}{r}\right)\right)} \, ds \right\|_{X'(0,1)}$$
(3.5.83)

for $f \in \mathcal{M}_+(0,1)$.

The reduction theorem takes a simpler form in the case of Gaussian measure.

Theorem 3.5.23 *Let $X(\mathbb{R}^n, \gamma_n)$ and $Y(\mathbb{R}^n, \gamma_n)$ be rearrangement-invariant spaces, and let $m \geq 1$. There exists a constant C_1 such that*

$$\|u\|_{Y(\mathbb{R}^n, \gamma_n)} \leq C_1 \|\nabla^m u\|_{X(\mathbb{R}^n, \gamma_n)}$$

for every $u \in V_\perp^m X(\mathbb{R}^n, \gamma_n)$ if and only if there exists a constant C_2 such that

$$\left\| \frac{1}{\left(1 + \log \frac{1}{s}\right)^{\frac{m-1}{2}}} \int_s^1 f(r) \frac{\left(\log \frac{r}{s}\right)^{m-1}}{r\left(1 + \log \frac{1}{r}\right)^{1/2}} \, dr \right\|_{Y(0,1)} \leq C_2 \|f\|_{X(0,1)}$$

for every $f \in X(0,1)$.

Given $n, m \in \mathbb{N}$, and a rearrangement-invariant function norm $\|\cdot\|_{X(0,1)}$, define $\|\cdot\|_{X_{m,G}(0,1)}$ as the rearrangement-invariant function norm whose associate function norm is given by

$$\|f\|_{X'_{m,G}(0,1)} = \left\| \frac{1}{r\left(\log \frac{2}{r}\right)^{1/2}} \int_0^r f^*(s) \frac{\left(\log \frac{r}{s}\right)^{m-1}}{\left(\log \frac{2}{s}\right)^{\frac{m-1}{2}}} \, ds \right\|_{X'(0,1)}$$
(3.5.84)

for $f \in \mathcal{M}_+(0,1)$.

Theorem 3.5.24 *Let $n \in \mathbb{N}$, $m \in \mathbb{N}$, and let $\|\cdot\|_{X(0,1)}$ be a rearrangement-invariant function norm. Then the functional $\|\cdot\|_{X'_{m,G}(0,1)}$, given by (3.5.84), is a rearrangement-invariant function norm, whose associate norm $\|\cdot\|_{X_{m,G}(0,1)}$ satisfies*

$$V^m X(\mathbb{R}^n, \gamma_n) \to X_{m,G}(\mathbb{R}^n, \gamma_n)$$
(3.5.85)

3 Reduction Principles   215

with norm independent of n, and

$$\|u\|_{X_{m,G}(\mathbb{R}^n,\gamma_n)} \leq C \|\nabla^m u\|_{X(\mathbb{R}^n,\gamma_n)} \qquad (3.5.86)$$

for some constant C independent of n, for every $u \in V^m_\perp X(\mathbb{R}^n, \gamma_n)$. *Moreover, the function norm* $\|\cdot\|_{X_{m,G}(0,1)}$ *is optimal in (3.5.85) and (3.5.86) among all rearrangement-invariant norms.*

We finish this section with an application of the results of this section to the particular case when $\mu_{\Phi,n}$ is a Boltzmann measure, and the norms are of Lorentz–Zygmund type.

Theorem 3.5.25 *Let* $n, m \in \mathbb{N}$, *let* $\beta \in [1, 2]$ *and let* $p, q \in [1, \infty]$ *and* $\alpha \in \mathbb{R}$ *be such that one of the conditions in (3.2.13) is satisfied. Then*

$$V^m L^{p,q;\alpha}(\mathbb{R}^n, \gamma_{n,\beta}) \to \begin{cases} L^{p,q;\alpha+\frac{m(\beta-1)}{\beta}}(\mathbb{R}^n, \gamma_{n,\beta}) & \text{if } p < \infty, \\ L^{\infty,q;\alpha-\frac{m}{\beta}}(\mathbb{R}^n, \gamma_{n,\beta}) & \text{if } p = \infty. \end{cases} \qquad (3.5.87)$$

Moreover, in both cases, the target space is optimal among all rearrangement-invariant spaces.

When $\beta = 2$, Theorem 3.5.25 yields the following sharp Sobolev type embeddings in Gauss space.

Theorem 3.5.26 *Let* $n, m \in \mathbb{N}$, *and let* $p, q \in [1, \infty]$ *and* $\alpha \in \mathbb{R}$ *be such that one of the conditions in (3.2.13) is satisfied. Then*

$$V^m L^{p,q;\alpha}(\mathbb{R}^n, \gamma_n) \to \begin{cases} L^{p,q;\alpha+\frac{m}{2}}(\mathbb{R}^n, \gamma_n) & \text{if } p < \infty, \\ L^{\infty,q;\alpha-\frac{m}{2}}(\mathbb{R}^n, \gamma_n) & \text{if } p = \infty. \end{cases}$$

Moreover, in both cases, the target space is optimal among all rearrangement-invariant spaces.

A further specialization of the indices p, q, α appearing in Theorem 3.5.26 leads to the following basic embeddings.

Corollary 3.5.27 *Let* $n, m \in \mathbb{N}$.

(i) *Assume that* $p \in [1, \infty)$. *Then*

$$V^m L^p(\mathbb{R}^n, \gamma_n) \to L^p(\log L)^{\frac{mp}{2}}(\mathbb{R}^n, \gamma_n),$$

and the target space is optimal among all rearrangement-invariant spaces.

(ii) *Assume that $\gamma > 0$. Then*

$$V^m \exp L^\gamma(\mathbb{R}^n, \gamma_n) \to \exp L^{\frac{2\gamma}{2+m\gamma}}(\mathbb{R}^n, \gamma_n),$$

and the target space is optimal among all rearrangement-invariant spaces.

(iii)

$$V^m L^\infty(\mathbb{R}^n, \gamma_n) \to \exp L^{\frac{2}{m}}(\mathbb{R}^n, \gamma_n),$$

and the target space is optimal among all rearrangement-invariant spaces.

3.5.4 The Ornstein–Uhlenbeck Operator

The Ornstein–Uhlenbeck operator was introduced in Sect. 3.1.2. In [43], norm estimates for functions in the Gauss space in terms of the Ornstein-Uhlenbeck operator were studied. This time the norms depend on global integrability properties of functions, measured, once again, by rearrangement-invariant function norms. The specific Sobolev inequalities studied are of the form

$$\|u - \mathrm{m}(u)\|_{Y(\mathbb{R}^n, \gamma_n)} \le c \|\mathcal{L}u\|_{X(\mathbb{R}^n, \gamma_n)} \quad (3.5.88)$$

for some constant c and for all functions $u \colon \mathbb{R}^n \to \mathbb{R}$ such that $\mathcal{L}u \in X(\mathbb{R}^n, \gamma_n)$, in which $\mathrm{m}(u)$ is either the mean value or a median of a function u. The theory is more complicated in this case but the resulting reduction principle is quite neat.

Theorem 3.5.28 (Reduction Principle for Ornstein-Uhlenbeck Embeddings)
Let $X(\mathbb{R}^n, \gamma_n)$ and $Y(\mathbb{R}^n, \gamma_n)$ be rearrangement-invariant spaces. The following statements are equivalent:

1. *There exists a constant $c_1 > 0$ such that*

$$\|u - \mathrm{med}(u)\|_{Y(\mathbb{R}^n, \gamma_n)} \le c_1 \|\mathcal{L}u\|_{X(\mathbb{R}^n, \gamma_n)}$$

for every $u \in W_\mathcal{L} X(\mathbb{R}^n, \gamma_n)$.

2. *There exists a constant $c_2 > 0$ such that*

$$\left\| \frac{1}{s\ell(s)} \int_0^s g(r)\,\mathrm{d}r + \int_s^1 \frac{g(r)}{r\ell(r)}\,\mathrm{d}r \right\|_{Y(0,1)} \le c_2 \|g\|_{X(0,1)} \quad (3.5.89)$$

for every nonnegative function $g \in X(0, 1)$.

Moreover, the constants c_1 and c_2 depend only on each other.

3 Reduction Principles

Here $W_\mathcal{L} L^2(\mathbb{R}^n, \gamma_n)$ consists of all functions $u \in W^{1,2}(\mathbb{R}^n, \gamma_n)$ such that there exists a function $f \in L^2_\perp(\mathbb{R}^n, \gamma_n)$ fulfilling

$$\int_{\mathbb{R}^n} \nabla u \cdot \nabla v \, d\gamma_n = - \int_{\mathbb{R}^n} v f \, d\gamma_n \quad (3.5.90)$$

for every $v \in W^{1,2}(\mathbb{R}^n, \gamma_n)$. We then set $\mathcal{L}u = f$ for $u \in W_\mathcal{L} L^2(\mathbb{R}^n, \gamma_n)$.

The usual machinery can be then applied to obtain formulas for optimal range and domain spaces when the partner space is fixed. A bunch of examples of optimal pairs of spaces is then obtained, in which, similarly as we have seen in the case of the first-order Gaussian Sobolev embeddings, a mixture of gain/loss results is obtained. We present some of them.

Example 3.5.29 The following embeddings hold:

$$\begin{cases} W_\mathcal{L} L(\log\log L)^{\alpha+1} \to L(\log\log L)^\alpha & \text{if } \alpha \geq 0 \\ W_\mathcal{L} L^1 (\log L)^\alpha \to L^1 (\log L)^\alpha & \text{if } \alpha \geq 0 \\ W_\mathcal{L} L^p (\log L)^\alpha \to L^p (\log L)^{\alpha+p} & \text{if } p \in (1, \infty) \text{ and } \alpha \in \mathbb{R} \\ W_\mathcal{L} \exp L^\beta \to \exp L^\beta & \text{if } \beta > 0 \\ W_\mathcal{L} \exp\exp L^{\beta+1} \to \exp\exp L^\beta & \text{if } \beta > 0 \\ W_\mathcal{L} L^\infty \to \exp\exp L, \end{cases} \quad (3.5.91)$$

where all the spaces are over $(\mathbb{R}^n, \gamma_n)$.

A distinctive feature of these examples is that all the domain spaces and the target spaces in Eq. (3.5.91) are optimal within the class of Orlicz spaces, but also, at the same time, within the (wider) class of rearrangement-invariant spaces. This property is in sharp contrast with Sobolev embeddings in Euclidean domains, including those for the Laplace operator, where the optimal target and domain rearrangement-invariant space is always better (namely, essentially smaller on the target side and essentially larger one the domain side) than that in the smaller class of Orlicz spaces.

Next, observe that the norm in the target space can either be stronger, equivalent or weaker than that in the domain space. This means that there can be a gain, or a draw or a loss in the degree of integrability of a function inherited from that of the Ornstein-Uhlenbeck operator. On the contrary, a function vanishing on the boundary of a domain with finite Lebesgue measure always enjoys stronger integrability properties than its Laplacian.

Finally, the embeddings above are worth being compared with standard second-order Gaussian Sobolev embeddings. In particular, one has that

$$\begin{cases} W^2 L^p (\log L)^\alpha \to L^p (\log L)^{\alpha+1} & \text{if } p \in [1, \infty) \text{ and } \alpha \geq 0 \\ W^2 \exp L^\beta \to \exp L^{\frac{\beta}{\beta+1}} & \text{if } \beta > 0 \\ W^2 L^\infty \to \exp L, \end{cases} \quad (3.5.92)$$

where all the spaces are over $(\mathbb{R}^n, \gamma_n)$, see [40, Theorems 7.8 and 7.13, Corollary 7.14]. The norms of the target spaces in the embeddings displayed in (3.5.92) are always weaker than those in the respective embeddings with the same domain norms in (3.5.91), save when $p = 1$ in the last one of (3.5.92), in which case the norm in the target space is stronger. This phenomenon can be explained by the fact that, because of a multiplying factor blowing up near infinity, the first-order term in the Ornstein-Uhlenbeck operator plays a dominant role with respect not only to the Laplacian, but also to the full Hessian of a function when norms sufficiently far from the $L^1(\mathbb{R}^n, \gamma_n)$ endpoint are taken. Viceversa, when getting close to this endpoint, at which no embedding into a rearrangement-invariant space for the Ornstein-Uhlenbeck operator holds, the impact of the missing second-order derivatives and the gap between the properties of the Laplacian and the Hessian become apparent.

3.5.5 Traces of Sobolev Functions

One important property enjoyed by functions from the Sobolev space $W^{m,p}(\mathbb{R}^n)$, $m \in \mathbb{N}$, $p \in [1, \infty]$, is that their restrictions, called traces, to lower dimensional spaces can be properly defined, provided that the dimension d of the relevant subspaces is not too small, depending on n, m and p. The trace of a function $u \in W^{m,p}(\mathbb{R}^n)$ turns out to be measurable with respect to the d-dimensional measure on the relevant subspaces, and also integrable to some power q, depending on n, m, p and d. Loosely speaking, increasing the values of m and p causes u to be more regular, and hence allows smaller values of d and larger values of q.

The situation is much easier when the *boundary* traces on regular domains are studied. In such case, one has $d = n - 1$.

The theory of boundary traces in Sobolev spaces has a number of applications, especially to boundary-value problems for partial differential equations, in particular when the Neumann problem is studied. The trace operator can be defined just by

$$\text{Tr } u = u_{|\partial\Omega}$$

for a continuous function u on $\overline{\Omega}$, and then extended to a bounded linear operator

$$\text{Tr} : W^{1,1}(\Omega) \to L^1(\partial\Omega),$$

where, $L^1(\partial\Omega)$ denotes the Lebesgue space of summable functions on $\partial\Omega$ with respect to the $(n-1)$–dimensional Hausdorff measure $\mathcal{H}^{n-1}$. There exist many powerful methods for proving embedding theorems for the trace operator Tr, usually however quite dependent on a concrete norms involved. For specific limiting situations, other (e.g. potential) methods have been used, but there does not seem to

3 Reduction Principles

exist a unified flexible approach that would cover the whole range of situations of interest in applications.

In [38], a method for obtaining sharp trace inequalities in a general context of rearrangement-invariant spaces was developed. The corresponding reduction principle reads as follows.

Theorem 3.5.30 (Reduction Principle for Boundary Traces) *Let $X(\Omega)$ and $Y(\partial\Omega)$ be rearrangement-invariant spaces. Then*

$$\left\| \int_{t^{n'}}^1 f(s) s^{\frac{m}{n}-1} ds \right\|_{\overline{Y}(0,1)} \leq C \|f\|_{\overline{X}(0,1)}, \quad f \in \mathcal{M}_+(0,1),$$

if and only if

$$\|\operatorname{Tr} u\|_{Y(\partial\Omega)} \leq C \|u\|_{W^m X(\Omega)}$$

for every $u \in W^m X(\Omega)$.

It is of interest to note that this time the corresponding Hardy-type operator in the reduction principle has an extra power in the lower bound. In applications of the reduction principle thus one can use the known boundedness properties of such operators [63].

An extension to traces on a lower-dimensional manifolds was obtained in [36].

Theorem 3.5.31 (Reduction Principle for Trace Embeddings) *Let Ω be a bounded open set with the cone property in $\mathbb{R}^n$, $n \geq 2$. Assume that $m \in \mathbb{N}$ and $d \in \mathbb{N}$ are such that $1 \leq d \leq n$ and $d \geq n - m$. Let $\|\cdot\|_{X(0,1)}$ and $\|\cdot\|_{Y(0,1)}$ be rearrangement-invariant function norms. Then the following facts are equivalent.*

(i) *The Sobolev trace embedding*

$$\operatorname{Tr}: W^m X(\Omega) \to Y(\Omega_d) \qquad (3.5.93)$$

holds.

(ii) *The inequality*

$$\left\| \int_{t^{\frac{n}{d}}}^1 f(s) s^{-1+\frac{m}{n}} ds \right\|_{Y(0,1)} \leq C_2 \|f\|_{X(0,1)} \qquad (3.5.94)$$

holds for some constant C_2 and for every nonnegative $f \in X(0,1)$.

Further extension to spaces endowed with Frostman measures was developed in [42] and [41], in which all values of d were considered. It turns out that the dimensional threshold appears here in consequence of which the cases when $d \geq n - m$ and $d < n - m$ need separate treatments. The latter case (the so-called *slowly-decaying* measures) even requires an additional one-dimensional inequality in the reduction principles. This result ignited an investigation of certain new scale of function spaces, see [127].

3.5.6 Miscellaneous Applications of Reduction Principles

Let us finally mention, without going into details, that reduction principles are of use in several further situations. These include, for example, compactness of Sobolev embeddings [80, 116, 117], fractional Sobolev embeddings [2], K-interpolation inequalities in limiting cases [4], embeddings on entire Euclidean space [1, 96, 97, 128], higher-order Sobolev-type embeddings on Carnot–Carathéodory spaces [60] and their compactness [61], compactness for Sobolev-type trace operators [23], compactness of Sobolev-type embeddings with measures [24], Sobolev inequalities in the hyperbolic space [99], boundedness of integral operators [55], Sobolev embeddings into Morrey, Campanato and Hölder spaces [25, 33], embeddings of Sobolev spaces involving symmetric gradients [15], optimal function spaces in weighted Sobolev embeddings with monomial weights [51], and more.

Acknowledgments The course was supported in part by Fondazione C.I.M.E. Roberto Conti, Goethe Universität Frankfurt am Main, Università degli studi di Firenze, INdAM, EMS, Springer, and the Czech Science Foundation.

References

1. Alberico, A., Cianchi, A., Pick, L., Slavíková, L.: Sharp Sobolev type embeddings on the entire Euclidean space. Commun. Pure Appl. Anal. **17**(5), 2011–2037 (2018)
2. Alberico, A., Cianchi, A., Pick, L., Slavíková, L.: Fractional Orlicz-Sobolev embeddings. J. Math. Pures Appl. **149**, 216–253 (2021)
3. Aubin, T.: Problèmes isopérimétriques et espaces de Sobolev. J. Differ. Geom. **11**(4), 573–598 (1976)
4. Baena-Miret, S., Gogatishvili, A., Mihula, Z., Pick, L.: Reduction principle for Gaussian K-inequality. J. Math. Anal. Appl. **516**(2), 126522, 23 (2022)
5. Baernstein, A., II.: A unified approach to symmetrization. In: Partial Differential Equations of Elliptic Type (Cortona, 1992), Symposium in Mathematics, vol. XXXV, pp. 47–91. Cambridge University Press, Cambridge (1994)
6. Barthe, F., Cattiaux, P., Roberto, C.: Interpolated inequalities between exponential and Gaussian, Orlicz hypercontractivity and isoperimetry. Rev. Mat. Iberoam. **22**(3), 993–1067 (2006)
7. Barthe, F., Cattiaux, P., Roberto, C.: Isoperimetry between exponential and Gaussian. Electron. J. Probab. **12**(44), 1212–1237 (2007)
8. Bathory, M.: Joint weak type interpolation on Lorentz-Karamata spaces. Math. Inequal. Appl. **21**(2), 385–419 (2018)
9. Bennett, C., Rudnick, K.: On Lorentz-Zygmund spaces. Dissertationes Math. (Rozprawy Mat.) 175:67 (1980)
10. Bennett, C., Sharpley, R.: Interpolation of Operators. Pure and Applied Mathematics, vol. 129. Academic Press, Boston (1988)

3 Reduction Principles

11. Bingham, N.H., Goldie, C.M., Teugels, J.L.: Regular Variation. Encyclopedia of Mathematics and its Applications, vol. 27. Cambridge University Press, Cambridge (1989)
12. Bobkov, S.G., Houdré, C.: Some connections between isoperimetric and Sobolev-type inequalities. Mem. Am. Math. Soc. **129**(616), viii+111 (1997)
13. Bojarski, B.: Remarks on Sobolev imbedding inequalities. In: Complex Analysis, Joensuu 1987. Lecture Notes in Mathematics, pp. 52–68. Springer, Berlin (1988)
14. Borell, C.: The Brunn-Minkowski inequality in Gauss space. Invent. Math. **30**(2), 207–216 (1975)
15. Breit, D., Cianchi, A.: Symmetric gradient Sobolev spaces endowed with rearrangement-invariant norms. Adv. Math. **391**, 107954 (2021)
16. Brézis, H., Wainger, S.: A note on limiting cases of Sobolev embeddings and convolution inequalities. Commun. Partial Differ. Equ. **5**(7), 773–789 (1980)
17. Brothers, J.E., Ziemer, W.P.: Minimal rearrangements of Sobolev functions. J. Reine Angew. Math. **384**, 153–179 (1988)
18. Brudnyĭ, J.A.: Rational approximation and imbedding theorems. Dokl. Akad. Nauk SSSR **247**(2), 269–272 (1979)
19. Buckley, S., Koskela, P.: Sobolev-Poincaré implies John. Math. Res. Lett. **2**(5), 577–593 (1995)
20. Caetano, A.M., Gogatishvili, A., Opic, B.: Embeddings and the growth envelope of Besov spaces involving only slowly varying smoothness. J. Approx. Theory **163**(10), 1373–1399 (2011)
21. Carlen, E.A., Kerce, C.: On the cases of equality in Bobkov's inequality and Gaussian rearrangement. Calc. Var. Partial Differ. Equ. **13**(1), 1–18 (2001)
22. Carro, M., Pick, L., Soria, J., Stepanov, V.D.: On embeddings between classical Lorentz spaces. Math. Inequal. Appl. **4**(3), 397–428 (2001)
23. Cavaliere, P., Mihula, Z.: Compactness for Sobolev-type trace operators. Nonlinear Anal. **183**, 42–69 (2019)
24. Cavaliere, P., Mihula, Z.: Compactness of Sobolev-type embeddings with measures. Commun. Contemp. Math. **24**(9), 2150036 (2022)
25. Cavaliere, P., Cianchi, A., Pick, L., Slavíková, L.: Higher-order Sobolev embeddings into spaces of Campanato and Morrey type. arxiv 2404.09702
26. Chong, K.M., Rice, N.M.: Equimeasurable Rearrangements of Functions. Queen's Papers in Pure and Applied Mathematics, vol. 28. Queen's University, Kingston (1971)
27. Cianchi, A.: A sharp embedding theorem for Orlicz-Sobolev spaces. Ind. Univ. Math. J. **45**(1), 39–65 (1996)
28. Cianchi, A.: Second-order derivatives and rearrangements. Duke Math. J. **105**(3), 355–385 (2000)
29. Cianchi, A.: Optimal Orlicz-Sobolev embeddings. Rev. Mat. Iberoamericana **20**(2), 427–474 (2004)
30. Cianchi, A.: Symmetrization and second-order Sobolev inequalities. Ann. Mat. Pura Appl. (4) **183**(1), 45–77 (2004)
31. Cianchi, A.: Higher-order Sobolev and Poincaré inequalities in Orlicz spaces. Forum Math. **18**(5), 745–767 (2006)
32. Cianchi, A., Pick, L.: Sobolev embeddings into BMO, VMO, and L_∞. Ark. Mat. **36**(2), 317–340 (1998)
33. Cianchi, A., Pick, L.: Sobolev embeddings into spaces of Campanato, Morrey, and Hölder type. J. Math. Anal. Appl. **282**(1), 128–150 (2003)
34. Cianchi, A., Pick, L.: Optimal Gaussian Sobolev embeddings. J. Funct. Anal. **256**(11), 3588–3642 (2009)
35. Cianchi, A., Pick, L.: An optimal endpoint trace embedding. Ann. Inst. Fourier (Grenoble) **60**(3), 939–951 (2010)
36. Cianchi, A., Pick, L.: Optimal Sobolev trace embeddings. Trans. Am. Math. Soc. **368**(12), 8349–8382 (2016)

37. Cianchi, A., Kerman, R., Opic, B., Pick, L.: A sharp rearrangement inequality for the fractional maximal operator. Stud. Math. **138**(3), 277–284 (2000)
38. Cianchi, A., Kerman, R., Pick, L.: Boundary trace inequalities and rearrangements. J. Anal. Math. **105**, 241–265 (2008)
39. Cianchi, A., Fusco, N., Maggi, F., Pratelli, A.: On the isoperimetric deficit in Gauss space. Am. J. Math. **133**(1), 131–186 (2011)
40. Cianchi, A., Pick, L., Slavíková, L.: Higher-order Sobolev embeddings and isoperimetric inequalities. Adv. Math. **273**, 568–650 (2015)
41. Cianchi, A., Pick, L., Slavíková, L.: Sobolev embeddings in Orlicz and Lorentz spaces with measures. J. Math. Anal. Appl. **485**(2), 123827 (2020)
42. Cianchi, A., Pick, L., Slavíková, L.: Sobolev embeddings, rearrangement-invariant spaces and Frostman measures. Ann. Inst. H. Poincaré Anal. Non Linéaire **37**(1), 105–144 (2020)
43. Cianchi, A., Musil, V., Pick, L.: Optimal Sobolev embeddings for the Ornstein-Uhlenbeck operator. J. Differ. Equ. **359**, 414–475 (2023)
44. Courant, R.: Beweis des Satzes, daßvon allen homogenen Membranen gegebenen Umfanges und gegebener Spannung die kreisförmige den tiefsten Grundton besitzt. Math. Z. **1**(2–3), 321–328 (1918)
45. Cwikel, M., Nilsson, P.: Interpolation of Marcinkiewicz spaces. Math. Scand. **56**(1), 29–42 (1985)
46. Cwikel, M., Pustylnik, E.: Sobolev type embeddings in the limiting case. J. Fourier Anal. Appl. **4**(4–5), 433–446 (1998)
47. Day, P.W.: Rearrangements of measurable functions. ProQuest LLC, Ann Arbor. Thesis (Ph.D.)–California Institute of Technology (1970)
48. Day, P.W.: Decreasing rearrangements and doubly stochastic operators. Trans. Am. Math. Soc. **178**, 383–392 (1973)
49. De Giorgi, E.: Sulla proprietà isoperimetrica dell'ipersfera, nella classe degli insiemi aventi frontiera orientata di misura finita. Atti Accad. Naz. Lincei Mem. Cl. Sci. Fis. Mat. Natur. Sez. Ia **5**, 33–44 (1958)
50. DeVore, R., Scherer, K.: Interpolation of linear operators on Sobolev spaces. Ann. Math. **109**(3), 583–599 (1979)
51. Drážný, L.: Optimal function spaces in weighted Sobolev embeddings with monomial weight. Charles University, Prague. Diploma Thesis (2023)
52. Edmunds, D.E., Evans, W.D.: Hardy Operators, Function Spaces and Embeddings. Springer Monographs in Mathematics. Springer, Berlin (2004)
53. Edmunds, D.E., Opic, B.: Alternative characterisations of Lorentz-Karamata spaces. Czechoslovak Math. J. **58**(2), 517–540 (2008)
54. Edmunds, D.E., Kerman, R., Pick, L.: Optimal Sobolev imbeddings involving rearrangement-invariant quasinorms. J. Funct. Anal. **170**(2), 307–355 (2000)
55. Edmunds, D.E., Mihula, Z., Musil, V., Pick, L.: Boundedness of classical operators on rearrangement-invariant spaces. J. Funct. Anal. **278**(4), 108341 (2020)
56. Evans, W.D., Opic, B., Pick, L.: Interpolation of operators on scales of generalized Lorentz-Zygmund spaces. Math. Nachr. **182**, 127–181 (1996)
57. Faber, G.: Beweis, dass unter allen homogenen Membranen von gleicher Fläche und gleicher Spannung die kreisförmige den tiefsten Grundton gibt. Sitzungsberichte Bayerische Akademie der Wissenschaften München. Mathematisch-Physikalische Klasse, pp. 169–172. Verlagd.Bayer.Akad.d.Wiss., Munchen (1923)
58. Federer, H., Fleming, W.H.: Normal and integral currents. Ann. Math. **72**, 458–520 (1960)
59. Fernández-Martínez, P., Signes, T.M.: An application of interpolation theory to renorming of Lorentz-Karamata type spaces. Ann. Acad. Sci. Fenn. Math. **39**(1), 97–107 (2014)
60. Franců, M.: Higher-order Sobolev-type embeddings on Carnot-Carathéodory spaces. Math. Nachr. **290**(7), 1033–1052 (2017)
61. Franců, M.: Higher-order compact embeddings of function spaces on Carnot-Carathéodory spaces. Banach J. Math. Anal. **12**(4), 970–994 (2018)

3 Reduction Principles

62. Gagliardo, E.: Proprietà di alcune classi di funzioni in più variabili. Ricerche Mat. **7**, 102–137 (1958)
63. Gogatishvili, A., Lang, J.: The generalized Hardy operator with kernel and variable integral limits in Banach function spaces. J. Inequal. Appl. **4**(1), 1–16 (1999)
64. Gogatishvili, A., Opic, B., Neves, J.S.: Optimality of embeddings of Bessel-potential-type spaces into Lorentz-Karamata spaces. Proc. R. Soc. Edinburgh Sect. A **134**(6), 1127–1147 (2004)
65. Gogatishvili, A., Opic, B., Pick, L.: Weighted inequalities for Hardy-type operators involving suprema. Collect. Math. **57**(3), 227–255 (2006)
66. Gogatishvili, A., Opic, B., Trebels, W.: Limiting reiteration for real interpolation with slowly varying functions. Math. Nachr. **278**(1–2), 86–107 (2005)
67. Gross, L.: Logarithmic Sobolev inequalities. Am. J. Math. **97**(4), 1061–1083 (1975)
68. Grothendieck, A.: Réarrangements de fonctions et inégalités de convexité dans les algèbres de von Neumann munies d'une trace. Sémin. N. Bourbaki. **113**(3), 127–139 (1956)
69. Gurka, P., Opic, B.: Sharp embeddings of Besov-type spaces. J. Comput. Appl. Math. **208**(1), 235–269 (2007)
70. Hajłasz, P., Koskela, P.: Isoperimetric inequalities and imbedding theorems in irregular domains. J. Lond. Math. Soc. **58**(2), 425–450 (1998)
71. Hansson, K.: Imbedding theorems of Sobolev type in potential theory. Math. Scand. **45**(1), 77–102 (1979)
72. Hempel, J.A., Morris, G.R., Trudinger, N.S.: On the sharpness of a limiting case of the Sobolev imbedding theorem. Bull. Austral. Math. Soc. **3**, 369–373 (1970)
73. Hildén, K.: Symmetrization of functions in Sobolev spaces and the isoperimetric inequality. Manuscript. Math. **18**(3), 215–235 (1976)
74. Ho, K.P.: Sublinear operators on radial rearrangement-invariant quasi-Banach function spaces. Acta Math. Hungar. **160**(1), 88–100 (2020)
75. Holmstedt, T.: Interpolation of quasi-normed spaces. Math. Scand. **26**, 177–199 (1970)
76. Karamata, J.: Sur un mode de croissance régulière. Théorèmes fondamentaux. Bull. Soc. Math. France **61**, 55–62 (1933)
77. Kerman, R.A.: Function spaces continuously paired by operators of convolution type. Can. Math. Bull. **22**(4), 499–507 (1979)
78. Kerman, R.A.: An integral extrapolation theorem with applications. Stud. Math. **76**(3), 183–195 (1983)
79. Kerman, R., Pick, L.: Optimal Sobolev imbeddings. Forum Math. **18**(4), 535–570 (2006)
80. Kerman, R., Pick, L.: Compactness of Sobolev imbeddings involving rearrangement-invariant norms. Stud. Math. **186**(2), 127–160 (2008)
81. Kilpeläinen, T., Malý, J.: Sobolev inequalities on sets with irregular boundaries. Z. Anal. Anwendungen **19**(2), 369–380 (2000)
82. Kolyada, V.I.: Rearrangements of functions, and embedding theorems. Uspekhi Mat. Nauk **44**(5), 61–95 (1989)
83. Krahn, E.: Über eine von Rayleigh formulierte Minimaleigenschaft des Kreises. Math. Ann. **94**(1), 97–100 (1925)
84. Krée, P.: Interpolation d'espaces vectoriels qui ne sont ni normés, ni complets. Applications. Ann. Inst. Fourier (Grenoble) **17**(fasc. 2), 137–174 (1968), 1967
85. Kreĭn, S.G., Petunīn, Y.I., Semënov, E.M.: Interpolation of linear operators. Translations of Mathematical Monographs, vol. 54. American Mathematical Society, Providence (1982). Translated from the Russian by J. Szűcs
86. Ledoux, M.: The Concentration of Measure Phenomenon. Mathematical Surveys and Monographs, vol. 89. American Mathematical Society, Providence (2001)
87. Ledoux, M.: Spectral gap, logarithmic Sobolev constant, and geometric bounds. In: Surveys in Differential Geometry. Vol. IX. Surveys in Differential Geometry, vol. 9, pp. 219–240. International Press, Somerville (2004)

88. Lindenstrauss, J., Tzafriri, L.: Classical Banach Spaces. II. Function Spaces. Ergebnisse der Mathematik und ihrer Grenzgebiete [Results in Mathematics and Related Areas], vol. 97. Springer, Berlin (1979)
89. Lorentz, G.G.: Bernstein Polynomials. Mathematical Expositions, vol. 8. University of Toronto Press, Toronto (1953)
90. Lorentz, G.G.: Approximation of Functions. Holt, Rinehart and Winston, New York (1966)
91. Luxemburg, W.A.J.: Rearrangement-invariant Banach function spaces. In: Proceedings of Symposium in Analysis. Queen's Papers in Pure and Appliled Mathematics, vol. 10 (1967)
92. Malý, J., Pick, L.: An elementary proof of sharp Sobolev embeddings. Proc. Am. Math. Soc. **130**(2), 555–563 (2002)
93. Maz'ya, V.G.: Classes of domains and imbedding theorems for function spaces. Soviet Math. Dokl. **1**, 882–885 (1960)
94. Maz'ya, V.G.: p-conductivity and theorems on imbedding certain functional spaces into a C-space. Dokl. Akad. Nauk SSSR **140**, 299–302 (1961)
95. Maz'ya, V.G.: Sobolev spaces with applications to elliptic partial differential equations. In: Grundlehren der Mathematischen Wissenschaften [Fundamental Principles of Mathematical Sciences], vol. 342. Springer, Heidelberg, augmented edition (2011)
96. Mihula, Z.: Embeddings of homogeneous Sobolev spaces on the entire space. Proc. R. Soc. Edinburgh Sect. A **151**(1), 296–328 (2021)
97. Mihula, Z.: Poincaré-Sobolev inequalities with rearrangement-invariant norms on the entire space. Math. Z. **298**(3–4), 1623–1640 (2021)
98. Mihula, Z.: Optimal behavior of weighted Hardy operators on rearrangement-invariant spaces. Math. Nachr. **296**(8), 3492–3538 (2023)
99. Mihula, Z.: Optimal Sobolev inequalities in the hyperbolic space (2023). arxiv2305.06797
100. Musil, V.: Optimal Orlicz domains in Sobolev embeddings into Marcinkiewicz spaces. J. Funct. Anal. **270**(7), 2653–2690 (2016)
101. Musil, V., Pick, L., Takáč, J.: Optimality problems in Orlicz spaces. Adv. Math. **432**, 109273 (2023)
102. Nelson, E.: The free Markoff field. J. Funct. Anal. **12**, 211–227 (1973)
103. Neves, J.S.: Lorentz-Karamata spaces, Bessel and Riesz potentials and embeddings. Dissertationes Math. (Rozprawy Mat.) **405**, 46 (2002)
104. Nirenberg, L.: On elliptic partial differential equations. Ann. Scuola Norm. Sup. Pisa Cl. Sci. **13**, 115–162 (1959)
105. Opic, B., Pick, L.: On generalized Lorentz-Zygmund spaces. Math. Inequal. Appl. **2**(3), 391–467 (1999)
106. Peetre, J.: Espaces d'interpolation et théorème de Soboleff. Ann. Inst. Fourier (Grenoble) **16**(fasc. 1), 279–317 (1966)
107. Peša, D.: Lorentz–Karamata spaces (2023). arxiv 2006.14455
108. Peša, D.: On the smoothness of slowly varying functions (2023). arxiv 2304.14148
109. Pick, L.: Optimal Sobolev Embeddings. Rudolph Lipschitz Vorlesung, vol. 43. Rheinische Friedrich–Wilhelms–Universität Bonn, Bonn (2001)
110. Pick, L., Kufner, A., John, O., Fučík, S.: Function Spaces. Vol. 1. De Gruyter Series in Nonlinear Analysis and Applications, vol. 14. Walter de Gruyter, Berlin, extended edition (2013)
111. Pohozhaev, S.I.: On the imbedding Sobolev theorem for $pl = n$ (in Russian). Dokl. Conf. Section Math. Moscow Power Inst. **165**, 158–170 (1965)
112. Rayleigh, Baron, J.W.S.: The Theory of Sound, 2nd edn. Dover Publications, New York (1945)
113. Ryff, J.V.: Orbits of L^1-functions under doubly stochastic transformations. Trans. Am. Math. Soc. **117**, 92–100 (1965)
114. Ryff, J.V.: Measure preserving transformations and rearrangements. J. Math. Anal. Appl. **31**, 449–458 (1970)
115. Sawyer, E.: Boundedness of classical operators on classical Lorentz spaces. Stud. Math. **96**(2), 145–158 (1990)

116. Slavíková, L.: Almost-compact embeddings. Math. Nachr. **285**(11–12), 1500–1516 (2012)
117. Slavíková, L.: Compactness of higher-order Sobolev embeddings. Publ. Mat. **59**(2), 373–448 (2015)
118. Sobolev, S.: On some estimates relating to families of functions having derivatives that are square integrable. Dokl. Akad. Nauk. SSSR **1**, 267–270 (1936)
119. Sobolev, S.: On a theorem in functional analysis. Sb. Math. **4**, 471–497 (1938)
120. Stein, E.M.: Editor's note: the differentiability of functions in $\mathbf{R}^n$. Ann. Math. **113**(2), 383–385 (1981)
121. Strichartz, R.S.: A note on Trudinger's extension of Sobolev's inequalities. Ind. Univ. Math. J. **21**, 841–842 (1971/1972)
122. Sudakov, V.N., Cirel'son, B.S.: Extremal properties of half-spaces for spherically invariant measures. Zap. Naučn. Sem. Leningrad. Otdel. Mat. Inst. Steklov. (LOMI) **41**, 14–24, 165 (1974). Problems in the theory of probability distributions, II
123. Talenti, G.: Best constant in Sobolev inequality. Ann. Mat. Pura Appl. **110**, 353–372 (1976)
124. Talenti, G.: An inequality between u^* and $|\operatorname{grad} u|^*$. In: General Inequalities, 6 (Oberwolfach, 1990). International Series of Numerical Mathematics, vol. 103, pp. 175–182. Birkhäuser, Basel (1992)
125. Talenti, G.: Inequalities in rearrangement invariant function spaces. In: Nonlinear Analysis, Function Spaces and Applications, vol. 5, pp. 177–230. Prometheus, Prague (1994)
126. Trudinger, N.S.: On imbeddings into Orlicz spaces and some applications. J. Math. Mech. **17**, 473–483 (1967)
127. Turčinová, H.: Basic functional properties of certain scale of rearrangement-invariant spaces. Math. Nachr. **296**(8), 3652–3675 (2023)
128. Vybíral, J.: Optimal Sobolev embeddings on $\mathbb{R}^n$. Publ. Mat. **51**(1), 17–44 (2007)
129. Yudovich, V.I.: Some estimates connected with integral operators and with solutions of elliptic equations (in Russian). Soviet Math. Dokl. **2**, 746–749 (1961)
130. Zygmund, A.: Trigonometric Series, 2nd edn., vols. I–II. Cambridge University Press, New York (1959)

Chapter 4
The Monge-Ampère Equation

Ovidiu Savin

4.1 Introduction

The Monge-Ampère equation appears naturally in different areas of mathematics such as geometry, fluid dynamics, calculus of variations etc. In these lecture notes we give an overview of the basic regularity theory for this equation and present the interior $C^{2,\alpha}$ and $W^{2,p}$ estimates of Caffarelli [2].

The Monge-Ampère equation in its simplest form can be written as

$$det\, D^2 u = 1, \qquad u : \Omega \to \mathbb{R} \quad \text{convex}. \tag{4.1.1}$$

It expresses the fact that the change of variables $x \mapsto \nabla u(x)$ preserves volume.

Affine Invariance A key feature of the Monge-Ampère equation is its affine invariance. It can be thought as an affine invariant version of the Laplace equation. Before this we discuss the invariance groups for two other basic elliptic PDEs.

1. Laplace equation

$$\triangle u = 0 \quad \left(\text{or} \quad \sum \lambda_i = 0, \quad div(\nabla u) = 0.\right)$$

It is invariant under the rigid motions of $\mathbb{R}^n$ (translations and rotations) together with all the α-homogenous rescalings

$$r^{-\alpha} u(rx), \qquad \alpha \in \mathbb{R}.$$

O. Savin (✉)
Department of Mathematics, Columbia University, New York, NY, USA
e-mail: savin@math.columbia.edu

2. Minimal surface equation

$$(1+|\nabla u|^2)\Delta u - u_i u_j u_{ij} = 0 \quad \left(\text{or} \quad \sum \kappa_i = 0, \quad \text{div}\frac{\nabla u}{(1+|\nabla u|^2)^{1/2}} = 0.\right)$$

This is a geometric equations for the graph of u in $\mathbb{R}^{n+1}$. It remains invariant under the rigid motions of $\mathbb{R}^{n+1}$, and the dilations in $\mathbb{R}^{n+1}$ i.e.

$$r^{-1}u(rx).$$

The Laplace equation appears as the linearization of the minimal surface equation. Indeed, pick a point $X_0 = (x_0, u(x_0))$ on the graph of a C^2 solution u. After a dilation and a rotation around X_0 we end up in the situation where u is a small perturbation of the trivial solution 0 in B_1. Its behavior is dictated by the linearized equation of the trivial solution, which is the Laplace equation. Thus the minimal surface equation is a geometric version of the Laplace equation.

3. The Monge-Ampère equation

$$det\, D^2 u = 1, \quad \text{with} \quad D^2 u > 0, \quad (\text{or} \quad \Pi \lambda_i = 1, \quad \lambda_i > 0.)$$

It is invariant under the group of affine transformations

$$u(Ax), \quad \text{with } A \text{ an } n \times n \text{ matrix with } det\, A = 1,$$

to which we add the standard transformations which preserve the second derivatives

$$r^{-2}u(rx), \quad \text{and} \quad u - l \quad \text{with } l \text{ a linear function.}$$

The Laplace equation appears in the linearization of this equation. Indeed, fix a point, say $(0, u(0))$ of a C^2 solution, and after a combination of the 3 transformations above we end up in the setting where u is a perturbation of the trivial solution

$$u_0 := \frac{1}{2}|x|^2.$$

The linearized equation for u_0 is given by

$$L_{u_0}(w) = \lim_{\varepsilon \to 0} \frac{1}{\varepsilon}\left(det\, D^2(u_0 + \varepsilon w) - det\, D^2 u_0\right) = \Delta w = 0.$$

Concavity and Ellipticity Denote by $\mathcal{S}^n$ the space of symmetric $n \times n$ matrices, and by $\mathcal{S}^n_+$ the cone of positive definite matrices.

4 The Monge-Ampère Equation

Notice that if $M \in \mathcal{S}_+^n$, then

$$(det\, M)^{\frac{1}{n}} = \frac{1}{n} \inf_{A \in \mathcal{A}} tr(A\, M),$$

with

$$\mathcal{A} := \{A \in \mathcal{S}_+^n \mid \det A = 1\}.$$

Thus

$$F(M) := (det\, M)^{\frac{1}{n}}$$

is an infimum of linear functions, hence F is a concave operator on $\mathcal{S}_+^n$. It is *elliptic* since it is increasing in its variable

$$F(M + N) > F(M) \qquad \text{if } N > 0.$$

The operator becomes *uniformly elliptic* when restricted to a compact set $K \subset\subset \mathcal{S}_+^n$:

$$\Lambda_K |N| \geq F(M + N) - F(M) \geq \lambda_K |N| \qquad \text{if } N > 0,$$

for some positive constants λ_K, Λ_K depending on the set K.

Motivation We present a few examples where the Monge-Ampère equation appears.

1. *Optimal transportation.*

   Given two probability densities $f(x)dx$ and $g(y)dy$, and a cost function $c(x, y)$ representing the cost of transporting x into y, we want to transport one density into the other by a measure preserving map $y = T(x)$ which minimizes the total cost functional

   $$\int c(x, T(x)) f(x) dx.$$

   The quadratic cost

   $$c(x, y) := |x - y|^2$$

   has very good analytical properties due to the following theorem.

Theorem 4.1.1 (Brenier) *The optimal map T is given by the gradient of a convex function u*

$$T(x) = \nabla u(x).$$

Remark 4.1.2 $\nabla u(x)$ denotes the set of subgradients of u at x (i.e. the slopes of the supporting planes at x), and it is single-valued a.e. in x. In the theorem above the set of points where ∇u is multi-valued can be ignored since it is a set of measure 0 with respect to the density $f(x)dx$. In the case when the x-density is not absolutely continuous with respect to the Lebesgue measure (as in a discrete setting), then the optimal map $x \to T(x)$ can be multi-valued and then the Theorem reads as

$$T(x) \subset \nabla u(x).$$

If $T \in C^1$ then, the change of coordinate formula

$$g(y)\,dy = g(T(x))|\det DT(x)|\,dx$$

implies that u satisfies the Monge-Ampère equation

$$\det D^2 u = \frac{f(x)}{g(\nabla u(x))}.$$

A similar analysis can be carried out for general cost functions $c(x, y)$.

2. *Prescribed curvature equations.*

Given a convex C^2 surface $\Sigma \subset \mathbb{R}^{n+1}$, the Gauss curvature $K(X)$ at $X \in \Sigma$ is the Jacobian of the Gauss map

$$X \mapsto \nu(X) \in \partial B_1,$$

hence

$$K(X) = \prod \kappa_i \quad \text{with } \kappa_i \text{ the principal curvatures.}$$

If $\Sigma = \{x_{n+1} = u(x)\}$ is the graph of a function $u : \Omega \subset \mathbb{R}^n \to \mathbb{R}$, then

$$\nu(X) = \frac{(-\nabla u, 1)}{(1 + |\nabla u|^2)^{\frac{1}{2}}} = \nu_{n+1}(-\nabla u, 1),$$

and, since the tangent planes at X and $\nu(X)$ are parallel, the Jacobian can be computed from the projections in the first n coordinates

$$K(X) = \det D_x(\nu_{n+1}\nabla u) = \nu_{n+1}^{n+2} \det D^2 u(x).$$

The *prescribed Gauss curvature equation* for the graph u reads

$$\det D^2 u(x) = K(x)(1 + |\nabla u|^2)^{\frac{n+2}{2}},$$

for some given function $K(x) \geq 0$.

4 The Monge-Ampère Equation

A related problem deals with that case when Σ is given by a spherical graph

$$\Sigma = \{X = x\varphi(x) | x \in \partial B_1\},$$

and we prescribe the Jacobian between the variables $x \in \partial B_1$ and $\nu(X) \in \partial B_1$,

$$|det\, D_x \nu(X)| = f(x),$$

with $f(x) \geq 0$ a given function of average 1 on the unit sphere. This reduces locally to an equation of the form

$$det\, D^2 u = g(x, u, \nabla u)$$

by writing Σ as the graph of u.

The *Minkowski problem* asks for the existence of a convex set with boundary Σ such that

$$K(X) = f(\nu(X)),$$

with $f(\nu) > 0$ a given function that satisfies the compatibility condition

$$\int_{\partial B_1} \frac{\nu_i}{f(\nu)} d\nu = 0.$$

Then the support function of Σ

$$u(Y) = \sup_{X \in \Sigma} Y \cdot X,$$

has the property

$$\nabla u(\nu(X)) = X,$$

and it satisfies the following Monge-Ampère equation on the unit sphere

$$det\, (\nabla^2 u + uI) = \frac{1}{f}.$$

The *Weyl problem* concerns the isometric embedding of two dimensional spheres $(\partial B_1, g)$ into the Euclidean space $\mathbb{R}^3$. When the curvature $K_g > 0$ then the problem reduces locally to a Monge-Ampère equation

$$det\, D^2 u = h(\nabla u).$$

3. *Fluid dynamics.*

The *semi-geostrophic equation* is an active scalar equation similar to the equation for vorticity in the Euler equation in 2D. The difference is that the potential is obtained by solving the Monge-Ampère equation instead of the Laplace equation.

Precisely, let $\omega(\cdot, t) \geq 0$ be a probability density defined on the periodic torus $T \subset \mathbb{R}^2$ which is transported along the vector field $\nabla^\perp u$, with u defined by the second equation below

$$\begin{cases} \omega_t + \nabla^\perp u \cdot \nabla \omega = 0, \\ det\,(D^2 u + I) = \omega. \end{cases}$$

4.2 An Overview of the Main Results

Here we list some of the main regularity results for the Monge-Ampère equation

$$det\, D^2 u = f(x), \qquad u : \Omega \to \mathbb{R} \text{ convex}, \qquad (4.2.1)$$

and we refer to the books by Gutiérrez [17] and Figalli [15] for more details.

1. **Alexandrov solutions**

Convex functions enjoy strong compactness properties. For example, given a sequence of convex functions (extend u by $+\infty$ outside Ω), the supergraphs in $\mathbb{R}^{n+1}$ converge (up to subsequences) on compact sets of $\mathbb{R}^{n+1}$, in the Hausdorff distance sense. It is convenient to have a more general notion of solution for (4.2.1) that does not require $u \in C^2$, and which has good compactness properties.

When $u \in C^2$, $det\, D^2 u$ is the Jacobian of the transformation

$$x \mapsto \nabla u.$$

In general, we can define $det\, D^2 u$ as the Borel measure μ defined in Ω which is pushed forward by the gradient map into the Lebesgue measure

$$\mu(A) := |\nabla u(A)|.$$

Here $\nabla u(A)$ denotes the set of slopes for all supporting planes at points in $A \subset \Omega$.

Exercise

(i) Show that $\nabla u(A)$ is a closed set if A is a closed set.
(ii) Prove that the set

$$\{y | y \in \nabla u(x_1) \cap \nabla u(x_2), \quad \text{for some distinct points } x_1 \neq x_2\}$$

has measure 0.
(iii) Prove that μ is a well defined Borel measure.

4 The Monge-Ampère Equation 233

The measure μ is called the Monge-Ampère measure of u, and is denoted by abuse of notation by $det\, D^2 u$. In this way the Eq. (4.2.1) is understood in the sense of equality of measures (Alexandrov's point of view) with the right hand side being the measure $f(x)dx$.

(a) *Viscosity solutions.*

Another good notion of solution with good compactness properties comes from the theory of viscosity solutions:

$$det\, D^2 u \le f \quad (\text{or } det\, D^2 u \ge f) \text{ means that}$$

for any convex $\varphi \in C^2$ that touches u by below (or by above) at some point $x_0 \in \Omega$ we have

$$det\, D^2 \varphi(x_0) \le f(x_0) \quad (\text{or} \quad det\, D^2 \varphi(x_0) \ge f(x_0)).$$

Here we require f to be continuous. It turns out that the two definitions are equivalent for continuous f's (exercise). However, the notion of Alexandrov solution is more general as it makes sense for measures.

(b) *Stability.*

The notion of Alexandrov solution is stable under limits: if

$$det\, D^2 u_n = \mu_n, \qquad u_n \to u, \qquad \mu_n \rightharpoonup \mu$$

then

$$det\, D^2 u = \mu.$$

Here $u_n \to u$ means that u_n converges uniformly to u on compact sets, and μ_n converges weakly in the sense of measures to μ.

(c) *Existence and uniqueness for the Dirichlet problem.*

Let Ω be a strictly convex bounded domain, φ a continuous function on $\partial\Omega$ and μ a bounded Borel measure on Ω. There exists a unique Alexandrov solution to

$$det\, D^2 u = \mu, \qquad u = \varphi \quad \text{on} \quad \partial\Omega.$$

The existence in the case when μ is supported on a finite set follows by Perron's method. The general case can be deduced by approximation from (b). The uniqueness part follows by a version of the maximum principle.

2. **Pogorelov's singular example** $(n \ge 3)$

We look for a function

$$u(x) = |x'|^\alpha h(x_n)$$

Fig. 4.1 Pogorelov's example

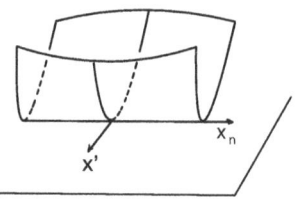

which solves (4.2.1) with constant $f \equiv 1$. It follows that

$$\det D^2 u = r^{n(\alpha-2)+2} h^{n-2} \left(c_1 h h'' - c_2 (h')^2 \right),$$

for some positive constants c_1, c_2 depending on n and α. Here $r = |x'|$ and h is evaluated at x_n. We choose $\alpha = 2 - \frac{2}{n}$, and h as a local solution to the ODE

$$h^{n-2} \left(c_1 h h'' - c_2 (h')^2 \right) = 1, \qquad h(0) = 1, \quad h'(0) = 0.$$

Then, near 0, u solves the Monge-Ampère equation in the Alexandrov sense with $f \equiv 1$, but u fails to be C^2 on the line $|x'| = 0$ (Fig. 4.1).

Exercise There is a non-C^1 singular solution in dimension $n \geq 3$ of the type

$$u(x) = |x'| + |x'|^\alpha h(x_n).$$

3. The Dirichlet problem for smooth data

Caffarelli-Nirenberg-Spruck [8] and Krylov showed that the Dirichlet problem with sufficiently smooth data has classical solutions:

Theorem 4.2.1 *There exists a unique solution $u \in C^2(\overline{\Omega})$ that solves*

$$\begin{cases} \det D^2 u = f & \text{in } \Omega, \\ u = \varphi & \text{on } \partial\Omega, \end{cases}$$

provided that $f \in C^2(\overline{\Omega})$, $f > 0$, $\partial\Omega \in C^4$ uniformly convex, and $\varphi \in C^4$.

The theorem follows from the method of continuity and the apriori estimate

$$\|u\|_{C^{2,\alpha}(\overline{\Omega})} \leq C(f, \partial\Omega, \varphi).$$

4. Pogorelov's interior estimate

Interior C^2 estimates for Eq. (4.2.1) are valid under the additional hypothesis that u is constant on $\partial\Omega$. Notice that this is consistent with the singular example.

4 The Monge-Ampère Equation

Fig. 4.2 The section $S_h(x_0)$

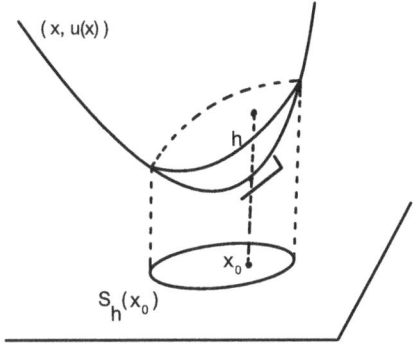

Theorem 4.2.2 *If u solves*

$$\begin{cases} \det D^2 u = f & \text{in } \Omega, \\ u = 0 & \text{on } \partial\Omega, \end{cases}$$

with $f \in C^2(\overline{\Omega})$, $f > 0$, then

$$|D^2 u(x)| \leq C\left(d_{\partial\Omega}(x), \operatorname{diam} \Omega, f\right),$$

where $d_{\partial\Omega}(x)$ denotes the distance from x to $\partial\Omega$.

5. Sections and John's Lemma

Since the Eq. (4.2.1) is invariant under the addition of linear functionals, Theorem 4.2.2 implies that singularities cannot occur at points of strict convexity. Indeed, let's define

$$S_h(x_0) := \{x \in \Omega \mid u(x) < u(x_0) + \nabla u(x_0) \cdot (x - x_0) + h\},$$

as the section at height h at x_0 (Fig. 4.2).

Then if u solves (4.2.1) with $f \in C^2$, $f > 0$, and

$$S_h(x_0) \subset\subset \Omega \quad \text{for some small } h > 0,$$

then, by Theorem 4.2.2, $u \in C^3$ near x_0. Moreover, by John's lemma, $S_h(x_0)$ has the shape of an ellipsoid up to a dilating factor, and can be made to look round after an affine transformation (Fig. 4.3).

Lemma 4.2.3 (John's Lemma) *Let $K \subset \mathbb{R}^n$ be a bounded convex set. There exists an ellipsoid E and $x_0 \in K$ such that*

$$x_0 + E \quad \subset \quad K \subset \quad x_0 + nE.$$

Fig. 4.3 John's lemma

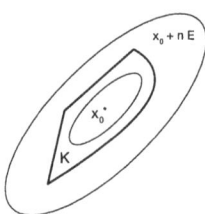

6. Caffarelli's interior regularity

Caffarelli [1–3] used the compactness modulo affine transformations of the family of solutions

$$det\, D^2 u = f \quad \text{in } \Omega, \quad u = 0 \quad \text{on } \partial\Omega, \quad \Lambda^{-1} \leq f \leq \Lambda,$$

to develop the Schauder and Calderon-Zygmund type estimates.

If u is a *strictly convex* solution of (4.2.1) then

$$\Lambda^{-1} \leq f \leq \Lambda \implies u \in C^{1,\alpha} \quad \text{for some} \quad \alpha(n, \Lambda) > 0,$$
$$f \in C^\alpha, \quad f > 0 \implies u \in C^{2,\alpha},$$
$$\Lambda^{-1} \leq f \leq \Lambda \implies u \in W^{2,1+\alpha} \quad \text{for some} \quad \alpha(n, \Lambda) > 0,$$
$$f \text{ continuous}, \quad f > 0 \implies u \in W^{2,p} \quad \text{for all } p < \infty.$$

The third implication is due to De Philippis and Figalli [12], see also [14, 24].

7. Partial regularity

Mooney [20] showed that solutions to (4.2.1) with $f \geq \Lambda^{-1}$ are strictly convex outside a set Σ, with

$$\mathcal{H}^{n-1}(\Sigma) = 0,$$

and the results of 6) apply in $\Omega \setminus \Sigma$. The Hausdorff dimension $n-1$ of Σ is optimal.

8. Boundary regularity

The Schauder and Calderon-Zygmund theory can be extended up to the boundary provided the boundary data separates quadratically from the tangent plane at a point $x_0 \in \partial\Omega$. Compactness of solutions is more delicate in this setting. One needs to show that sections centered at a boundary point x_0 are comparable to half-ellipsoids centered at x_0:

$$(x_0 + E) \cap \Omega \subset S_h(x_0) \subset x_0 + C(n, \Lambda)E.$$

The difference between interior and boundary regularity can be seen also from the behavior of global solutions. While in $\mathbb{R}^n$ global solutions are rigid (Calabi's theorem)

$$det\, D^2 u = 1 \quad \text{in } \mathbb{R}^n \quad \Longrightarrow \quad u \text{ is a quadratic polynomial,}$$

in $\mathbb{R}^n_+$ the family of global solutions

$$det\, D^2 u = 1 \quad \text{in } \mathbb{R}^n_+, \qquad u(x', 0) = \frac{1}{2}|x'|^2,$$

has non-quadratic solutions like

$$u(x_1, x_2) = \frac{1}{2}\frac{x_1^2}{1+x_2} + \frac{1}{2}x_2^2 + \frac{1}{6}x_2^3, \quad \text{in } \mathbb{R}^2_+.$$

9. **Degenerate Monge-Ampère equations**

If $f = 0$ or $f = \infty$ at some points in $\overline{\Omega}$ then we deal with a degenerate Monge-Ampère equation. Most of the previous results do not apply. We list a few such equations of interest and some references.

1.
$$det\, D^2 u \equiv 0 \quad \text{in } \Omega,$$

then u is the convex envelope of the boundary data.

2. Point degeneracy for f:

$$det\, D^2 u = |x|^2.$$

It is known that $u \in C^2$ (but not necessarily C^3) near the origin in dimension $n = 2$, see [11, 16]. It remains an open question to extend this result to general dimensions $n \geq 3$ (for strictly convex solutions u.)

3. Boundary degeneracy for f:

$$det\, D^2 u = f(x) \sim d^\alpha_{\partial\Omega}(x).$$

The boundary regularity of solutions was obtained in [22]. See also [19] where the smoothness of the eigenfunction for the Dirichlet problem was established.

4. The affine hyperbolic sphere equation

$$det\, D^2 u = |u|^{-(n+2)} \quad \text{in } \Omega, \ u = 0 \text{ on } \partial\Omega.$$

This problem appears in the celebrated work of Cheng and Yau [10].

5. The obstacle problem for the Monge-Ampère equation

$$det\ D^2u = \chi_{\{u>0\}}.$$

The equation is relevant in the study of isolated singularities for the Monge-Ampère equation. We refer to [18, 21] for a treatment of this free boundary problem.

10. The second boundary problem

The optimal transportation problem between two bounded convex sets Ω_1 and Ω_2 of the same measure reduces to

$$\begin{cases} det\ D^2u = 1 & \text{in } \Omega_1, \\ \nabla u(\Omega_1) = \Omega_2. \end{cases}$$

The main question concerns the global regularity of u.

Caffarelli [4, 5] showed that $u \in C^{1,\alpha}(\overline{\Omega}_1)$ for some $\alpha(n) > 0$.

Caffarelli [6] and Urbas [25] proved independently that $u \in C^{2,\alpha}(\overline{\Omega}_1)$ under stronger regularity assumptions: $\partial\Omega_i \in C^2$ and uniformly convex. These hypotheses were relaxed further to $\partial\Omega_i \in C^{1,1}$ by Chen-Liu-Wang [9].

In [23] we showed that $u \in W^{2,p}(\overline{\Omega}_1)$ for all $p < \infty$ in dimension $n = 2$.

11. The optimal transport problem for general costs

In the optimal transport problem for general costs $c(x, y)$, Brenier theorem implies that there exists a potential u which is $-c$ *convex*, in the sense that if x is transported into y then u has a tangent function $-c(\cdot, y) + const.$ by below at x. The point y is recovered from u by the relation

$$\nabla u(x) = -c_x(x, y) \implies y = T_c(x, \nabla u(x)).$$

After differentiating formally the left equality we find

$$D^2u + c_{xx}(x, y) = -c_{xy}(x, y)D_x y,$$

which means that at points where $u \in C^2$ it satisfies the Monge-Ampère type equation

$$det\left(D^2u + c_{xx}(x, y)\right) = det(-c_{xy}(x, y))\frac{f(x)}{g(y)}, \qquad y = T_c(x, \nabla u(x)).$$

Ma, Trudinger and Wang realized that the analysis for the standard Monge-Ampère equation can be carried out in this setting if the cost c satisfies the so-called MTW condition involving 4th order derivatives of c. On the other hand Loeper showed that the MTW condition is necessary for C^1 regularity.

De Philippis and Figalli [13] obtained the partial regularity of the transport map under mild regularity assumptions on c.

4.3 The C^2 Estimates

Here we provide the details for the global $C^{2,\alpha}$ estimates and interior C^2 estimates mentioned in the subsections 3) and 4) above.

Theorem 4.3.1 ($C^{2,\alpha}$ *Estimates*) *Assume $u \in C^4(\overline{\Omega})$ is convex and solves*

$$\begin{cases} \det D^2 u = f & \text{in } \Omega, \\ u = \varphi & \text{on } \partial\Omega, \end{cases}$$

with $f \in C^2(\overline{\Omega})$, $f > 0$, $\partial\Omega \in C^4$ uniformly convex, and $\varphi \in C^4$. Then

$$\|u\|_{C^{2,\alpha}(\overline{\Omega})} \leq C(f, \partial\Omega, \varphi).$$

Proof Since $F(M) = (\det M)^{\frac{1}{n}}$ is a concave operator on $\mathcal{S}_+^n$ it suffices to prove only the C^2 estimate

$$\|D^2 u\|_{L^\infty} \leq C,$$

with C depending on the data. Then $D^2 u \geq cI$ for some $c > 0$, and F becomes uniformly elliptic on the range of $D^2 u$. The $C^{2,\alpha}$ estimate is a consequence of the Evans-Krylov theorem, see [7].

Assume for simplicity that

$$f \equiv 1 \quad \text{and} \quad \Omega = B_1(e_n), \quad \text{so that } 0 \in \partial\Omega.$$

After subtracting a linear function we may assume further that

$$\nabla \varphi(0) = 0.$$

Step 1: $|u|, |\nabla u| \leq C$.
By maximum principle

$$-Cx_n + |x|^2 \leq u \leq Cx_n,$$

which gives $|u| \leq C$, $|\nabla u(0)| \leq C$. Similarly we can bound $|\nabla u|$ at all points on $\partial\Omega$, which by convexity gives the desired bound for ∇u in $\overline{\Omega}$.

Step 2: $|u_{ij}(0)| \leq C$ with $i, j < n$. The tangential second derivatives are bounded since

$$u_{ij}(0) = \varphi_{ij}(0) - u_n(0)\delta_{ij} \qquad i, j < n.$$

Step 3: $|u_{in}(0)| \leq C, i < n.$
For the mixed derivative $u_{in}(0)$ we use the tangential derivative

$$u_\tau := (1 - x_n)u_i + x_i u_n$$

which solves the linearized equation

$$Lu_\tau = 0 \quad \text{in } \Omega, \qquad u_\tau = \varphi_\tau \quad \text{on } \partial\Omega$$

where

$$Lv := U^{ij} v_{ij},$$

and U^{ij} is the cofactor matrix of $D^2 u$. Now compare u_τ with the explicit barriers

$$l \pm |x|^2, \qquad l \text{ linear,}$$

to find as in Step 1 that $|u_{in}(0)| = |\partial_{x_n} u_\tau(0)| \leq C$.

Step 4: $u_{nn}(0) \leq C$.
This follows from the equation once we prove that $D^2_{x'} u(0) \geq cI$. After subtracting the tangent plane of u at 0 we may assume we are in the situation

$$\nabla u(0) = 0, \qquad u = \tilde{\varphi}(x') \quad \text{on } \partial\Omega$$

for some function $\tilde{\varphi} \in C^4$. We want to show that $D^2_{x'} \tilde{\varphi}(0) \geq cI$.
Otherwise, if say $\tilde{\varphi}_{11}(0) = 0$, we find $\tilde{\varphi}(te_1) = O(t^4)$ which implies that for all small $h > 0$, the section $S_h(0) = \{u < h\}$ contains an ellipsoid E of axis

$$(ch^{1/4}, ch^{1/2}, .., ch^{1/2}, ch^{1/2}).$$

Then the solution to $\det D^2 v = 1$, $v = h$ on ∂E, is a quadratic polynomial which is strictly negative at the center of E. On the other hand $u \leq v$ in E by maximum principle and we contradict that $u \geq 0$.

Step 5. $|D^2 u| \leq C$ in $\overline{\Omega}$.
By Steps 2–4 the inequality holds on $\partial\Omega$. The concavity of F implies the pure second derivatives u_{ee} are subsolutions for the linearized operator $Lu_{ee} \geq 0$, and achieve the maximum on $\partial\Omega$. Thus $0 \leq u_{ee} \leq C$ in Ω. $\square$

4 The Monge-Ampère Equation

Next we provide an important estimate for u in terms of its Monge-Ampère measure in Ω.

Theorem 4.3.2 (Alexandrov's Estimate) *Assume that $\Omega \subset B_1$, and*

$$\det D^2 u = \mu \quad \text{in } \Omega, \qquad u = 0 \quad \text{on } \partial\Omega$$

Then

$$|u(x)| \le C_0 (d_{\partial\Omega}(x) \mu(\Omega))^{\frac{1}{n}},$$

with C_0 depending only on n.

Proof Let v be the function whose graph is the cone in $\mathbb{R}^{n+1}$ generated by the vertex $(x, u(x))$ and the cross-section $\partial\Omega \times \{0\}$. Then

$$\nabla v(x) \subset \nabla u(\Omega),$$

and, since $\Omega \subset B_1$, $\nabla v(x)$ contains an ellipsoid of axes

$$c_n \left(\frac{|u(x)|}{d_{\partial\Omega}(x)}, |u(x)|, ..., |u(x)| \right).$$

Looking at the measures of the two sets we find

$$c \frac{|u(x)|^n}{d_{\partial\Omega}(x)} \le \mu(\Omega).$$

□

Next we prove Pogorelov's interior estimate.

Theorem 4.3.3 *If u solves*

$$\begin{cases} \det D^2 u = 1 & \text{in } \Omega, \\ u = 0 & \text{on } \partial\Omega, \end{cases}$$

then

$$|D^2 u(x)| \le C(\delta, \Omega) \quad \text{in} \quad \{u < -\delta\}.$$

Remark 4.3.4 The result is true for more general RHS, $f \in C^2(\overline{\Omega})$, $f > 0$, with the constant C depending on f.

Proof Differentiating $\log(\det D^2 u) = 0$ in the unit e direction twice we obtain

$$Lu_e = 0, \qquad Lu_{ee} = u^{ik} u^{jl} u_{eij} u_{ekl}$$

where L denotes the linearized operator

$$Lv := u^{ij} v_{ij}$$

and u^{ij} are the entries of the inverse matrix $(D^2 u)^{-1}$. Here we used that

$$\partial_{a_{kl}} a^{ij} = -a^{ik} a^{jl}.$$

Let M be the maximum value of

$$u_{11} |u| e^{\frac{1}{2} u_1^2},$$

which is achieved at some point $x_0 \in \Omega$. Let

$$w = \log u_{11} + \log |u| + \frac{1}{2} u_1^2,$$

and after an affine deformation which leaves the e_1 direction invariant we may assume that $D^2 u$ is diagonal at x_0. At this point we have

$$w_i = 0, \qquad Lw \leq 0,$$

hence, by using the formulas for Lu_1, Lu_{11} above,

$$\frac{u_{11i}}{u_{11}} + \frac{u_i}{u} = 0, \quad \text{if } i \neq 1,$$

$$\sum_{i \neq 1} \frac{u_{1ij}^2}{u_{11} u_{ii} u_{jj}} + \frac{n}{u} - \frac{1}{u_{ii}} \frac{u_i^2}{u^2} + u_{11} \leq 0.$$

We find

$$\frac{n}{u} - \frac{1}{u_{11}} \frac{u_1^2}{u^2} + u_{11} \leq 0,$$

or

$$M^2 - n e^{\frac{1}{2} u_1^2} M \leq u_1^2 e^{u_1^2},$$

which shows that M is bounded above by a constant depending only on n and $\|u_1\|_{L^\infty}$. Now the result follows from the bound on $|\nabla u| \leq C(\delta, \Omega)$ in $\{u < -\delta\}$ which is a consequence of the Alexandrov estimate. $\square$

4.4 Compactness and Strict Convexity

Next we establish the compactness of strictly convex solutions of

$$det\ D^2 u = f, \qquad \Lambda^{-1} \le f \le \Lambda,$$

modulo affine transformations. We first prove John's lemma (see Lemma 4.2.3):

John's Lemma *Let $K \subset \mathbb{R}^n$ be a bounded convex set. There exists an ellipsoid E and $x_0 \in K$ such that*

$$x_0 + E \quad \subset \quad K \subset \quad x_0 + nE.$$

We say that K has the shape of E and write $K \sim E$.

Proof Let E be the ellipsoid of largest volume included in K. After an affine transformation we may assume that $K = B_1$ and need to prove that $K \subset B_n$. Otherwise we show that an ellipsoid of strictly larger volume is included in K.

Indeed, we claim that the ellipsoid of the same volume with $|B_1|$

$$(1+\varepsilon)^{\frac{2}{n-1}}|x'|^2 + (1+\varepsilon)^{-2}(x_n - \varepsilon)^2 = 1,$$

goes outside B_1 only in a neighborhood of $\partial B_1 \cap \{x_n \ge \frac{1}{n}\}$. Hence if $te_n \in K$ with $t > n$, then a slight deformation of the ellipsoid above is in K and has volume greater than $|B_1|$, a contradiction.

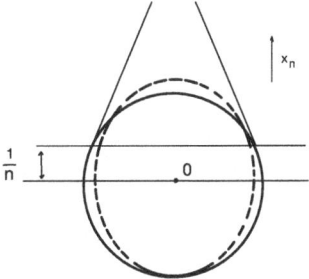

To prove the claim it suffices to look at the sign of the derivative in ε of the left hand side above (and use that $|x'|^2 + x_n^2 = 1$) to find

$$\frac{2}{n-1}(1 - x_n^2) - 2x_n^2 - 2x_n = \frac{2}{n-1}(1 + x_n)(1 - nx_n).$$

□

Recall that the section of center x_0 and height $h > 0$, is defined as

$$S_h(x_0) := \{u - l_{x_0} < h\}, \qquad l_{x_0}(x) := u(x_0) + \nabla u(x_0) \cdot (x - x_0).$$

Theorem 4.4.1 (Compactness) *Assume that*

$$\det D^2 u = f \quad \text{in } \Omega, \qquad \Lambda^{-1} \le f \le \Lambda,$$

and that

$$S_h(x_0) \subset\subset \Omega.$$

Then there exists a matrix A_h, $\det A_h = 1$ such that

$$c h^{\frac{1}{2}} A_h B_1 \subset S_h(x_0) - x_0 \subset C h^{\frac{1}{2}} A_h B_1,$$

for some positive constants c, C depending on n and Λ.

The theorem can be stated in the following way: the rescaled function

$$\tilde{u}(x) := \frac{1}{h} \left(u - l_{x_0} - h \right)(x_0 + h^{\frac{1}{2}} A_h x),$$

solves

$$\det D^2 \tilde{u} = \tilde{f}, \qquad \tilde{f} \in [\Lambda^{-1}, \Lambda], \tag{4.4.1}$$

with

$$\tilde{f}(x) := f(x_0 + h^{\frac{1}{2}} A_h x)$$

Moreover, the section $\tilde{S}_1(0)$ of center 0 and height 1 for $\tilde{u}$ is given by the level set

$$\tilde{S}_1(0) = \{\tilde{u} < 0\}, \tag{4.4.2}$$

and it satisfies the inclusions Fig. 4.4

$$c B_1 \subset \tilde{S}_1(0) \subset C B_1. \tag{4.4.3}$$

Proof By John's lemma we know that there exists y_0 such that $S_h(x_0) - y_0$ contains an ellipsoid E and is included in nE. Let A be a linear transformation such that $A B_1 = E$. Then the rescaled function

$$v(x) = |\det A|^{-\frac{2}{n}} (u - l_{x_0} - h)(y_0 + Ax)$$

4 The Monge-Ampère Equation

Fig. 4.4 Section rescaling

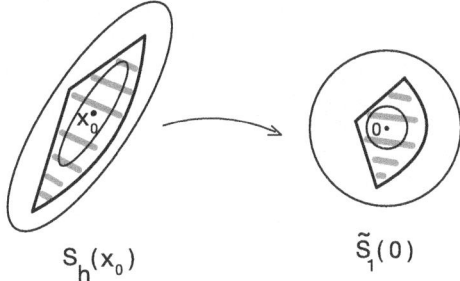

$S_h(x_0)$ $\tilde{S}_1(0)$

satisfies

$$\Lambda^{-1} \leq det\, D^2 v \leq \Lambda, \qquad B_1 \subset \{v < 0\} \subset B_n, \tag{4.4.4}$$

and the minimum of v occurs at z_0 defined by $x_0 = y_0 + A z_0$ and

$$v(z_0) = -h |det\, A|^{-\frac{2}{n}}.$$

By maximum principle, (4.4.4) implies that

$$c \leq |v(z_0)| \leq C \quad \Longrightarrow \quad c h^{\frac{n}{2}} \leq |det\, A| \leq C h^{\frac{n}{2}}.$$

Moreover, by Alexandrov estimate

$$B_{c'}(z_0) \subset \{v < 0\}.$$

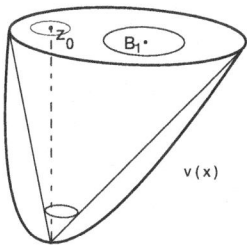

The theorem is proved since

$$S_h(x_0) - x_0 = A(\{v < 0\} - z_0).$$

□

Corollary 4.4.2 *Under the hypotheses of Theorem 4.4.1 we have that $S_h(x_0)$ is balanced with respect to x_0, i.e. for any line passing through x_0 that exits $S_h(x_0)$ at points y and z the ratio $|y - x_0|/|z - x_0|$ is bounded above and below by constants depending on n and Λ.*

We want to understand the situation when $S_h(x_0)$ is not compactly supported in Ω for any $h > 0$. In the Pogorelov's singular example, singularities lie on a line segment with end points on the boundary. The next proposition shows that the set where u is not strictly convex is generated from $\partial\Omega$.

Proposition 4.4.3 *Assume*

$$\Lambda^{-1} \leq \det D^2 u \leq \Lambda \quad \text{in } \Omega,$$

and let l_{x_0} be a supporting plane for u at some point x_0. Then either $\{u = l_{x_0}\}$ is equal to x_0 or it is a convex set supported on $\partial\Omega$.

Proof Assume by contradiction that $u \geq 0$, and $\{u = 0\}$ is a convex set that has an extremal point inside Ω, say for simplicity

$$0 \in \{u = 0\} \subset \{x_n < 0\} \cup \{0\}.$$

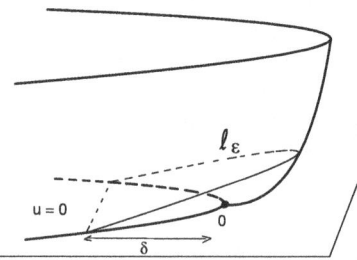

We look at the sections

$$S_\varepsilon := \{u < l_\varepsilon\}, \quad l_\varepsilon(x) = \varepsilon(x_n + \delta),$$

for some fixed $\delta > 0$ small. As $\varepsilon \to 0$ we find

$$S_\varepsilon \quad \to \quad \{u = 0\} \cap \{x_n \geq -\delta\},$$

(in the Hausdorff distance sense) and therefore $S_\varepsilon \subset\subset \Omega$.

Moreover, the minimum of $u - l_\varepsilon$ occurs at a point

$$x_\varepsilon \in S_\varepsilon \cap \{x_n \geq 0\},$$

which is the center of the section S_ε. Then $x_\varepsilon \to 0$ as $\varepsilon \to 0$, and S_ε cannot be balanced with respect to x_ε for small ε. We reached a contradiction. □

4.5 Interior $C^{1,\alpha}$ and $C^{2,\alpha}$ Estimates

We use the results of the previous section to derive interior estimates for strictly convex solutions. We focus in the interior of a section $S_h(x_0) \subset\subset \Omega$, which after normalization can be taken such that $x_0 = 0$, $h = 1$ and $S_1(0) \sim B_1$.

Theorem 4.5.1 (Interior $C^{1,\alpha}$ Estimates) *Assume that $u(0) = 0$,*

$$\Lambda^{-1} \le \det D^2 u \le \Lambda \quad \text{in } S_1(0),$$

and $S_1(0)$ is normalized such that

$$B_{c_0} \subset S_1(0) \subset B_{C_0},$$

for some c_0, C_0 depending on n and Λ. Then

$$\|u\|_{C^{1,\alpha}(S_{1/2}(0))} \le C,$$

with $\alpha > 0$, C depending on n and Λ.

Proof We claim that there exists $\delta(n, \Lambda) > 0$ such that

$$(\frac{1}{2} + \delta) S_1(0) \quad \subset \quad S_{1/2}(0) \quad \subset \quad (1 - \delta) S_1(0).$$

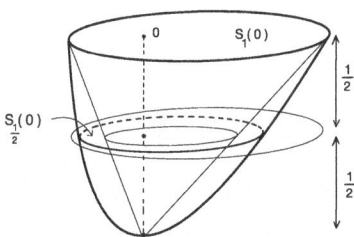

By compactness, we only need to prove the existence of some $\delta > 0$ for a given function u. Indeed, if the left inclusion does not hold for any $\delta > 0$ then the graph of u contains a line segment and we contradict Proposition 4.4.3. The right inclusion holds by Alexandrov's estimate.

By iterating the inclusion above we find

$$(\frac{1}{2} + \delta)^m B_{c_0} \quad \subset \quad S_{2^{-m}}(0) \quad \subset \quad (1-\delta)^m B_{C_0},$$

which implies

$$c|x|^M \le u \le C|x|^{1+\alpha}. \tag{4.5.1}$$

This shows that u is pointwise $C^{1,\alpha}$ at 0, and it also has a polynomial modulus of convexity at 0.

□

The sections $S_h(x) \subset\subset \Omega$ of a function u with bounded determinant play the same role as Euclidean balls for harmonic functions. They satisfy a similar engulfing property (which follows as in the proof above by compactness):

There exist t small, depending on n and Λ such that if $S_h(x) \subset\subset \Omega$ then

$$y \in S_{th}(x) \implies S_{th}(x) \subset S_{h/2}(y) \subset S_h(x).$$

As a consequence we have a Vitali type covering lemma.

Lemma 4.5.2 (Vitali Covering Lemma) *Assume that $u \in C(\Omega)$ is strictly convex and satisfies*

$$\Lambda^{-1} \leq \det D^2 u \leq \Lambda.$$

Let

$$\{S_{h_x}(x)\}_{x \in K}, \qquad K \subset\subset \Omega$$

be a collection of sections compactly included in Ω. There exists a finite cover of K

$$K \subset \cup S_{h_i}(x_i) \qquad \text{with } S_{ch_i}(x_i) \text{ disjoint,}$$

and $c > 0$ depending only on n and Λ.

The proof is the same as in the Euclidean case. We first cover K with a finite number of sections from the collection. Then we pick a maximal subcover from this collection inductively so that the $S_{ch_i}(x_i)$'s are disjoint from one-another and the h_i are chosen in decreasing order. We leave the details to the reader.

A Homogenous Example Let $u(x, y)$ be a function of 2 variables which satisfies

$$u(x, y) = \lambda^{-1} u(\lambda^\beta x, \lambda^{1-\beta} y), \qquad \forall \lambda > 0, \qquad \beta \in (0, 1),$$

i.e. u is homogenous of degree $1/\beta$ in x and $1/(1 - \beta)$ in y. Then u is uniquely determined from its values on the unit circle, and $\det D^2 u$ is constant on the curves

$$\{(\lambda^\beta x, \lambda^{1-\beta} y), \quad \lambda > 0\}.$$

It is not difficult to show that for any $\beta \in (0, 1)$ there exists such a convex function with determinant bounded away from 0 and ∞. This shows that the interior $C^{1,\alpha}$ estimate is optimal in the sense that $\alpha \to 0$ as $\Lambda \to \infty$.

4 The Monge-Ampère Equation

Theorem 4.5.3 (Interior $C^{2,\alpha}$ Estimates) *Assume that $u(0) = 0$,*

$$\det D^2 u = f \quad \text{in } S_1(0), \quad \Lambda^{-1} \leq f \leq \Lambda,$$

and $S_1(0)$ is normalized such that

$$B_{c_0} \subset S_1(0) \subset B_{C_0},$$

for some c_0, C_0 depending on n and Λ.
If $f \in C^\alpha$, then $u \in C^{2,\alpha}$ and

$$\|u\|_{C^{2,\alpha}(S_{1/2}(0))} \leq C(n, \Lambda, \|f\|_{C^\alpha}).$$

We only sketch the proof of this estimate.

Step 1. There exists $h_0 > 0$ small depending on n, Λ, $\|f\|_{C^\alpha}$ such that the rescaled solution

$$\tilde{u} = \frac{1}{h_0} u(A_0 x),$$

satisfies (see (4.4.1))

$$\det D^2 \tilde{u} = \tilde{f} \quad \text{in } \tilde{S}_1, \quad B_{c_0} \subset \tilde{S}_1(0) \subset B_{C_0},$$

and

$$[\tilde{f}]_{C^\alpha} \leq \varepsilon_0,$$

with $\varepsilon_0 = \varepsilon_0(n, \alpha)$ sufficiently small.
This follows from Theorem 4.4.1 and then we use

$$[\tilde{f}]_{C^\alpha} \leq |A_0|^\alpha [f]_{C^\alpha}.$$

Notice that (4.5.1) implies that $|A_0| \to 0$ as $h_0 \to 0$.

Step 2. By Step 1, it suffices to show the estimate in the setting

$$\|f - 1\|_{C^\alpha} \leq \varepsilon_0, \qquad f(0) = 1,$$

with ε_0 sufficiently small. Then there exists $h_1 > 0$ such that the rescaled solution

$$\tilde{u} = \frac{1}{h_1} u(A_1 x),$$

satisfies

$$\det D^2 \tilde{u} = \tilde{f} \quad \text{in } \tilde{S}_1, \qquad \|\tilde{u} - \frac{1}{2}|x|^2\|_{L^\infty(\tilde{S}_1(0))} \le \varepsilon_1$$

and

$$[\tilde{f}]_{C^\alpha} \le \varepsilon_1,$$

with $\varepsilon_1 = \varepsilon_1(n, \alpha)$ sufficiently small.

This is essentially the same as Step 1, except we also need to show that $\tilde{u}$ is well-approximated by a quadratic polynomial.

Let u_0 be the solution to the Dirichlet problem in $S_1(0)$ with right hand side 1 and same boundary data as u. Then

$$\|u - u_0\|_{L^\infty} \le C\varepsilon_0,$$

by maximum principle. Since $u_0 \in C^3$ it follows that

$$\|u_0 - P\| \le C|x|^3, \qquad \det D^2 P = 1,$$

with P a quadratic polynomial with bounded coefficients. The claim follows from the inequalities above by first choosing h_1 depending on ε_1, and then choosing ε_0 small depending on h_1, ε_1.

Step 3. After a rescaling we may assume that u satisfies the conclusion of Step 2 for some small ε_1 depending on n and α.

Now we are in a perturbative setting, and we can use the Campanato's method to show inductively that u is approximated by quadratic polynomials in a $C^{2,\alpha}$ fashion at all smaller scales, i.e.

$$\|u - P_k\|_{L^\infty(B_{r_k})} \le C\varepsilon_1 r_k^{2+\alpha} \qquad r_k = \rho^k, \quad k = 0, 1, 2, ..$$

with P_k quadratic polynomials.

Applications *Regularity in the Minkowski's Problem.* Assume that on the boundary Σ of a closed convex set in $K \subset \mathbb{R}^{n+1}$ the Gauss curvature is given as a positive smooth function depending on the position and the normal. If Σ is given by the graph of a C^2 function u locally, then the equation can be written in the form

$$\det D^2 u = g(x, u, \nabla u)$$

with $g > 0$, g smooth in its variables. We want to show that Σ must be smooth.

In order to apply the interior $C^{1,\alpha}$ and $C^{2,\alpha}$ estimates above we need first to show that Σ is a strictly convex surface. Assume by contradiction that this is not true. Then we can find a point $x_0 \in \Sigma$ which is extremal for K and it is the end point of a line segment lying in Σ. Here by "x_0 extremal for K" we understand that there exist hyperplanes which cut K in two parts, with one part included in an arbitrarily small neighborhoods of x_0. In such a region, Σ is given by the graph of a Lipschitz function u defined on a convex set, which is linear on its boundary. Moreover, $\det D^2 u$ and the Gauss curvature are comparable, hence u has determinant bounded by above and below. By Theorem 4.5.1, u is strictly convex in the interior which means Σ is strictly convex near x_0, and we reached a contradiction.

The same argument shows that $\Sigma \in C^{1,\alpha}$ locally. Then the right hand side above is Hölder continuous, which gives $u \in C^{2,\alpha}$ by Theorem 4.5.3. Now the standard Schauder estimates apply and we obtain $u \in C^\infty$.

Classification of Global Solutions We show that the only global solutions to

$$\det D^2 u = 1 \quad \text{in} \quad \mathbb{R}^n,$$

are the quadratic polynomials (Calabi's theorem).

Assume for simplicity that $u(0) = 0$, $u \geq 0$. We claim that the convex set $\{u = 0\}$ is a single point, the origin. Indeed, if this set contains a whole line say in the e_1 direction then u must be constant in this direction, i.e. ∇u belongs to a hyperplane. Then the equation cannot hold in the Alexandrov sense and we reach a contradiction. Thus $\{u = 0\}$ does not contains whole line. If $\{u = 0\}$ is not a singleton then it must contain an extremal point. However, this contradicts Proposition 4.4.3 and we proved the claim.

Since $\{u = 0\} = 0$, the sections S_h are compact, and $u \in C^2$ by Pogorelov's interior estimate Theorem 4.3.3. The rescaled function (see (4.4.1))

$$\tilde{u}(x) = \frac{1}{h} u(h^{1/2} A_h x),$$

satisfies the same equation in $\tilde{S}_1(0) \sim B_1$. Then $|D^2 \tilde{u}(0)| \leq C$ implies $|A_h^{-1}| \leq M$ for some constant M depending on $|D^2 u(0)|$ but independent of h. We have

$$\|D^3 u\|_{L^\infty(S_{h/2})} \leq h^{-\frac{1}{2}} M^3 \|D^3 \tilde{u}\|_{L^\infty(\tilde{S}_{1/2})} \leq C M^3 h^{-\frac{1}{2}}.$$

We let $h \to \infty$ and obtain $D^3 u = 0$ which gives the desired conclusion.

4.6 Interior $W^{2,p}$ Estimates

First we discuss the interior $W^{2,p}$ estimates in the case when f has small oscillation.

Theorem 4.6.1 (Interior $W^{2,p}$ Estimates) *Let $p \in (1, \infty)$, and assume $u(0) = 0$,*

$$\det D^2 u = f \quad \text{in } S_1(0), \qquad \|f - 1\|_{L^\infty} \leq \varepsilon_0(p),$$

with $\varepsilon_0(p)$ sufficiently small, and $S_1(0)$ normalized such that

$$B_{c_0} \subset S_1(0) \subset B_{C_0},$$

for some c_0, C_0 depending on n. Then $u \in W^{2,p}_{loc}$ and

$$\|u\|_{W^{2,p}(S_{1/2}(0))} \leq C(n).$$

If $f > 0$ is continuous and u is strictly convex, then we can always localize at an interior point $x_0 \in \Omega$ and rescale a section $S_h(x_0) \subset\subset \Omega$ with h small, to satisfy the hypotheses of the theorem above. As a consequence we have the following corollary.

Corollary 4.6.2 *If u is a strictly convex solution of*

$$\det D^2 u = f > 0 \quad \text{in} \quad \Omega,$$

and f is continuous, then $u \in W^{2,p}_{loc}(\Omega)$ for any $p < \infty$.

Before we prove Theorem 4.6.1 we need the following result.

Lemma 4.6.3 *Assume that $u \in C^2$ satisfies*

$$\det D^2 u = f \quad \text{in } S_1(0),$$

with

$$\|f - 1\|_{L^\infty(S_1(0))} \leq \varepsilon, \qquad \|u - \frac{1}{2}|x|^2\|_{L^\infty(S_1(0))} \leq \varepsilon.$$

Then

$$\left|\{|D^2 u - I| \geq \frac{1}{10}\} \cap S_{1/2}(0)\right| \leq C\varepsilon^{1/2}.$$

Proof The hypotheses imply that

$$(1 - C\varepsilon) B_{\sqrt{2}} \subset S_1(0) \subset (1 + C\varepsilon) B_{\sqrt{2}}.$$

Denote by

$$P := \frac{1}{2}|x|^2,$$

and the convexity of u and $\|u - P\|_{L^\infty} \leq \varepsilon$ implies

$$|\nabla u - \nabla P| \leq C\varepsilon^{1/2} \quad \text{in} \quad B_1.$$

4 The Monge-Ampère Equation

We have
$$\int_{B_1} \Delta(u - P)\, dx = \int_{\partial B_1} u_\nu - P_\nu\, d\sigma = O(\varepsilon^{1/2}).$$

The arithmetic geometric mean inequality gives
$$\Delta u \geq n\,(\det D^2 u)^{\frac{1}{n}} = n\,(\det D^2 P)^{\frac{1}{n}} + O(\varepsilon) = \Delta P + O(\varepsilon),$$

hence
$$\int_{B_1} \Delta u - n\,(\det D^2 u)^{\frac{1}{n}}\, dx = O(\varepsilon^{1/2}),$$

which gives the desired conclusion. $\square$

Proof of Theorem 4.6.1 Without loss of generality we assume that $u \in C^2$. Then the general result follows by approximation.

As in Step 2 of the proof of Theorem 4.5.3 we know that for each $x_0 \in S_{1/2}(0)$, the function u in the section $S_h(x_0)$ with $h \leq c$, c sufficiently small, can be normalized by an affine transformation T_h,

$$T_h:\quad S_h(x_0) - x_0 \;\longmapsto\; \tilde{S}_1(0), \qquad T_h = h^{-\frac{1}{2}} A_h^{-1}, \qquad \det A_h = 1,$$

so that the rescaled function $\tilde{u}$ satisfies the hypotheses of Lemma 4.6.3 for some ε sufficiently small. Notice that

$$\lim_{h \to 0} |A_h^{-1}|^2 \sim |D^2 u(x_0)|, \qquad |A_{2h}^{-1} A_h| \leq 1 + C\varepsilon. \tag{4.6.1}$$

For M large we denote by K_M the compact set
$$K_M := \{|D^2 u(x)| \geq M\} \cap \overline{S_{1/2}(0)},$$

and we show that its measure, $|K_M|$, decays fast as $M \to \infty$. Precisely, we claim that we have

$$|K_{2M}| \leq C\varepsilon^{1/2}|K_M|. \tag{4.6.2}$$

This gives the desired L^p bound for $D^2 u$ by choosing $\varepsilon = \varepsilon(p)$ small.

For each $x \in K_{2M}$ we choose a section $S_{2h}(x)$ such that
$$|A_{2h}^{-1}|^2 \in [\tfrac{4}{3}M, \tfrac{5}{3}M].$$

The existence of such a section with $h \leq c$ follows by (4.6.1).

Notice that after rescaling we have the inclusion,

$$\{|D^2\tilde{u} - I| \le \frac{1}{10}\} \cap \tilde{S}_{1/2}(0) \quad \subset \quad T_{2h}\left(\{M < |D^2u| < 2M\} \cap S_h(x)\right).$$

We apply Vitali's covering lemma for the collection of sections $S_h(x)$, $x \in K_{2M}$ and obtain

$$K_{2M} \subset \cup S_{h_i}(x_i), \quad \text{with } S_{\delta h_i}(x_i) \text{ disjoint.}$$

By Lemma 4.6.3 applied in $S_{2h_i}(x_i)$ (after rescaling) we find that

$$|K_{2M} \cap S_{h_i}(x_i)| \le C\varepsilon^{1/2}|S_{h_i}(x_i)|,$$

and

$$c|S_{h_i}(x_i)| \le \frac{1}{2}|S_{\delta h_i}(x_i) \cap S_{1/2}(0)| \le |S_{\delta h_i} \cap K_M|.$$

In the first part of the last inequality we used the engulfing properties of sections. Now (4.6.2) follows by adding these inequalities for all i. □

Convex functions have the property that the mass of $|D^2u|$ is locally bounded since

$$\int_B |D^2u|\,dx \le \int_B \Delta u\,dx = \int_{\partial B} u_\nu\,d\sigma, \qquad (4.6.3)$$

and the last integral is bounded since u is locally Lipschitz. In the next theorem we improve this result for functions u that have Monge-Ampère measure bounded away from 0 and ∞, and show that $D^2u \in L^{1+\varepsilon}$ for some small $\varepsilon > 0$.

Theorem 4.6.4 (Interior $W^{2,1+\varepsilon}$ Estimates) *Assume $u(0) = 0$,*

$$\Lambda^{-1} \le det\,D^2u \le \Lambda \quad in\ S_1(0),$$

and $S_1(0)$ normalized such that
for some c_0, C_0 depending on n and Λ. Then $u \in W^{2,1+\varepsilon}_{loc}$ for some $\varepsilon = \varepsilon(n, \Lambda) > 0$ small and

$$\|u\|_{W^{2,+\varepsilon}(S_{1/2}(0))} \le C.$$

Proof As before we denote by

$$K_M := \{|D^2u| \ge M\} \cap \overline{S_{1/2}(0)}.$$

4 The Monge-Ampère Equation

It suffices to show that

$$\int_{K_{C_0 M}} |D^2 u| \, dx \le C \int_{K_M \setminus K_{C_0 M}} |D^2 u| \, dx. \qquad (4.6.4)$$

for some large universal constants C_0, C. Indeed, then

$$\int_{K_{C_0 M}} |D^2 u| \, dx \le \frac{C}{C+1} \int_{K_M} |D^2 u| \, dx,$$

which gives the geometric decay of the truncated integrals,

$$\int_{K_M} |D^2 u| \, dx \le C' M^{-2\varepsilon},$$

and the $L^{1+\varepsilon}$ bound of $|D^2 u|$ follows easily.

We prove (4.6.4) by arguing similarly as in the proof of Theorem 4.6.1. For each $x \in K_{C_0 M}$ we consider the section $S_{2h}(x)$ such that

$$|A_{2h}^{-1}|^2 \in [C_0^{\frac{1}{3}} M, C_0^{\frac{2}{3}} M]. \qquad (4.6.5)$$

We claim that

$$\int_{S_h(x)} |D^2 u| \, dx \le CM |S_h(x)|, \qquad (4.6.6)$$

and

$$|S_{\delta h}(x) \cap (K_M \setminus K_{C_0 M})| \ge c |S_h(x)|. \qquad (4.6.7)$$

The last inequality implies

$$\int_{S_{\delta h}(x) \cap (K_M \setminus K_{C_0 M})} |D^2 u| \, dx \ge c' M |S_h(x)|.$$

Then (4.6.4) follows by taking a Vitalli cover of $K_{C_0 M}$ with sections $S_{h_i}(x_i)$ and then summing up the integral inequalities above over i.

In order to prove (4.6.6) and (4.6.7) it suffices to show that,

$$\int_{\tilde{S}_{1/2}(0)} |D^2 \tilde{u}| \, dx \le C, \qquad (4.6.8)$$

where $\tilde{u}$ denotes the renormalized solution of u in $S_{2h}(x)$. This implies that we also have

$$|\{|D^2\tilde{u}| \leq C_1\} \cap \tilde{S}_{c\delta}(y)| \geq c'', \quad \forall y \in \tilde{S}_{1/2}(0), \tag{4.6.9}$$

for some C_1 large universal. Then we use that

$$D^2 u = (A_{2h}^{-1})^T D^2\tilde{u} \, A_{2h}^{-1},$$

together with (4.6.5) to obtain

$$|D^2 u| \leq CM|D^2\tilde{u}|.$$

Moreover, since $det\, D^2\tilde{u}$ is bounded, we find

$$|D^2\tilde{u}| \leq C_1 \implies c_1 I \leq D^2\tilde{u} \leq C_1 I \implies M < |D^2 u| < C_0 M,$$

provided that we first choose C_0 large depending on c_1, C_1. Now (4.6.6) and (4.6.7) can be deduced from (4.6.8), (4.6.9).

It remains to prove (4.6.6). However, this follows directly from the computation (4.6.3). Indeed, we replace B by $\tilde{S}_{1/2}(0)$, and recall that $\tilde{u}$ has bounded Lipschitz norm in $\tilde{S}_{1/2}(0)$.

□

References

1. Caffarelli, L.: A localization property of viscosity solutions to the Monge-Ampère equation and their strict convexity. Ann. Math. **131**(1), 129–134 (1990)
2. Caffarelli, L.: An interior $W^{2,p}$ estimates for solutions of the Monge-Ampère equation. Ann. Math. **131**(1), 135–150 (1990)
3. Caffarelli, L.: Some regularity properties of solutions of Monge-Ampère equation. Commun. Pure Appl. Math. **44**(8–9), 965–969 (1991)
4. Caffarelli, L.: The regularity of mappings with a convex potential. J. Am. Math. Soc. **5**(1), 99–104 (1992)
5. Caffarelli, L.: Boundary regularity of maps with convex potentials. Commun. Pure Appl. Math. **45**(9), 1141–1151 (1992)
6. Caffarelli, L.: Boundary regularity of maps with convex potentials II. Ann. Math. **144**(3), 453–496 (1996)
7. Caffarelli, L., Cabrè, X.: Fully Nonlinear Elliptic Equations, vol. 43, College Publications, American Mathematical Society, Providence (1995)
8. Caffarelli L., Nirenberg, L., Spruck, J.: The Dirichlet problem for nonlinear second-order elliptic equations. I. Monge-Ampère equation. Commun. Pure Appl. Math. **37**(3), 369–402 (1984)
9. Chen, S., Liu, J., Wang, X.J.: Global regularity for the Monge-Ampère equation with natural boundary condition (2018). arXiv:1802.07518

10. Cheng, S.Y., Yau, S.T.: On the regularity of the Monge-Ampère equation $det(\partial^2 u/\partial x_i \partial x_j) = F(x, u)$. Commun. Pure Appl. Math. **30**, 41–68 (1977)
11. Daskalopoulos, P., Savin, O.: On Monge-Ampère equations with homogenous right-hand sides. Commun. Pure Appl. Math. **62**(5), 639–676 (2009)
12. De Philippis, G., Figalli, A.: $W^{2,1}$ regularity for solutions of the Monge-Ampère equation. Invent. Math. **192**(1), 55–69 (2013)
13. De Philippis, G., Figalli, A.: Partial regularity for optimal transport maps. Publ. Math. Inst. Hautes Études Sci. **121**, 81–112 (2015)
14. De Philippis, G., Figalli, A., Savin, O.: A note on interior $W^{2,1+\varepsilon}$ estimates for the Monge-Ampère equation. Math. Ann. **357**(1), 11–22 (2013)
15. Figalli, A.: The Monge-Ampère Equation and its Applications. Zurich Lectures in Advanced Mathematics. European Mathematical Society, Zürich (2017)
16. Guan, P.: Regularity of a class of quasilinear degenerate elliptic equations. Adv. Math. **132**(1), 2445 (1997)
17. Gutiérrez, C.: The Monge-Ampère Equation. Progress in Nonlinear Differential Equations and Their Applications, vol. 89. Birkhäuser, Cham (2016)
18. Huang, G., Tang, L., Wang, X.J.: Regularity of free boundary for the Monge-Ampère obstacle problem. arXiv:2111.10575
19. Le, N.Q., Savin, O.: Schauder estimates for degenerate Monge-Ampère equations and smoothness of the eigenfunctions. Invent. Math. **207**(1), 389–423 (2017)
20. Mooney, C.: Partial regularity for singular solutions to the Monge-Ampère equation. Commun. Pure Appl. Math. **68**(6), 1066–1084 (2014)
21. Savin, O.: The obstacle problem for Monge-Ampère equation. Calc. Var. Partial Differ. Equ. **22**(3), 303–320 (2005)
22. Savin, O.: A localization theorem and boundary regularity for a class of degenerate Monge-Ampère equations. J. Differ. Equ. **256**(2), 327–388 (2014)
23. Savin, O., Yu, H.: Regularity of optimal transport between planar convex domains (2018). arXiv:1806.06252
24. Schmidt, T.: $W^{2,1+\varepsilon}$ estimates for the Monge-Ampère equation. Adv. Math. **240**, 672–689 (2013)
25. Urbas, J.: On the second boundary value problem for equations of Monge-Ampère type. J. Reine Angew. Math. **487**, 115–124 (1997)

Chapter 5
Injective Ellipticity, Cancelling Operators, and Endpoint Gagliardo-Nirenberg-Sobolev Inequalities for Vector Fields

Jean Van Schaftingen

Abstract Although Ornstein's nonestimate entails the impossibility to control in general all the L^1-norm of derivatives of a function by the L^1-norm of a constant coefficient homogeneous vector differential operator, the corresponding endpoint Sobolev inequality has been known to hold in many cases: the gradient of scalar functions (Gagliardo and Nirenberg), the deformation operator (Korn-Sobolev inequality by M.J. Strauss), and the Hodge complex (Bourgain and Brezis). The class of differential operators for which estimates holds can be characterized by a cancelling condition. The proof of the estimates rely on a duality estimate for L^1-vector fields lying in the kernel of a cocancelling differential operator, combined with classical linear algebra and harmonic analysis techniques. This characterization unifies classes of known Sobolev inequalities and extends to fractional Sobolev and Hardy inequalities. A similar weaker condition introduced by Raiță characterizes the operators for which there is an L^∞-estimate on lower-order derivatives.

5.1 Introduction

The central question in the present notes is to determine for which constant coefficient homogeneous vector differential operator $A(D)$ an estimate of the form

$$\|D^\ell u\|_{L^q(\mathbb{R}^n)} \le \|A(D)u\|_{L^1(\mathbb{R}^n)} \tag{5.1}$$

holds for every vector field $u \in C_c^\infty(\mathbb{R}^n, V)$. After reviewing how classical harmonic analysis, including Fourier analysis and singular integral theory settle the case where $L^p(\mathbb{R}^n)$ with $p \in (1, +\infty)$ replaces $L^1(\mathbb{R}^n)$ in the right-hand

---

J. Van Schaftingen (✉)
Université Catholique de Louvain (UCLouvain), Institut de Recherche en Mathématique et Physique (IRMP), Louvain-la-Neuve, Belgium
e-mail: Jean.VanSchaftingen@UCLouvain.be

© The Author(s), under exclusive license to Springer Nature Switzerland AG 2024
A. Cianchi et al. (eds.), *Geometric and Analytic Aspects of Functional Variational Principles*, C.I.M.E. Foundation Subseries 2348,
https://doi.org/10.1007/978-3-031-67601-7_5

side of (5.1) and how Ornstein's nonestimate dooms this approach for (5.1) in Sect. 5.2, we present the cancellation condition characterizing Bourgain-Brezis endpoint Sobolev estimates in Sect. 5.3 and the corresponding duality estimates for cocancelling operators in Sect. 5.4. Finally, in Sect. 5.5 we connect the duality estimates with functions of bounded mean oscillation, we study the corresponding fractional Sobolev and Hardy inequalities, and we present Raiţă's characterization of operators for which uniform estimates holds.

The original material covered in these notes corresponds essentially to the references [12, 53, 71, 72, 74], complemented by the survey article [75] and the more informal lecture notes [76].

5.2 Estimates for Vector Elliptic Operators

The central topic of the present lecture notes is *constant coefficients vector differential operators*.

Definition 5.1 Given $n \in \mathbb{N} \setminus \{0\}$ and finite-dimensional (real) vector spaces V and E, $A(\mathrm{D})$ is a *homogeneous constant coefficient differential operator of order* $k \in \mathbb{N} \setminus \{0\}$ *from V to E on $\mathbb{R}^n$* whenever there exists $A \in \mathrm{Lin}(\mathrm{Lin}_{\mathrm{sym}}^k(\mathbb{R}^n, V), E)$ such that for every $u \in C^\infty(\mathbb{R}^n, V)$ and every $x \in \mathbb{R}^n$, one has $A(\mathrm{D})u(x) = A[\mathrm{D}^k u(x)]$.

In other words, $A(\mathrm{D})$ is a homogeneous constant coefficient differential operator of order $k \in \mathbb{N} \setminus \{0\}$ from V to E on $\mathbb{R}^n$ when there exists a linear map $A : \mathrm{Lin}_{\mathrm{sym}}^k(\mathbb{R}^n, V) \to E$ such that the differential operator $A(\mathrm{D})$ can be represented at any point $x \in \mathbb{R}^n$ as the linear map A applied to the total k-th order derivative $\mathrm{D}^k u : \mathbb{R}^n \to \mathrm{Lin}_{\mathrm{sym}}^k(\mathbb{R}^n, V)$, seen as a function from $\mathbb{R}^n$ to symmetric k-linear maps from $(\mathbb{R}^n)^k$ to V.

In terms of partial derivatives, $A(\mathrm{D})$ is a homogeneous constant coefficient differential operator of order $k \in \mathbb{N} \setminus \{0\}$ from V to E on $\mathbb{R}^n$ if and only if for each multiindex $\alpha = (\alpha_1, \ldots, \alpha_n) \in \mathbb{N}^n$ satisfying $|\alpha| := \alpha_1 + \cdots + \alpha_n = k$ there exists a linear map $A_\alpha \in \mathrm{Lin}(V, E)$, such that for every $u \in C^\infty(\mathbb{R}^n, V)$ and for every $x \in \mathbb{R}^n$, we have

$$A(\mathrm{D})u(x) = \sum_{\substack{\alpha \in \mathbb{N}^n \\ |\alpha|=k}} A_\alpha[\partial^\alpha u(x)] . \tag{5.2}$$

Alternatively, the representation (5.2) can be written as

$$A(\mathrm{D})u(x) = \sum_{\substack{\alpha \in \mathbb{N}^n \\ |\alpha|=k}} \partial^\alpha (A_\alpha[u])(x) ,$$

5 Injective Ellipticity, Cancelling Operators and Endpoint Sobolev Inequalities

since A_α does not depend on the variable x and commutes thus with the partial derivative ∂^α. Our question in the present notes will be to determine when, whether and how a k-th order differential operator $A(\mathrm{D})u$ is a good substitute for the total k-th order derivative $\mathrm{D}^k u$.

5.2.1 Injective Ellipticity

We first examine what the Fourier theory tells us about u in terms of $A(\mathrm{D})u$. Given a Schwartz test function $u \in \mathcal{S}(\mathbb{R}^n, V)$, its Fourier transform $\mathcal{F}u : \mathbb{R}^n \to V + iV$ (here $V + iV$ is understood as the complexification of the linear space V) is defined for every $\xi \in \mathbb{R}^n$ as

$$(\mathcal{F}u)(\xi) := \int_{\mathbb{R}^n} e^{-2\pi i \langle \xi, x \rangle} u(x) \, \mathrm{d}x \ .$$

By classical properties of the Fourier transform, we have for every $\xi \in \mathbb{R}^n$

$$\mathcal{F}(A(\mathrm{D})u)(\xi) = A\big((2\pi i)^k \mathcal{F}u(\xi) \otimes \xi^{\otimes k}\big) = (2\pi i)^k A(\xi)[\mathcal{F}u(\xi)] \ , \tag{5.3}$$

where $A(\xi) \in \mathrm{Lin}(V+iV, E+iE)$ also denotes the complexification of the operator $A(\xi)$, which is defined for every $v \in V$ by

$$A(\xi)[v] := A(v \otimes \xi^{\otimes k}) \tag{5.4}$$

and is the symbol of the differential operator $A(\mathrm{D})$. Equivalently, the identity (5.4) can be rewritten by the representation formula (5.2) as

$$A(\xi) = \sum_{\substack{\alpha \in \mathbb{N}^n \\ |\alpha|=k}} \xi^\alpha A_\alpha \in \mathrm{Lin}(V, E) \ , \tag{5.5}$$

where for every $\alpha = (\alpha_1, \ldots, \alpha_n) \in \mathbb{N}^n$ and $\xi = (\xi_1, \ldots, \xi_n) \in \mathbb{R}^n$, we have written $\xi^\alpha := \xi_1^{\alpha_1} \cdots \xi_n^{\alpha_n} \in \mathbb{R}$. By the Parseval identity for the Fourier transform and by the identity (5.3), we observe that

$$\int_{\mathbb{R}^n} |A(\mathrm{D})u|^2 = (2\pi)^{2k} \int_{\mathbb{R}^n} |A(\xi)[\mathcal{F}u(\xi)]|^2 \, \mathrm{d}\xi \ . \tag{5.6}$$

In order to use the right-hand side of the inequality (5.6) to control derivatives of u, we will use the *injective ellipticity* condition, which will guarantee that the integrand in the right-hand side integral of (5.6) does not vanish at $\xi \in \mathbb{R}^n \setminus \{0\}$ unless $\mathcal{F}u(\xi)$ does.

Definition 5.2 (Injective Ellipticity[1]**)** Given $n \in \mathbb{N} \setminus \{0\}$ and finite-dimensional vector spaces V and E, a homogeneous constant coefficient differential operator $A(\mathrm{D})$ of order $k \in \mathbb{N} \setminus \{0\}$ from V to E on $\mathbb{R}^n$ is *injectively elliptic* whenever for every $\xi \in \mathbb{R}^n \setminus \{0\}$, one has $\ker A(\xi) = \{0\}$.

In other words, the operator $A(\mathrm{D})$ is injectively elliptic if and only if for every $\xi \in \mathbb{R}^n$ and $v \in V$, the condition $A(\xi)[v] = 0$ implies either $\xi = 0$ or $v = 0$.

In the one-dimensional case $n = 1$, injectively elliptic operators are those operators that can be written as an injective function of the derivative; this allows then one to recover the full derivative D^k from $A(\mathrm{D})$.

Proposition 5.1 (One-Dimensional Injectively Elliptic Operators) *Given finite-dimensional vector spaces V and E, a homogeneous constant coefficient differential operator $A(\mathrm{D})$ of order $k \in \mathbb{N} \setminus \{0\}$ from V to E on $\mathbb{R}$ is injectively elliptic if and only if there exists $A \in \mathrm{Lin}(V, E)$ such that $\ker A = \{0\}$ and for every $u \in C^\infty(\mathbb{R}, V)$,*

$$A(\mathrm{D})u = A(u^{(k)}) \,.$$

Let us now review a few examples of injectively elliptic operators.

Example 5.1 (Total Derivative) The *k-th order total derivative* $A(\mathrm{D}) = \mathrm{D}^k$ is injectively elliptic as a differential operator from $\mathbb{R}$ to $\mathrm{Lin}^k_{\mathrm{sym}}(\mathbb{R}^n, \mathbb{R})$ on $\mathbb{R}$. Indeed, if $\xi \in \mathbb{R}^n$ and $v \in \mathbb{R}$ satisfy $A(\xi)[v] = \xi^{\otimes k} v = 0$, then either $\xi = 0$ or $v = 0$.

Example 5.2 (Laplacian) The *Laplacian* $A(\mathrm{D}) := \Delta$ is injectively elliptic as a differential operator from $\mathbb{R}$ to $\mathbb{R}$ on $\mathbb{R}^n$. Indeed, if $\xi \in \mathbb{R}^n$ and $v \in \mathbb{R}$ satisfy the condition $A(\xi)[v] = |\xi|^2 v = 0$, then either $\xi = 0$ or $v = 0$.

Example 5.3 (Cauchy-Riemann Operator) The *Cauchy-Riemann operator* $A(\mathrm{D}) = \bar{\partial}$, which is the differential operator from $\mathbb{C}$ to $\mathbb{C}$ on $\mathbb{R}^2$ defined for each function $u \in C^\infty(\mathbb{R}^2, \mathbb{C})$ by $A(\mathrm{D})u = (\partial_1 u + i\partial_2 u)/2$ is injectively elliptic. Indeed, for every $\xi = (\xi_1, \xi_2) \in \mathbb{R}^2$ and $v \in \mathbb{C}$, if $A(\xi)v = (\xi_1 + i\xi_2)v/2 = 0$ then either $\xi = 0$ or $v = 0$.

Example 5.4 (Hodge Complex) The *Hodge complex operator* $A(\mathrm{D}) = (d, d^*)$ on differential forms from $\bigwedge^m \mathbb{R}^n$ to $\bigwedge^{m+1} \mathbb{R}^n \times \bigwedge^{m-1} \mathbb{R}^n$, with $m \in \{1, \ldots, n-1\}$ is injectively elliptic. Here d and d^* denote respectively the exterior differential and codifferential. Indeed, for every $\xi \in \mathbb{R}^n$ and $v \in \bigwedge^m \mathbb{R}^n$, if $A(\xi)[v] = (\xi \wedge v, \xi \lrcorner v) = 0$, then one has the Lagrange identity $|A(\xi)[v]|^2 = |\xi \wedge v|^2 + |\xi \lrcorner v|^2 = |\xi|^2 |v|^2$ and thus either $\xi = 0$ or $v = 0$.

---

[1] The definition of injective ellipticity (Definition 5.2) appeared for overdetermined differential operators [40, Th. 1], [61, Def. 1.7.1], [20, Def. 3.4], [59], and [56, Def. 1.1] (when dim $V = 1$, see also [2, §7], and [3, Def. 6.3]); it is also called *right ellipticity* [26, Def. 26], which might seem somehow inconsistent with the *left*-invertibility of the operator it is equivalent to; in the study of Sobolev inequalities it is standard [74, Def. 1.1]. The terminology *injectively elliptic* is borrowed from [39, §4] in the context of pseudo-differential operators (see also [5, Rem. 18.2 (d) (2a)]).

5 Injective Ellipticity, Cancelling Operators and Endpoint Sobolev Inequalities   263

Example 5.5 (Symmetric Derivative) The *symmetric derivative*, also known as *deformation operator*, defined for every $u \in C^\infty(\mathbb{R}^n, \mathbb{R}^n)$ and every $x \in \mathbb{R}^n$ by $A(\mathrm{D})u(x) := \mathrm{D}_{\mathrm{sym}} u(x) := (\mathrm{D}u(x) + \mathrm{D}u(x)^*)/2 \in \mathrm{Lin}(\mathbb{R}^n, \mathbb{R}^n)$ is injectively elliptic. Indeed if $\xi \in \mathbb{R}^n$ and $v \in \mathbb{R}^n$ satisfy $A(\xi)[v] = (v \otimes \xi + \xi \otimes v)/2 = 0$, then $|A(\xi)[v]|^2 = (|v|^2 |\xi|^2 + \langle \xi, v \rangle^2)/2 = 0$, and hence either $\xi = 0$ or $v = 0$.

Example 5.6 (Trace-Free Symmetric Derivative) The operator $A(\mathrm{D})u := \mathrm{D}_{\mathrm{sym}} u + \lambda \operatorname{div} u \operatorname{id}$ is injectively elliptic if and only if $n \geq 2$ or $n = 1$ and $\lambda \neq -1$. When $\lambda = -1/n$, the operator $A(\mathrm{D})$ is the *trace-free symmetric derivative* which is injectively elliptic if and only if $n \geq 2$. To prove the injective ellipticity, we have for every $\xi \in \mathbb{R}^n$ and $v \in \mathbb{R}^n$, $A(\xi)[v] = (v \otimes \xi + \xi \otimes v)/2 + \lambda \langle \xi, v \rangle \operatorname{id}$ and thus $|A(\xi)[v]|^2 = |\xi|^2 |v|^2/2 + (1/2 + 2\lambda + n\lambda^2) \langle \xi, v \rangle^2$. We note that $1 + 2\lambda + n\lambda^2 = (1+\lambda)^2 + (n-1)\lambda^2 > 0$ if and only if either $n = 1$ and $\lambda \neq -1$ or $n \geq 2$; in these cases if $A(\xi)[v] = 0$ we have either $\xi = 0$ or $v = 0$, and the operator $A(\mathrm{D})$ is injectively elliptic.

Injective ellipticity turns out to be always a necessary condition to obtain L^p estimates on $\mathrm{D}^k u$.

Theorem 5.1 (Neccessity of the Injective Ellipticity for Equivalence of Norms) *Let $n \in \mathbb{N} \setminus \{0\}$, let V and E be finite-dimensional vector spaces, let $A(\mathrm{D})$ be a homogeneous constant coefficient differential operator of order $k \in \mathbb{N} \setminus \{0\}$ from V to E on $\mathbb{R}^n$, and let $p \in [1, +\infty)$. If there exists a constant $C \in (0, +\infty)$ such that for every $u \in C_c^\infty(\mathbb{R}^n, V)$,*

$$\int_{\mathbb{R}^n} |\mathrm{D}^k u|^p \leq C \int_{\mathbb{R}^n} |A(\mathrm{D})[u]|^p \,, \tag{5.7}$$

then the operator $A(\mathrm{D})$ is injectively elliptic.

Proof We fix a function $\eta \in C_c^\infty(\mathbb{R}^n, \mathbb{R})$ such that $\eta = 1$ on the unit ball $B_1(0) \subseteq \mathbb{R}^n$, and we define for every $v \in V$ and $\xi \in \mathbb{R}^n$, the function $u_{\xi,v} : \mathbb{R}^n \to V$ for each $x \in \mathbb{R}^n$ by

$$u_{\xi,v}(x) := v \sin(\langle \xi, x \rangle) \eta(x) \,.$$

We compute, since $\eta = 1$ on $B_1(0)$,

$$\int_{\mathbb{R}^n} |\mathrm{D}^k u_{\xi,v}|^p \,\mathrm{d}x \geq \int_{B_1(0)} |\mathrm{D}^k u_{\xi,v}|^p = \int_{B_1(0)} |v|^p |\xi|^{kp} |\sin^{(k)}(\langle \xi, x \rangle)|^p \,\mathrm{d}x$$

$$\geq C_1 |v|^p |\xi|^{kp} \,, \tag{5.8}$$

if $|\xi| \geq 1$, for some constant $C_1 \in (0, +\infty)$. On the other hand, we have by Lemma 5.1 below

$$\int_{\mathbb{R}^n} |A(D)u_{\xi,v}|^p \, dx \leq C_2 \int_{\mathbb{R}^n} |\eta|^p |A(\xi)[v]|^p + |v|^p \sum_{j=1}^{k} |D^j \eta|^p |\xi|^{(k-j)p} \,. \tag{5.9}$$

When $|\xi|$ is large enough we deduce from our assumption (5.7) and from the estimates (5.8) and (5.9) that

$$|v|^p |\xi|^{kp} \leq C_3 |A(\xi)[v]|^p \,. \tag{5.10}$$

By linearity of A, the inequality (5.10) holds for every $\xi \in \mathbb{R}^n$ and the operator $A(D)$ is injectively elliptic in view of Definition 5.2. □

In the proof of Theorem 5.1, we used the fact that commutators between $A(D)$ and a multiplication only involve lower-order derivatives of u:

Lemma 5.1 (Generalized Leibnitz Rule) *Let $n \in \mathbb{N} \setminus \{0\}$, let V and E be finite-dimensional vector spaces, and let $A(D)$ be a homogeneous constant coefficient differential operator of order $k \in \mathbb{N} \setminus \{0\}$ from V to E on $\mathbb{R}^n$. For every $u \in C^\infty(\mathbb{R}^n, V)$ and every $\varphi \in C^\infty(\mathbb{R}^n, E)$, one has*

$$A(D)(\varphi u) - \varphi A(D)u = \sum_{j=1}^{k} \binom{k}{j} A\big(\mathrm{Sym}(D^j \varphi \otimes D^{k-j} u)\big) \,. \tag{5.11}$$

Here, Sym denotes the symmetrization of tensors.

Proof of Lemma 5.1 From the generalized Leibnitz formula, we have

$$D^k(\varphi u) = \sum_{j=0}^{k} \binom{k}{j} \mathrm{Sym}(D^j \varphi \otimes D^{k-j} u) \,,$$

and the conclusion (5.11) follows. □

5.2.2 Fourier Analysis and L^2 Estimates

The injective ellipticity is necessary and sufficient for having an L^2 estimate of the k-th order total derivative.

5 Injective Ellipticity, Cancelling Operators and Endpoint Sobolev Inequalities

Theorem 5.2 (L^2 Estimate for Injectively Elliptic Operators) *Let $n \in \mathbb{N} \setminus \{0\}$, let V and E be finite-dimensional vector spaces, and let $A(D)$ be a homogeneous constant coefficient differential operator of order $k \in \mathbb{N} \setminus \{0\}$ from V to E on $\mathbb{R}^n$. If $A(D)$ is injectively elliptic, then there exists a constant $C \in (0, +\infty)$ such that for every $u \in C_c^\infty(\mathbb{R}^n, V)$,*

$$\int_{\mathbb{R}^n} |D^k u|^2 \le C \int_{\mathbb{R}^n} |A(D)[u]|^2 .$$

The proof of Theorem 5.2 will rely on the following quantitative reformulation of injective ellipticity.

Lemma 5.2 (Quantitative Characterization of Injective Ellipticity) *Let $n \in \mathbb{N} \setminus \{0\}$, let V and E be finite-dimensional vector spaces, and let $A(D)$ be a homogeneous constant coefficient differential operator of order $k \in \mathbb{N} \setminus \{0\}$ from V to E on $\mathbb{R}^n$. The operator $A(D)$ is injectively elliptic if and only if there exists a constant $C \in (0, +\infty)$ such that for every $\xi \in \mathbb{R}^n$ and every $v \in V$,*

$$|\xi|^k |v| \le C |A(\xi)[v]| . \tag{5.12}$$

Proof If the inequality (5.12) is satisfied and if $\xi \in \mathbb{R}^n \setminus \{0\}$, then the inequality (5.12) implies that the mapping $A(\xi)$ is injective on V.

Conversely, if the operator $A(D)$ is injectively elliptic, then the function $(\xi, v) \in K \mapsto |A(\xi)[v]| \in \mathbb{R}$ is continuous and positive on the compact set $K := \{(\xi, v) \in \mathbb{R}^n \times V \mid |\xi| = |v| = 1\}$, hence it is bounded from below by a positive constant on that set. By homogeneity and linearity, we have for every $\xi \in \mathbb{R}^n \setminus \{0\}$ and $v \in V \setminus \{0\}$

$$|A(\xi)[v]| = |A(v \otimes \xi^{\otimes k})| = |\xi|^k |v| \left| A\left(\tfrac{v}{|v|} \otimes (\tfrac{\xi}{|\xi|})^{\otimes k}\right) \right| = |\xi|^k |v| \left| A\left(\tfrac{\xi}{|\xi|}\right)\left[\tfrac{v}{|v|}\right] \right|$$

and the conclusion (5.12) follows. □

We are now in position to prove the L^2 estimate on $D^k u$ for injectively elliptic operators of Theorem 5.2.

Proof of Theorem 5.2 By Parseval's identity for the Fourier transform and by Lemma 5.2, we have for some constant $C_4 \in (0, +\infty)$

$$\int_{\mathbb{R}^n} |D^k u|^2 = (2\pi)^{2k} \int_{\mathbb{R}^n} |u(\xi)|^2 |\xi|^{2k} \, d\xi$$

$$\le C_4^2 (2\pi)^{2k} \int_{\mathbb{R}^n} |A(\xi)[u(\xi)]|^2 \, d\xi = C_4^2 \int_{\mathbb{R}^n} |A(D)[u]|^2 ,$$

and the conclusion follows. □

5.2.3 Representation Kernel

In order to extend Theorem 5.2 to L^p, with $p \in (1, +\infty) \setminus \{2\}$, we will rely on a representation formula of u by $A(D)u$. In order to construct such a formula, one can first observe on the Fourier frequency domain that since the symbol $A(\xi)$ is injective on V for every $\xi \in \mathbb{R}^n \setminus \{0\}$, it should be possible to recover u from $A(D)u$ by the identity (5.3). In order to implement this, we start from the observation that for every $\xi \in \mathbb{R}^n \setminus \{0\}$, the operator $A(\xi)^* \circ A(\xi) : V \to V$ is invertible, where $A(\xi)^* : E \to V$ denotes the adjoint operator of $A(\xi)$,[2] so that $A(\xi)$ has a well-defined pseudo-inverse

$$A(\xi)^\dagger := \left(A(\xi)^* \circ A(\xi)\right)^{-1} \circ A(\xi)^* : E \to V, \qquad (5.13)$$

—this is the classical construction for least-square solutions of linear systems through normal equations—and the symbol $A(\xi)^\dagger$ satisfies

$$A(\xi)^\dagger \circ A(\xi) = \mathrm{id}_V .$$

The next natural step would be to define a representation kernel (or Green function) $G_A : \mathbb{R}^n \setminus \{0\} \to \mathrm{Lin}(E, V)$ through the requirement that for every $\xi \in \mathbb{R}^n \setminus \{0\}$

$$\mathcal{F}(G_A * u)(\xi) = (\mathcal{F}G_A)(\xi)[\mathcal{F}(u)(\xi)] = (2\pi i)^k A(\xi)^\dagger [\mathcal{F}(A(D)u)(\xi)] ,$$

and therefore, with a suitable inversion formula, to define

$$G_A := \frac{1}{(2\pi i)^k} \mathcal{F}^{-1} A^\dagger , \qquad (5.14)$$

The problem is that in general $A^\dagger$ in the right-hand side of (5.14) is not even a distribution on $\mathbb{R}^n$ when $k \geq n$. Through a careful analysis, the next proposition bypasses this mathematical obstacle to construct a representation kernel.

Proposition 5.2 (Representation Kernel of an Injectively Elliptic Operator[3])
Let $n \in \mathbb{N} \setminus \{0\}$, let V and E be finite-dimensional vector spaces, and let $A(D)$ be a homogeneous constant coefficient differential operator of order $k \in \mathbb{N} \setminus \{0\}

---

[2] The definition of the adjoint $A(\xi)^*$ of $A(\xi)$ and hence of its pseudo-inverse $A(\xi)^\dagger$ in (5.13) depends on the choice of Euclidean structures on the spaces V and E. These structures allow to choose consistently a left-inverse and the final results will always be independent on this arbitrary choice of Euclidean structures.

[3] The construction of the representation kernel in Proposition 5.2 for injectively elliptic operators is due to [12, Lem. 2.1] and [53]. We give here a self-contained construction of the Green function, based on the extension of the homogeneous distribution $A(\xi)^\dagger$ from $\mathbb{R}^n \setminus \{0\}$ to homogeneous distributions on $\mathbb{R}^n$ [41, Th. 3.2.3 & 3.2.4], the preservation of the smoothness in the extension and its temperate character [41, Th. 7.1.18], and the homogeneity of the resulting Fourier transform [41, Th. 7.1.16].

5 Injective Ellipticity, Cancelling Operators and Endpoint Sobolev Inequalities 267

from V to E on $\mathbb{R}^n$. If the operator $A(D)$ is injectively elliptic, then there exists a function $G_A \in C^\infty(\mathbb{R}^n \setminus \{0\}, \text{Lin}(E, V)) \cap L^1_{\text{loc}}(\mathbb{R}^n \setminus \{0\}, \text{Lin}(E, V))$, such that

(i) for every $u \in \mathcal{S}(\mathbb{R}^n, V)$,

$$u(x) = \int_{\mathbb{R}^n} G_A(x-y)[A(D)u(y)]\,dy,\quad (5.15)$$

(ii) for every $x \in \mathbb{R}^n$ and every $\lambda \in \mathbb{R} \setminus \{0\}$,

$$G_A(\lambda x) = \lambda^{k-n}\bigl(G_A(x) - \ln|\lambda|\,P_A(x)\bigr),$$

where the function $P_A : \mathbb{R}^n \to \text{Lin}(E, V)$ is a homogeneous polynomial of degree $k - n$ when $k \geq n$ and is 0 when $k < n$,

(iii) if $\ell \in \mathbb{N}$ satisfies $\ell > k - n$, then

$$\int_{\mathbb{S}^{n-1}} D^\ell G_A(x)\,dx = 0,$$

(iv) given $f \in \mathcal{S}(\mathbb{R}^n, E)$ such that for every $\xi \in \mathbb{R}^n$, $\mathcal{F}f(\xi) \in A(\xi)[V + iV]$, the function $u \in C^\infty(\mathbb{R}^n, V)$ defined for each $x \in \mathbb{R}^n$ by

$$u(x) := \int_{\mathbb{R}^n} G_A(x-y)[f(y)]\,dy,\quad (5.16)$$

satisfies $A(D)u = f$,

(v) for every $e \in \bigcap_{\xi \in \mathbb{R}^n \setminus \{0\}} A(\xi)[V]$ we have $A(D)G_A[e] = e\delta_0$ on $\mathbb{R}^n$ in the sense of distributions,

(vi) if $k \geq n$, then for every $x \in \mathbb{R}^n$ and for every $e \in \bigcap_{\xi \in \mathbb{R}^n \setminus \{0\}} A(\xi)[V]$, we have

$$P_A(x)[e] = \int_{\mathbb{S}^{n-1}} \frac{\langle \xi, x \rangle^{k-n} A(\xi)^{-1}[e]}{(k-n)!(2\pi i)^n}\,d\xi.$$

Proof Fixing an even function $\psi \in C_c^\infty(\mathbb{R}^n, \mathbb{R})$ such that $\psi = 0$ in a neighbourhood of 0 and such that for every $x \in \mathbb{R}^n \setminus \{0\}$,

$$\int_0^{+\infty} \frac{\psi(x/t)}{t}\,dt = 1,\quad (5.17)$$

we define the function $H_A : \mathbb{R}^n \to \text{Lin}(E + iE, V + iV)$ for every $x \in \mathbb{R}^n$ by

$$H_A(x) := \int_{\mathbb{R}^n} \frac{e^{2\pi i \langle \xi, x \rangle}}{(2\pi i)^k}\psi(\xi)A(\xi)^\dagger\,d\xi,\quad (5.18)$$

where $A(\xi)^\dagger \in \text{Lin}(E, V)$ is the left-inverse of $A(\xi)$ that was defined in (5.13), which is well-defined since the operator $A(D)$ is injectively elliptic (Definition 5.2).

Since ψ is an even function, for every $x \in \mathbb{R}^n$ we have by (5.18), when $k \in \mathbb{N} \setminus \{0\}$ is an even number,

$$H_A(x) = \int_{\mathbb{R}^n} \frac{\cos(2\pi \langle \xi, x \rangle)}{(2\pi)^k (-1)^{k/2}} \psi(\xi) A(\xi)^\dagger \, d\xi , \tag{5.19}$$

since $\psi A^\dagger$ is then an even function, and, when $k \in \mathbb{N} \setminus \{0\}$ is an odd number,

$$H_A(x) = \int_{\mathbb{R}^n} \frac{\sin(2\pi \langle \xi, x \rangle)}{(2\pi)^k (-1)^{(k-1)/2}} \psi(\xi) A(\xi)^\dagger \, d\xi , \tag{5.20}$$

since $\psi A^\dagger$ is then an odd function. It follows from (5.19) or (5.20) that $H_A(x) \in \mathrm{Lin}(E, V)$ and that for every $x \in \mathbb{R}^n$,

$$H_A(-x) = (-1)^k H_A(x) . \tag{5.21}$$

By properties of Fourier transforms, we also have $H_A \in \mathcal{S}(\mathbb{R}^n, \mathrm{Lin}(E, V))$. In particular, for every $x \in \mathbb{R}^n \setminus \{0\}$, we have $t \in \mathbb{R} \mapsto H_A(tx) \in \mathcal{S}(\mathbb{R}, \mathrm{Lin}(E, V))$.

When $k < n$, we define the function $G_A : \mathbb{R}^n \setminus \{0\} \to \mathrm{Lin}(E, V)$ for each $x \in \mathbb{R}^n \setminus \{0\}$ by

$$G_A(x) := \int_0^{+\infty} H_A(tx) \, t^{n-k-1} \, dt . \tag{5.22}$$

If $u \in \mathcal{S}(\mathbb{R}^n, V)$, then we have by (5.22)

$$\int_{\mathbb{R}^n} G_A(x - y)[A(D)u(y)] \, dy$$

$$= \int_{\mathbb{R}^n} \int_0^{+\infty} H_A(t(x - y)) \, t^{n-k-1} \, dt [A(D)u(y)] \, dy \tag{5.23}$$

$$= \int_0^{+\infty} \int_{\mathbb{R}^n} H_A(t(x - y))[A(D)u(y)] \, dy \, t^{n-k-1} \, dt .$$

Next, we compute by the definition of H_A in (5.18)

$$\int_{\mathbb{R}^n} H_A(t(x - y))[A(D)u(y)] \, dy$$

$$= \int_{\mathbb{R}^n} \int_{\mathbb{R}^n} \frac{e^{2\pi i \langle \xi, t(x-y) \rangle}}{(2\pi i)^k} \psi(\xi) A(\xi)^\dagger [A(D)u(y)] \, d\xi \, dy$$

$$= \int_{\mathbb{R}^n} \int_{\mathbb{R}^n} \frac{e^{2\pi i \langle t\xi, x-y \rangle}}{(2\pi i)^k} \psi(\xi) A(\xi)^\dagger [A(D)u(y)] \, dy \, d\xi,$$

5 Injective Ellipticity, Cancelling Operators and Endpoint Sobolev Inequalities

so that by a change of variable $\zeta = t\xi$

$$\int_{\mathbb{R}^n} H_A(t(x-y))[A(D)u(y)]\,dy$$

$$= \int_{\mathbb{R}^n} e^{2\pi i \langle t\xi, x\rangle} \psi(\xi) A(\xi)^\dagger [A(t\xi)[\mathcal{F}u(t\xi)]]\,d\xi \qquad (5.24)$$

$$= t^{k-n} \int_{\mathbb{R}^n} e^{2\pi i \langle \zeta, x\rangle} \psi(\zeta/t) \mathcal{F}u(\zeta)\,d\zeta\,.$$

Hence by (5.23) and (5.24), we get by (5.17) and the Fourier inversion formula.

$$\int_{\mathbb{R}^n} G_A(x-y)[A(D)u(y)]\,dy = \int_0^{+\infty}\!\!\int_{\mathbb{R}^n} e^{2\pi i \langle \zeta, x\rangle} \psi(\zeta/t) \mathcal{F}u(\zeta)\,d\zeta\,\frac{dt}{t}$$

$$= \int_{\mathbb{R}^n} \mathcal{F}u(\zeta)\,d\zeta = u(x)\,, \qquad (5.25)$$

We have thus proved that (i) holds when $k < n$.

Differentiating (5.22), we have for every $\ell \in \mathbb{N}$ that $G_A \in C^\ell(\mathbb{R}^n \setminus \{0\}, \mathrm{Lin}(E, V))$, with for every $x \in \mathbb{R}^n \setminus \{0\}$,

$$D^\ell G_A(x) = \int_0^{+\infty} D^\ell H_A(tx)\,t^{n+\ell-k-1}\,dt\,. \qquad (5.26)$$

For every $\lambda \in (0, +\infty)$, we have by (5.22) and by the change of variable $s = \lambda t$,

$$G(\lambda x) = \int_0^{+\infty} H_A(t\lambda x)\,t^{n-k-1}\,dt = \frac{1}{\lambda^{n-k}} \int_0^{+\infty} H_A(sx)\,s^{n-k-1}\,ds = \frac{G_A(x)}{\lambda^{n-k}}\,, \qquad (5.27)$$

so that the assertion (ii) holds for $k < n$ in view of (5.21) and (5.27).

When $k \geq n$, we define the function $G_A : \mathbb{R}^n \setminus \{0\} \to \mathrm{Lin}(E, V)$ for each $x \in \mathbb{R}^n \setminus \{0\}$ by

$$G_A(x) := \int_0^{+\infty} \frac{\langle D^{k-n+1} H_A(tx), x^{\otimes k-n+1}\rangle}{(k-n)!} \ln \frac{1}{t}\,dt$$

$$+ \frac{\langle D^{k-n} H_A(0), x^{\otimes k-n}\rangle}{(k-n)!} \sum_{j=1}^{k-n} \frac{1}{j}\,. \qquad (5.28)$$

Assuming that $x = 0$, we first have by (5.28)

$$\int_{\mathbb{R}^n} G_A(x-y)[A(D)u(y)]\,dy$$

$$= \int_0^{+\infty} \int_{\mathbb{R}^n} \frac{\langle D^{k-n+1}H_A(-ty), (-y)^{\otimes k-n+1}\rangle[A(D)u(y)]}{(k-n)!}\,dy \ln\frac{1}{t}\,dt$$

$$+ \int_{\mathbb{R}^n} \frac{\langle D^{k-n}H_A(0), (-y)^{\otimes k-n}\rangle[A(D)u(y)]}{(k-n)!}\,dy \sum_{j=1}^{k-n}\frac{1}{j}\,.$$

(5.29)

In order to compute out the first-term in the right-hand side of (5.29), we first work out the inner integral for every $t \in (0, +\infty)$ in view of (5.18) as

$$\int_{\mathbb{R}^n} \langle D^{k-n+1}H_A(-ty), (-y)^{\otimes k-n+1}\rangle[A(D)u(y)]\,dy$$

$$= \int_{\mathbb{R}^n}\int_{\mathbb{R}^n} \frac{e^{-2\pi i\langle\xi,ty\rangle}}{(2\pi i)^{n-1}}(-\langle\xi,y\rangle)^{k-n+1}\psi(\xi)A(\xi)^\dagger[A(D)u(y)]\,dy\,d\xi$$

$$= \int_{\mathbb{R}^n}\int_{\mathbb{R}^n} \frac{e^{-2\pi i\langle t\xi,y\rangle}}{(2\pi i)^{n-1}}\psi(\xi)A(\xi)^\dagger[(-\langle\xi,y\rangle)^{k-n+1}A(D)u(y)]\,dy\,d\xi$$

$$= \int_{\mathbb{R}^n} \frac{\psi(\xi)A(\xi)^\dagger[D^{k-n+1}\mathcal{F}(A(D)u)(t\xi)[\xi^{\otimes k-n+1}]]}{(2\pi i)^k}\,d\xi\,.$$

(5.30)

By $k-n+1$ successive integration by parts we have for every $\xi \in \mathbb{R}^n$

$$\int_0^{+\infty} \frac{D^{k-n+1}\mathcal{F}(A(D)u)(t\xi)[\xi^{\otimes k-n+1}]}{(k-n)!} \ln\frac{1}{t}\,dt$$

$$= \int_0^{+\infty} \frac{\mathcal{F}(A(D)u)(t\xi)}{t^{k-n+1}}\,dt$$

$$= \int_0^{+\infty} \frac{(2\pi i)^k A(t\xi)\mathcal{F}u(t\xi)}{t^{k-n+1}}\,dt\,,$$

(5.31)

5 Injective Ellipticity, Cancelling Operators and Endpoint Sobolev Inequalities 271

and thus by (5.30) and (5.31), we get by a change of variable $\zeta = t\xi$

$$\int_{\mathbb{R}^n} \langle D^{k-n+1} H_A(-ty), (-y)^{\otimes k-n+1}\rangle [A(D)u(y)]\, dy$$

$$= \int_{\mathbb{R}^n} \int_0^{+\infty} \frac{\psi(\xi) A(t\xi)^\dagger [A(\xi)[\mathcal{F}u(t\xi)]]}{t^{k-n+1}}\, dt\, d\xi$$

$$= \int_{\mathbb{R}^n} \int_0^{+\infty} \frac{\psi(\zeta/t) \mathcal{F}u(\zeta)}{t}\, dt\, d\zeta = u(0)\,,$$

(5.32)

in view of (5.17). It remains to evaluate the second term in the right-hand side of (5.29): by integration by parts we have

$$\int_{\mathbb{R}^n} \frac{\langle D^{k-n} H_A(0), (-y)^{\otimes k-n}\rangle [A(D)u(y)]}{(k-n)!}\, dy = 0\,,$$ (5.33)

since $A(D)$ is a homogeneous differential operator of order $k \in \mathbb{N} \setminus \{0\}$. Combining (5.29), (5.32) and (5.33), we have proved (i) when $k \geq n$.

Next, we claim that for every $\ell \in \{0, \ldots, k-n\}$ and $x \in \mathbb{R}^n$, we have

$$D^\ell G_A(x) = \int_0^{+\infty} \frac{\langle D^{k-n+1} H_A(tx), x^{\otimes k-n-\ell+1}\rangle}{(k-n-\ell)!} \ln\frac{1}{t}\, dt$$

$$+ \frac{\langle D^{k-n} H_A(0), x^{\otimes k-n-\ell}\rangle}{(k-n-\ell)!} \sum_{j=1}^{k-n-\ell} \frac{1}{j}\,.$$ (5.34)

Indeed, for $\ell = 0$, (5.34) reduces to (5.28); if we assume that (5.34) holds for some $\ell \in \{0, \ldots, k-n-1\}$ and if we differentiate (5.34) we obtain

$$D^{\ell+1} G_A(x)$$

$$= \int_0^{+\infty} \left(\frac{\langle D^{k-n+1} H_A(tx), x^{\otimes k-n-\ell}\rangle}{(k-n-\ell-1)!} + \frac{\langle D^{k-n+1} H_A(tx), x^{\otimes k-n-\ell}\rangle}{(k-n-\ell)!}\right.$$

$$\left. + \frac{\langle D^{k-n+2} H_A(tx), x^{\otimes k-n-\ell+1}\rangle t}{(k-n-\ell)!}\right) \ln\frac{1}{t}\, dt$$

$$+ \frac{\langle D^{k-n} H_A(0), x^{\otimes k-n-\ell-1}\rangle}{(k-n-\ell-1)!} \sum_{j=1}^{k-n-\ell} \frac{1}{j}\,;$$

(5.35)

integrating by parts the last term in the integrand on the right-hand side of (5.35), we have

$$\int_0^{+\infty} \langle D^{k-n+2} H_A(tx), x^{\otimes k-n-\ell+1} \rangle t \ln \frac{1}{t} \, dt$$
$$= \int_0^{+\infty} \langle D^{k-n+1} H_A(tx), x^{\otimes k-n-\ell} \rangle \left(1 - \ln \frac{1}{t}\right) dt$$
$$= -\int_0^{+\infty} \langle D^{k-n+1} H_A(tx), x^{\otimes k-n-\ell} \rangle \ln \frac{1}{t} \, dt - \langle D^{k-n} H_A(0), x^{\otimes k-n-\ell-1} \rangle,$$
(5.36)

and it follows then from (5.35) and (5.36) that

$$D^{\ell+1} G_A(x) = \int_0^{+\infty} \frac{\langle D^{k-n+1} H_A(tx), x^{\otimes k-n-\ell} \rangle}{(k-n-\ell-1)!} \ln \frac{1}{t} \, dt$$
$$+ \frac{\langle D^{k-n} H_A(0), x^{\otimes k-n-\ell-1} \rangle}{(k-n-\ell-1)!} \sum_{j=1}^{k-n-\ell-1} \frac{1}{j},$$

and thus (5.34) holds for each $\ell \in \{0, \ldots, n-k\}$ by induction.

Differentiating (5.34) with $\ell = k - n$, it also follows by integration by parts that

$$D^{k-n+1} G_A(x) = \int_0^{+\infty} \left(D^{k-n+1} H_A(tx) + \langle D^{k-n+2} H_A(tx), x \rangle t \right) \ln \frac{1}{t} \, dt$$
$$= \int_0^{+\infty} D^{k-n+1} H_A(tx) \, dt,$$

and thus (5.26) holds for $\ell > k - n$ also when $k \geq n$.

For each $x \in \mathbb{R}^n \setminus \{0\}$ and $\lambda \in (0, +\infty)$, we have by (5.28) and by the change of variable $t = s/\lambda$,

$$G_A(\lambda x) = \int_0^{+\infty} \frac{\langle D^{k-n+1} H_A(t\lambda x), (\lambda x)^{\otimes k-n+1} \rangle}{(k-n)!} \ln \frac{1}{t} \, dt$$
$$+ \frac{\langle D^{k-n} H_A(0), (\lambda x)^{\otimes k-n} \rangle}{(k-n)!} \sum_{j=1}^{k-n} \frac{1}{j}$$
$$= \lambda^{k-n} \left(G_A(x) + \ln \lambda \int_0^{+\infty} \frac{\langle D^{k-n+1} H_A(sx), x^{\otimes k-n+1} \rangle}{(k-n)!} \, ds \right)$$
$$= \lambda^{k-n} \left(G_A(x) - \ln \lambda \, P_A(x) \right),$$

5 Injective Ellipticity, Cancelling Operators and Endpoint Sobolev Inequalities   273

where, in view of (5.18), the function $P_A : \mathbb{R}^n \to \mathrm{Lin}(E + iE, V + iV)$ is defined for every $x \in \mathbb{R}^n$ by

$$
\begin{aligned}
P_A(x) &:= -\int_0^{+\infty} \frac{\langle D^{k-n+1} H_A(tx), x^{\otimes k-n+1}\rangle}{(k-n)!} \, dt \\
&= -\int_0^{+\infty} \int_{\mathbb{R}^n} \frac{\langle \xi, x\rangle^{k-n+1} e^{2\pi i t \langle \xi, x\rangle} \psi(\xi) A(\xi)^\dagger}{(k-n)!(2\pi i)^{n-1}} \, d\xi \, dt \\
&= -\int_0^{+\infty} \int_0^{+\infty} \int_{\mathbb{S}^{n-1}} \frac{\langle \zeta, x\rangle^{k-n+1} e^{2\pi i t r \langle \zeta, x\rangle} \psi(r\zeta) A(\zeta)^\dagger}{(k-n)!(2\pi i)^{n-1}} \, d\zeta \, dr \, dt \\
&= -\int_{\mathbb{S}^{n-1}} \int_0^{+\infty} \int_0^{+\infty} \frac{\langle \zeta, x\rangle^{k-n} e^{2\pi i s r} \psi(r\zeta) A(\zeta)^\dagger}{(k-n)!(2\pi i)^{n-1}} \, dr \, ds \, d\zeta \, ,
\end{aligned}
\tag{5.37}
$$

by the change of variables $\xi = r\zeta$ and $s = t\langle \zeta, x\rangle$. The innermost integrals can be computed by the Riemann-Lebesgue lemma as

$$
\begin{aligned}
\int_0^{+\infty} \int_0^{+\infty} e^{2\pi i s r} \psi(r\xi) \, dr \, ds &= \lim_{S\to +\infty} \int_0^S \int_0^{+\infty} e^{2\pi i s r} \psi(r\xi) \, dr \, ds \\
&= \lim_{S\to \infty} \int_0^{+\infty} \frac{e^{2\pi i S r} - 1}{2\pi i r} \psi(r\xi) \, dr \\
&= -\int_0^{+\infty} \frac{\psi(r\xi)}{2\pi i r} \, dr = -\frac{1}{2\pi i} \, ,
\end{aligned}
\tag{5.38}
$$

in view of (5.17), and therefore for every $x \in \mathbb{R}^n$, by (5.37) and (5.38) we have

$$
P_A(x) = \int_{\mathbb{S}^{n-1}} \frac{\langle \xi, x\rangle^{k-n} A(\xi)^\dagger}{(k-n)!(2\pi i)^n} \, d\xi \, .
\tag{5.39}
$$

Since for every $\xi \in \mathbb{R}^n$ and $x \in \mathbb{R}^n$, $\langle \xi, -x\rangle^{k-n} A(-\xi)^\dagger = (-1)^n \langle \xi, x\rangle^{k-n} A(\xi)^\dagger$, we have $P_A = 0$ when n is odd. When n is even, we have for every $x \in \mathbb{R}^n$, $P_A(x) \in \mathrm{Lin}(E, V)$, and we have thus proved (ii) when $k \geq n$.

If $e \in \bigcap_{\xi \in \mathbb{R}^n \setminus \{0\}} A(\xi)[V]$, then for every $\xi \in \mathbb{R}^n$ we have $A(\xi)^\dagger[e] = A(\xi)^{-1}[e]$ and the assertion (vi) follows immediately from the identity (5.39).

We consider now a function $f \in \mathcal{S}(\mathbb{R}^n, E)$ which satisfies for every $\xi \in \mathbb{R}^n$ the condition that $\mathcal{F}f(\xi) \in A(\xi)[V + iV]$. If we let the function $u : \mathbb{R}^n \to V$ be defined by (5.16), then we have by the differentiation formula (5.26), since $k > k - n$,

$$
A(D)u(x) = \int_0^{+\infty} \int_{\mathbb{R}^n} A(D)H_A(ty)[f(y-x)] t^{n-1} \, dt \, dy \, .
\tag{5.40}
$$

We compute by the definition of H_A in (5.18)

$$\int_{\mathbb{R}^n} A(D)H_A(ty)[f(x-y)]\,\mathrm{d}y$$
$$= \int_{\mathbb{R}^n}\int_{\mathbb{R}^n} A(\xi)[A(\xi)^\dagger[e^{2\pi i\langle t\xi,y\rangle}f(x-y)]]\psi(\xi)\,\mathrm{d}\xi\,\mathrm{d}y$$
$$= \int_{\mathbb{R}^n}\int_{\mathbb{R}^n} A(\xi)[A(\xi)^\dagger[e^{2\pi i\langle t\xi,y\rangle}f(x-y)]]\psi(\xi)\,\mathrm{d}y\,\mathrm{d}\xi$$
$$= \int_{\mathbb{R}^n} A(t\xi)[A(t\xi)^\dagger[e^{2\pi i\langle t\xi,x\rangle}\mathcal{F}f(t\xi)]]\psi(\xi)\,\mathrm{d}\xi$$
$$= \int_{\mathbb{R}^n} e^{2\pi i\langle t\xi,x\rangle}\mathcal{F}f(t\xi)\psi(\xi)\,\mathrm{d}\xi$$
$$= \frac{1}{t^n}\int_{\mathbb{R}^n} e^{2\pi i\langle \zeta,x\rangle}\mathcal{F}f(\zeta)\psi(\zeta/t)\,\mathrm{d}\zeta\,.$$

(5.41)

and it follows thus from (5.17), (5.40) and (5.41) that

$$A(D)u(x) = \int_0^{+\infty}\int_{\mathbb{R}^n} e^{2\pi i\langle \zeta,x\rangle}\mathcal{F}f(\zeta)\frac{\psi(\zeta/t)}{t}\,\mathrm{d}\zeta\,\mathrm{d}t$$
$$= \int_{\mathbb{R}^n} e^{2\pi i\langle \zeta,x\rangle}\mathcal{F}f(\zeta)\,\mathrm{d}\zeta = f(x)\,.$$

We have thus proved the assertion (iv).

In order to prove the assertion (v), given

$$e \in \bigcap_{\xi\in\mathbb{R}^n\setminus\{0\}} A(\xi)[V]\,,$$

we fix a function $\eta \in C_c^\infty(\mathbb{R}^n,\mathbb{R})$ such that $\int_{\mathbb{R}^n}\eta = 1$ and for every $\delta \in (0,+\infty)$, we consider the function $f_\delta \in C_c^\infty(\mathbb{R}^n, E)$ defined for each $x \in \mathbb{R}^n$ by $f_\delta(x) := e\eta(x/\delta)/\delta^n$, so that $\int_{\mathbb{R}^n} f_\delta = e$. We define $u_\delta : \mathbb{R}^n \to V$ for each $x \in \mathbb{R}^n$ by

$$u_\delta(x) := \int_{\mathbb{R}^n} G_A(x-y)[f_\delta(y)]\,\mathrm{d}y\,.$$

For every $\xi \in \mathbb{R}^n$, we have

$$\mathcal{F}f_\delta(\xi) = e\mathcal{F}\eta(\delta\xi) \in \mathbb{C}e \subseteq \bigcap_{\xi\in\mathbb{R}^n\setminus\{0\}} A(\xi)[V + iV]\,.$$

By (iv), we have $A(D)u_\delta = f_\delta$ in $\mathbb{R}^n$. Since $G_A \in L^1_{\text{loc}}(\mathbb{R}^n, \text{Lin}(E, V))$, we have $u_\delta \to G_A[e]$ in the sense of distributions as $\delta \to 0$. Since $f_\delta \to e\delta_0$ in the sense of distributions as $\delta \to 0$, we conclude that $A(D)G_A[e] = e\delta_0$ in $\mathbb{R}^n$ in the sense of distributions and we have proved the assertion (v).

We finally prove (iii). For $\ell > n - k$, if $\varphi \in C_c^\infty(\mathbb{R}^n \setminus \{0\}, \mathbb{R})$ and if for every $x \in \mathbb{R}^n$, $\int_0^{+\infty} \varphi(x/t) \, t^{\ell-k-1} \, dt = 1$, then by (5.26) we have

$$\begin{aligned}
\int_{\mathbb{S}^{n-1}} D^\ell G_A(x) \, dx &= \int_{\mathbb{S}^{n-1}} \varphi(xr) D^\ell G_A(x) \, r^{k-\ell-1} \, dx \\
&= \int_{\mathbb{R}^n} \varphi(z) D^\ell G_A(z) \, dz \\
&= \int_{\mathbb{R}^n} \int_0^{+\infty} \varphi(z) D^\ell H_A(tz) \, t^{n+\ell-k-1} \, dt \, dz \\
&= \int_{\mathbb{R}^n} \int_0^{+\infty} \varphi(z/t) D^\ell H_A(z) \, t^{\ell-k-1} \, dz \, dt \\
&= \int_{\mathbb{R}^n} D^\ell H_A(z) \, dz = 0 \, ;
\end{aligned}$$

this proves (iii). □

5.2.4 Compatibility Conditions

The condition $\mathcal{F}f(\xi) \in A(\xi)[V + iV]$ in Proposition 5.2 (iv) can be reformulated as being in the kernel of some differential operator by the next proposition.

Proposition 5.3 (Compatibility Conditions) *Let* $n \in \mathbb{N} \setminus \{0\}$, *let V and E be finite-dimensional vector spaces, and let* $A(D)$ *be a homogeneous constant coefficient differential operator of order* $k \in \mathbb{N} \setminus \{0\}$ *from V to E on* $\mathbb{R}^n$. *If* $A(D)$ *is injectively elliptic, then there exists a homogeneous constant coefficient differential operator* $L(D)$ *from E to F on* $\mathbb{R}^n$ *such that for every* $\xi \in \mathbb{R}^n \setminus \{0\}$,

$$A(\xi)[V] = \ker L(\xi) \, . \tag{5.42}$$

The operator $L(D)$ arising in Proposition 5.3 describes the *compatibility conditions* of the operator $A(D)$. It describes the obstructions to the solution of systems of the form $A(D)u = f$ and the structure of the range of the differential operator $A(D)$.

Such conditions are well-known for the gradient operator, whose image are curl-free vector fields: one has indeed for every $i, j \in \{1, \ldots, n\}$,

$$\partial_j(\partial_i u) = \partial_i(\partial_j u) \, .$$

Similarly for the *Hodge complex*, where $A(D) = (d, d^*)$, one has $(d, d^*)A(D) = (d^2, d^{*2}) = 0$. A more subtle but still classical setting are the compatibility conditions for the symmetric derivative or deformation operator known in linear elasticity as the *Saint-Venant compatibility conditions*:[4] for every $i, j, k, \ell \in \{1, \ldots, n\}$, we have

$$\partial_{k\ell}(\partial_i u^j + \partial_j u^i) + \partial_{ij}(\partial_k u^\ell + \partial_\ell u^k) = \partial_{kj}(\partial_i u^\ell + \partial_\ell u^i) + \partial_{i\ell}(\partial_k u^j + \partial_j u^k).$$
(5.43)

Proof of Proposition 5.3 We define the operator $L(D)$ by setting for every $\xi \in \mathbb{R}^n \setminus \{0\}$,

$$L(\xi) := \det\left(A(\xi)^* \circ A(\xi)\right)\left(\mathrm{id}_E - A(\xi) \circ (A(\xi)^* \circ A(\xi))^{-1} \circ A(\xi)^*\right) \quad (5.44)$$

and $L(0) := 0$. We observe that for every $\xi \in \mathbb{R}^n \setminus \{0\}$,

$$\ker L(\xi) = \{e \in E \mid A(\xi) \circ (A(\xi)^* \circ A(\xi))^{-1} \circ A(\xi)^*[e] = e\} = A(\xi)[V],$$

so that (5.42) holds. Moreover, since $A(\xi)^* \circ A(\xi)$ is a polynomial in ξ, then

$$\det(A(\xi)^* \circ A(\xi))(A(\xi)^* \circ A(\xi))^{-1}$$

is also a polynomial in ξ and thus $L(\xi)$ is a polynomial in ξ. □

Remark 5.1 The proof of Proposition 5.3 yields an operator $L(D)$ of order $2k \dim V$, which is much more than what is needed in typical examples.

5.2.5 Singular Integrals and L^p Estimates

We are now is position to obtain L^p estimates for injectively elliptic operators.

Theorem 5.3 (L^p **Estimate for Injectively Elliptic Operators**) *Let $n \in \mathbb{N} \setminus \{0\}$, let V and E be finite-dimensional vector spaces, let $A(D)$ be a homogeneous constant coefficient differential operator of order $k \in \mathbb{N} \setminus \{0\}$ from V to E on $\mathbb{R}^n$, and let $p \in (1, +\infty)$. If $A(D)$ is injectively elliptic, then there exists a constant $C \in (0, +\infty)$ such that for every $u \in C_c^\infty(\mathbb{R}^n, V)$,*

$$\int_{\mathbb{R}^n} |\mathrm{D}^k u|^p \le C \int_{\mathbb{R}^n} |A(\mathrm{D})[u]|^p. \quad (5.45)$$

Our main tool will be the Calderón-Zygmund singular integral theorem.

---

[4] For a discussion of Saint-Venant compatibility conditions in physical context see for example [66, Ch. 9] and in the mathematical theory for smooth functions see [23, §6.18].

5 Injective Ellipticity, Cancelling Operators and Endpoint Sobolev Inequalities

Theorem 5.4 (Singular Integral Theorem[5]) *Let $n \in \mathbb{N} \setminus \{0\}$ and let V and E be finite-dimensional vector spaces. If $K \in C^1(\mathbb{R}^n \setminus \{0\}, \mathrm{Lin}(E, F))$ is homogeneous of degree $-n$ and if*

$$\int_{\mathbb{S}^{n-1}} K = 0 ,$$

then for every $p \in (1, +\infty)$, there exists a constant $C \in (0, +\infty)$ such that if we define for each $\varepsilon \in (0, +\infty)$ the operator $\mathcal{K}_\varepsilon$ for every $f \in L^p(\mathbb{R}^n, E)$ by

$$(\mathcal{K}_\varepsilon f)(x) := \int_{\mathbb{R}^n \setminus B_\varepsilon(0)} K(h)[f(x - h)] \, \mathrm{d}h ,$$

then $\mathcal{K}_\varepsilon$ is well-defined and we have for every $f \in L^p(\mathbb{R}^n, E)$,

$$\int_{\mathbb{R}^n} |\mathcal{K}_\varepsilon f|^p \le C \int_{\mathbb{R}^n} |f|^p .$$

Moreover, $\mathcal{K}_\varepsilon f$ converges in $L^p(\mathbb{R}^n, F)$ as $\varepsilon \to 0$.

In order to apply Theorem 5.4 and prove Theorem 5.3, we will rely on the following singular representation formula for $\mathrm{D}^k u$ that follows from the construction of representation kernels (Proposition 5.2).

Lemma 5.3 (Singular Integral Representation Formula) *Under the assumptions of Proposition 5.2 and with $G_A : \mathbb{R}^n \setminus \{0\} \to \mathrm{Lin}(E, V)$ given by the same proposition, for every $u \in C_c^\infty(\mathbb{R}^n, V)$ and every $x \in \mathbb{R}^n$, one has*

$$\begin{aligned}\mathrm{D}^k u(x) = &\int_{\mathbb{S}^{n-1}} \mathrm{D}^{k-1} G_A(z)[A(\mathrm{D})u(x)] \otimes z \, \mathrm{d}z \\ &+ \lim_{\varepsilon \to 0} \int_{\mathbb{R}^n \setminus B_\varepsilon(0)} \mathrm{D}^k G_A(h)[A(\mathrm{D})u(x - h)] \, \mathrm{d}h .\end{aligned} \quad (5.46)$$

Proof The first $k - 1$ derivatives of G_A are locally integrable in view of Proposition 5.2 (ii) and thus differentiating $k - 1$ times (5.15), we get for every $x \in \mathbb{R}^n$

$$\mathrm{D}^{k-1} u(x) = \int_{\mathbb{R}^n} \mathrm{D}^{k-1} G_A(x - y)[A(\mathrm{D})u(y)] \, \mathrm{d}y . \quad (5.47)$$

---

[5] The L^p theory of singular integrals of Theorem 5.4 originates in the work of Alberto Calderón and Antoni Zygmund [19] (see also [62, Ch. II Th. 3]).

We fix a function $\eta \in C_c^\infty(\mathbb{R}^n, \mathbb{R})$ such that $\eta = 1$ on $B_{1/2}(0)$ and $\eta = 0$ on $\mathbb{R}^n \setminus B_1(0)$. If $R > 0$ is large enough so that $\operatorname{supp} u \subseteq B_R(0)$, for every $\varepsilon \in (0, R/2)$, we rewrite (5.47) in view of Proposition 5.42 (iii) as

$$D^{k-1}u(x) = \int_{\mathbb{R}^n} \eta\left(\tfrac{|x-y|}{R}\right)\left(1 - \eta\left(\tfrac{|x-y|}{\varepsilon}\right)\right)D^{k-1}G_A(x-y)[A(D)u(y) - A(D)u(x)]\,dy$$

$$+ \int_{\mathbb{R}^n} \eta\left(\tfrac{|h|}{\varepsilon}\right)D^{k-1}G_A(h)[A(D)u(x-h)]\,dh\,. \tag{5.48}$$

Differentiating both integrals in the right-hand side of (5.48), we get

$$D^k u(x)$$

$$= \int_{\mathbb{R}^n} \eta\left(\tfrac{|x-y|}{R}\right)\left(1 - \eta\left(\tfrac{|x-y|}{\varepsilon}\right)\right)D^k G_A(x-y)[A(D)u(y) - A(D)u(x)]\,dy$$

$$+ \int_{\mathbb{R}^n} \left(\tfrac{1}{R}\eta'\left(\tfrac{|x-y|}{R}\right) - \tfrac{1}{\varepsilon}\eta'\left(\tfrac{|x-y|}{\varepsilon}\right)\right)D^{k-1}G_A(x-y)[A(D)u(y) - A(D)u(x)] \otimes \tfrac{x-y}{|x-y|}\,dy$$

$$+ \int_{\mathbb{R}^n} \eta\left(\tfrac{|h|}{\varepsilon}\right)D^{k-1}G_A(h)[DA(D)u(x-h)]\,dh\,. \tag{5.49}$$

Letting $\varepsilon \to 0$ in (5.49), the conclusion (5.46) follows from Proposition 5.2 (iii), Lebesgue's dominated convergence theorem and a direct computation.

Proof of Theorem 5.3 This follows from Theorem 5.4 and Lemma 5.3. □

5.2.6 Failure of the Endpoint Estimates

The restriction that $1 < p < \infty$ in the estimate (5.45) of Theorem 5.3 is essential. Indeed, when $p = \infty$, Karel DE LEEUW and Hazleton MIRKIL have proved that there is no nontrivial estimate [28, 29].

Theorem 5.5 (L^∞ **Nonestimate**) *Let $n \in \mathbb{N} \setminus \{0\}$, let V, E and F be finite-dimensional vector spaces, let $A(D)$ be a homogeneous constant coefficient differential operator of order $k \in \mathbb{N} \setminus \{0\}$ from V to E on $\mathbb{R}^n$, and let $B(D)$ be a be a homogeneous constant coefficient differential operator of order k from V to F on $\mathbb{R}^n$. If there exists a constant $C \in (0, +\infty)$ such that for every $u \in C_c^\infty(\mathbb{R}^n, V)$ one has*

$$\sup_{\mathbb{R}^n} |B(D)u| \le C \sup_{\mathbb{R}^n} |A(D)u|, \tag{5.50}$$

then there exists $L \in \operatorname{Lin}(E, F)$ such that $B(D) = L \circ A(D)$.

5 Injective Ellipticity, Cancelling Operators and Endpoint Sobolev Inequalities

Proof Let assume for contradiction that there does not exist $L \in \mathrm{Lin}(V, E)$ such that $B(\mathrm{D}) = L \circ A(\mathrm{D})$. Then by duality, there exists a homogeneous polynomial $P : \mathbb{R}^n \to V$ of degree k such that $A(\mathrm{D})P = 0$ and $B(\mathrm{D})P = e \in E \setminus \{0\}$ on $\mathbb{R}^n$.

We choose a function $\eta \in C_c^\infty(\mathbb{R}^n, \mathbb{R})$ such that $\eta = 1$ on $B_1(0)$. We define for every $\varepsilon \in (0, +\infty)$ the function $u_\varepsilon : \mathbb{R}^n \to V$ for each $x \in \mathbb{R}^n$ by

$$u_\varepsilon(x) := \eta(x) P(x) \ln \frac{1}{|x|^2 + \varepsilon^2} .$$

In view of Lemma 5.1, we have for every $x \in \mathbb{R}^n$ and $\varepsilon \in (0, 1)$

$$|A(\mathrm{D})u_\varepsilon(x)| \le C_5 \left(\left| A(\mathrm{D})(\eta P)(x) \ln \frac{1}{|x|^2 + \varepsilon^2} \right| + \sum_{j=1}^k \frac{|\mathrm{D}^{k-j}(\eta P)(x)|}{(|x|^2 + \varepsilon^2)^{j/2}} \right)$$

$$\le C_9 \left(1 + \sum_{j=1}^k \frac{|\mathrm{D}^{k-j}(\eta P)(x)|}{|x|^j} \right) \le C_{10} ,$$

whereas $|B(\mathrm{D})u_\varepsilon(0)| = |e| \ln 1/\varepsilon$; the contradiction follows from (5.50) as $\varepsilon \to 0$.
$\square$

For $p = 1$, Donald S. ORNSTEIN has proved there is no nontrivial estimate [52].[6]

Theorem 5.6 (L^1 **Nonestimate**) *Let $n \in \mathbb{N} \setminus \{0\}$, let V, E and F be finite-dimensional vector spaces, let $A(\mathrm{D})$ be a homogeneous constant coefficient differential operator of order $k \in \mathbb{N} \setminus \{0\}$ from V to E on $\mathbb{R}^n$, and let $B(\mathrm{D})$ be a be a homogeneous constant coefficient differential operator of order k from V to F on $\mathbb{R}^n$. If there exists a constant $C \in (0, +\infty)$ such that for every $u \in C_c^\infty(\mathbb{R}^n, V)$ one has*

$$\int_{\mathbb{R}^n} |B(\mathrm{D})u| \le C \int_{\mathbb{R}^n} |A(\mathrm{D})u| ,$$

then there exists $L \in \mathrm{Lin}(E, F)$ such that $B(\mathrm{D}) = L \circ A(\mathrm{D})$.

The proof of Theorem 5.6 is also based on the existence of a homogeneous polynomial $P : \mathbb{R}^n \to V$ of degree k such that $A(\mathrm{D})P = 0$ and $B(\mathrm{D})P = e \in E \setminus \{0\}$ on $\mathbb{R}^n$. However, the construction of a family of compactly supported functions for which the inequality fails is much more delicate and relies on techniques of convex integration.

---

[6] Although Ornstein's original paper does not cover explicitly the vector case in Theorem 5.6 its proofs seems to do, and more recent approaches explicitly do [42, 43] (for nonestimates through convex integration see also [25]).

5.3 Cancelling Operators

Since in view of Ornstein's L^1 nonestimate a vector differential operator cannot be we cannot controlled nontrivially in L^1 by another vector differential operator, one can consider whether lower-order derivatives can be controlled through Sobolev-type embeddings.

5.3.1 Sobolev Embeddings and Injective Ellipticity

The classical Sobolev embedding theorem on the Euclidean space states that the integrability of a derivative implies some higher-integrability of lower-order derivatives.

Theorem 5.7 (Sobolev Embedding Theorem[7]) *Let $n \in \mathbb{N} \setminus \{0\}$, let V be a finite-dimensional vector space, let $k \in \mathbb{N} \setminus \{0\}$ and $\ell \in \mathbb{N}$ satisfy $0 < k - \ell < n$, and let $p \in [1, \frac{n}{k-\ell})$. There exists a constant $C \in (0, +\infty)$ such that for every $u \in C_c^\infty(\mathbb{R}^n, V)$,*

$$\left(\int_{\mathbb{R}^n} |\mathrm{D}^\ell u|^{\frac{np}{n-(k-\ell)p}} \right)^{1 - \frac{(k-\ell)p}{n}} \le C \int_{\mathbb{R}^n} |\mathrm{D}^k u|^p .$$

When $1 < p < n/(k - \ell)$, as a consequence of the Sobolev embedding theorem (Theorem 5.7) and of Theorem 5.3, if $A(\mathrm{D})$ is a homogeneous constant coefficient differential operator of order $k \in \mathbb{N} \setminus \{0\}$ from V to E on $\mathbb{R}^n$ which is injectively elliptic, then there exists a constant $C \in (0, +\infty)$ such that for every $u \in C_c^\infty(\mathbb{R}^n, V)$, we have the inequality

$$\left(\int_{\mathbb{R}^n} |\mathrm{D}^\ell u|^{\frac{np}{n-(k-\ell)p}} \right)^{1 - \frac{(k-\ell)p}{n}} \le C \int_{\mathbb{R}^n} |A(\mathrm{D})[u]|^p . \tag{5.51}$$

The injective ellipticity of $A(\mathrm{D})$ turns out to be a necessary condition for (5.51) to hold when $\ell = k - 1$ and $1 \le p < n$.

Theorem 5.8 (Necessity of the Injective Ellipticity for Sobolev Embeddings[8]) *Let $n \in \mathbb{N} \setminus \{0\}$, let V and E be finite-dimensional vector spaces, let $A(\mathrm{D})$ be a homogeneous constant coefficient differential operator of order $k \in \mathbb{N} \setminus \{0\}$ from V to E on $\mathbb{R}^n$, let $B(\mathrm{D})$ be a homogeneous constant coefficient differential operator*

---

[7] The Sobolev embedding theorem (Theorem 5.7) is due to Sergei Lvovich SOBOLEV when $p \in (1, m/(k - \ell))$ [58] and to Emilio Gagliardo [35] and simultaneuously to Louis Nirenberg [51] in the endpoint case $p = 1$.
[8] The necessity of injective ellipticity for Sobolev estimates (Theorem 5.8) follows from well-known techniques; an explicit proof appears for example in [74, Prop. 5.1].

of order $k-1$ from V to E on $\mathbb{R}^n$, and let $p \in [1, n)$. If there exists a constant $C \in (0, +\infty)$ such that for every $u \in C_c^\infty(\mathbb{R}^n, V)$,

$$\left(\int_{\mathbb{R}^n} |B(D)u|^{\frac{np}{n-p}}\right)^{1-\frac{p}{n}} \le C \int_{\mathbb{R}^n} |A(D)[u]|^p, \tag{5.52}$$

then for every $\xi \in \mathbb{R}^n \setminus \{0\}$, one has $\ker B(\xi) \subseteq \ker A(\xi)$.

Proof Let $\xi \in \mathbb{R}^n \setminus \{0\}$ and $v \in V$. We take functions $\theta \in C_c^\infty(\mathbb{R}, \mathbb{R}) \setminus \{0\}$ and $\eta \in C_c^\infty(\mathbb{R}^n, \mathbb{R})$ such that $\eta(0) = 1$. For each $\lambda \in \mathbb{R}$, we define the function $u_\lambda : \mathbb{R}^n \to V$ for each $x \in \mathbb{R}^n$ by

$$u_\lambda(x) := \theta(\lambda \langle \xi, x \rangle) \eta(x) v .$$

By Lemma 5.1 and by Minkowski's inequality, we have

$$\left| \left(\int_{\mathbb{R}^n} |A(D)u_\lambda|^p\right)^{\frac{1}{p}} - \left(\lambda^{kp} |A(\xi)[v]| \int_{\mathbb{R}^n} |\theta^{(k)}(\lambda\langle\xi,x\rangle)|^p |\eta(x)|^p \, dx\right)^{\frac{1}{p}} \right|$$
$$\le C_{11} \sum_{j=1}^{k} \left(\lambda^{(k-j)p} \int_{\mathbb{R}^n} |\theta^{(k-j)}(\lambda\langle\xi,x\rangle)|^p |D^j \eta(x)|^p \, dx \right)^{\frac{1}{p}} .$$

and thus if $\lambda \ge 1$,

$$\left(\int_{\mathbb{R}^n} |A(D)u_\lambda|^p\right)^{\frac{1}{p}} \le \lambda^{k-\frac{1}{p}} |A(\xi)[v]| \left(\int_{\mathbb{R}} |\theta^{(k)}|^p\right)^{\frac{1}{p}} \left(\int_{\xi^\perp} |\eta|^p\right)^{\frac{1}{p}} + C_{15} \lambda^{k-1-\frac{1}{p}} . \tag{5.53}$$

Hence if $v \in \ker A(\xi)$, the inequality (5.53) implies

$$\limsup_{\lambda \to \infty} \frac{1}{\lambda^{k-\frac{1}{p}}} \left(\int_{\mathbb{R}^n} |A(D)u_\lambda|^p\right)^{\frac{1}{p}} < \infty . \tag{5.54}$$

By a similar reasoning on the operator $B(D)$, we have

$$\lim_{\lambda \to \infty} \frac{1}{\lambda^{k-1-\frac{1}{p}+\frac{1}{n}}} \left(\int_{\mathbb{R}^n} |B(D)u_\lambda|^{\frac{np}{n-p}}\right)^{\frac{1}{p}-\frac{1}{n}} = |B(\xi)[v]| \int_{\mathbb{R}} |\theta^{(k-1)}|^p \int_{\xi^\perp} |\eta|^p , \tag{5.55}$$

and thus by (5.52), (5.54) and (5.55), we have $|B(\xi)[v]| = 0$, that is, $v \in \ker B(\xi)$. $\square$

Remark 5.2 Ellipticity is not necessary for estimates of lower-order derivatives: Indeed, one has for example for every $u \in C_c^\infty(\mathbb{R}^{2n}, \mathbb{R})$,

$$\left(\int_{\mathbb{R}^{2n}} |u|^2\right)^{\frac{1}{2}} \leq C \int_{\mathbb{R}^{2n}} |\partial_1 \cdots \partial_n u| + |\partial_{n+1} \cdots \partial_{2n} u| \,.$$

5.3.2 Sobolev Embeddings and Cancelling Operators

In the case $p = 1$, injective ellipticity is still not sufficient to have a Sobolev estimate.

Example 5.7 We define for each $\lambda \in (0, +\infty)$ the function $u_\lambda : \mathbb{R}^n \to \mathbb{R}^n$ by setting for each $x \in \mathbb{R}^n$

$$u_\lambda(x) := \frac{x}{|x|^n} \eta(|x|)(1 - \eta(|x|/\lambda)) \,,$$

with $\eta \in C^\infty((0, +\infty), \mathbb{R})$ such that $\eta = 1$ on $(0, 1)$ and $\eta = 0$ on $(2, +\infty)$. We have then $\operatorname{curl} u_\lambda = 0$ on $\mathbb{R}^n$,

$$\int_{\mathbb{R}^n} |\operatorname{div} u_\lambda| \leq C_{16}$$

and

$$\lim_{\lambda \to +\infty} \int_{B_1(0)} |u_\lambda|^{\frac{n}{n-1}} \geq \int_{B_1(0)} \frac{dx}{|x|^n} = +\infty \,.$$

When $p = 1$, by Theorem 5.6 the Calderón-Zygmund theory of singular integrals (Theorem 5.3) fails, but one can still ask whether there are some endpoint Sobolev estimates for vector differential operators.

We will show that such estimates under an additional cancellation condition defined as follows.

Definition 5.3 (Cancelling Operator[9]) Let $n \in \mathbb{N} \setminus \{0\}$ and let V and E be finite-dimensional vector spaces. A homogeneous constant coefficient differential operator $A(D)$ of order $k \in \mathbb{N} \setminus \{0\}$ from V to E on $\mathbb{R}^n$ is *cancelling* whenever

$$\bigcap_{\xi \in \mathbb{R}^n \setminus \{0\}} A(\xi)[V] = \{0\} \,.$$

---

[9] The definition of cancelling operators (Definition 5.3) is due to the author [74].

5 Injective Ellipticity, Cancelling Operators and Endpoint Sobolev Inequalities   283

When $n = 1$, an operator $A(D)$ is cancelling if and only if $A(D) = 0$; hence this notion only makes sense in higher dimensions $n \geq 2$.

Formally, the cancellation condition means that $A(D)u = e\delta_0$ does not have a solution unless $e = 0$, as this would imply that $(2\pi i)^k A(\xi)\mathcal{F}u(\xi) = \mathcal{F}(A(D)u)(\xi) = \mathcal{F}(e\delta_0) = e$; this will be proved in Proposition 5.6.

The cancellation property characterizes when endpoint Sobolev inequalities hold.

Theorem 5.9 (Endpoint Sobolev Inequality for Cancelling Operators[10]**)** *Let $n \in \mathbb{N} \setminus \{0\}$, let V and E be finite-dimensional vector spaces, and let $A(D)$ be a homogeneous constant coefficient differential operator of order $k \in \mathbb{N} \setminus \{0\}$ from V to E on $\mathbb{R}^n$. Assume that $A(D)$ is injectively elliptic and that $\ell \in \mathbb{N} \setminus \{0\}$ satisfies $0 < k - \ell < n$. There exists a constant $C \in (0, +\infty)$ such that for every $u \in C_c^\infty(\mathbb{R}^n, V)$*

$$\left(\int_{\mathbb{R}^n} |D^\ell u|^{\frac{n}{n-(k-\ell)}} \right)^{1-\frac{k-\ell}{n}} \leq C \int_{\mathbb{R}^n} |A(D)[u]|, \qquad (5.56)$$

if and only if the operator $A(D)$ is cancelling.

It should be noted that Theorem 5.9 imposes strong boundary conditions on u—requiring in fact compact support—for the estimate (5.56) to hold. Working with weaker boundary conditions would introduce additionnal restrictions on the admissible class of operators [17, 36, 37].

We close this section by the proof of the necessity of the cancellation condition in Theorem 5.9. Roughly speaking one would like to take $u(x) := G_A(x)[e]$ for some $e \in \bigcap_{\xi \in \mathbb{R}^n \setminus \{0\}} A(\xi)[V]$ so that $A(D)u = \delta e_0$ (see Proposition 5.2 (v)), but unfortunately this function u is neither smooth nor compactly supported. The next lemma provides a careful approximation of such a function.

Lemma 5.4 (Regularization of the Representation Kernel[11]**)** *Let $n \in \mathbb{N} \setminus \{0\}$, let V and E be finite-dimensional vector spaces, and let $A(D)$ be a homogeneous constant coefficient differential operator of order $k \in \mathbb{N} \setminus \{0\}$ from V to E on $\mathbb{R}^n$. If $A(D)$ is injectively elliptic and if $e \in \bigcap_{\xi \in \mathbb{R}^n \setminus \{0\}} A(\xi)[V]$, then there exists a sequence $(u_j)_{j \in \mathbb{N}}$ in $C_c^\infty(\mathbb{R}^n, V)$ such that*

(i) for every $j \in \mathbb{N}$, $\operatorname{supp} u_j \subset B_2(0)$,
(ii) $\lim_{j \to \infty} \int_{\mathbb{R}^n} |A(D)u_j| \leq C|e|$,
(iii) $u_j \to G_A[e]$ in $C_{\mathrm{loc}}^\infty(B_1(0) \setminus \{0\}, E)$ as $j \to \infty$.

---

[10] The necessity and sufficiency of cancellation for vector Sobolev estimates is due to the author [74, Prop. 4.6 and 5.5].

[11] The statement of Lemma 5.4 is due to the author [74]*Prop. 5.5. We give here a proof based on the properties of the representation kernel instead of a direct construction in Fourier space as in [74].

Proof of Lemma 5.4 We take a function $\varrho \in C_c^\infty(\mathbb{R}^n, \mathbb{R})$ such that $\int_{\mathbb{R}^n} \varrho = 1$. We define first for every $\lambda \in (0, +\infty)$ the function $v_\lambda : \mathbb{R}^n \to V$ by setting for every $x \in \mathbb{R}^n$

$$v_\lambda(x) := \lambda^n \int_{\mathbb{R}^n} G_A(x-y)[\varrho(\lambda y)e] \, dy \,.$$

By our assumption, for every $\xi \in \mathbb{R}^n$, we have

$$\mathcal{F}(\varrho e)(\xi) = (\mathcal{F}\varrho(\xi))e \in \mathbb{C}e \subseteq A(\xi)[V + iV]$$

and thus by Proposition 5.2 (iv), for every $x \in \mathbb{R}^n$, $A(D)v_\lambda(x) = \lambda^n \varrho(\lambda x)$. Moreover, as $\lambda \to +\infty$, we have $v_\lambda \to G_A[e]$ in $C_{\text{loc}}^\infty(\mathbb{R}^n \setminus \{0\}, E)$. We take now a function $\eta \in C^\infty(\mathbb{R}^n, \mathbb{R})$ such that $\eta = 0$ on $\mathbb{R}^n \setminus B_2(0)$ and $\eta = 1$ on $B_1(0)$ and we set $u_j := \eta v_{\lambda_j}$ for some sequence $(\lambda_j)_{j \in \mathbb{N}}$ in $(0, +\infty)$ going to $+\infty$. □

Proof of the Necessity of Cancellation in Theorem 5.9 We take some vector $e \in \bigcap_{\xi \in \mathbb{R}^n \setminus \{0\}} A(\xi)[V]$ and we let the sequence $(u_j)_{j \in \mathbb{N}}$ be given by Lemma 5.4 for this vector e. By assumption (5.56), we have for every $j \in \mathbb{N}$,

$$\left(\int_{\mathbb{R}^n} |D^\ell u_j|^{\frac{n}{n-(k-\ell)}} \right)^{1-\frac{k-\ell}{n}} \le C_{20} \int_{\mathbb{R}^n} |A(D)u_j| \,.$$

It follows thus by Fatou's lemma that

$$\left(\int_{B_1(0)} |D^\ell G_A[e]|^{\frac{n}{n-(k-\ell)}} \right)^{1-\frac{k-\ell}{n}} \le C_{21}|e| < +\infty \,. \tag{5.57}$$

Since $\ell > n - k$, the function $D^\ell G_A[e] : \mathbb{R}^n \setminus \{0\} \to V$ is homogeneous of degree $n - (k - \ell)$ (by Proposition 5.2 (ii)); (5.57) then implies that $G_A[e] = 0$ on $\mathbb{R}^n$ and thus that $e = 0$ in view of Proposition 5.2 (v). □

5.3.3 Examples of Cancelling Operators

We review now several examples of cancelling operators.

Example 5.8 (Derivative) The derivative $A(D) = D^k$ is cancelling on $\mathbb{R}^n$ if and only if $n \ge 2$. Indeed, for every $\xi \in \mathbb{R}^n$, $v, w \in \mathbb{R}^n$ and $\eta_1, \ldots, \eta_k \in \mathbb{R}^n$, we have

$$\langle A(\xi)[v], w \otimes \eta_1 \otimes \cdots \otimes \eta_k \rangle = \langle v \otimes \xi^{\otimes k}, w \otimes \eta_1 \otimes \cdots \otimes \eta_k \rangle$$
$$= \langle v, w \rangle \langle \eta_1, \xi \rangle \cdots \langle \eta_k, \xi \rangle \,.$$

Since $n \geq 2$, for every $\eta_1 \in \mathbb{R}^n$, there exists $\xi \in \mathbb{R}^n \setminus \{0\}$ such that $\langle \eta_1, \xi \rangle = 0$, and thus

$$\bigcap_{\xi \in \mathbb{R}^n \setminus \{0\}} A(\xi)[V] \subseteq \bigcap_{\substack{\eta_1,\ldots,\eta_k \in \mathbb{R}^n \\ w \in V}} (w \otimes \eta_1 \otimes \cdots \otimes \eta_k)^\perp = \{0\} \ .$$

Example 5.9 (Hodge Complex) The operator $A(\mathrm{D}) = (d, d^*)$ acting on $C^\infty(\mathbb{R}^n, \bigwedge^m \mathbb{R}^n)$ is cancelling if and only if $m \notin \{1, n-1\}$. Here, d and d^* are respectively the exterior differential and codifferential. Indeed, if $(v, v_*) \in \bigcap_{\xi \in \mathbb{R}^n \setminus \{0\}} A(\xi)[V]$, then for every $\xi \in \mathbb{R}^n$, $\xi \wedge v = 0$ and $\xi \lrcorner v_* = 0$. Since $m \notin \{1, n-1\}$, this implies that $v = 0$ and $v_* = 0$. As a consequence of Theorem 5.9, one gets Jean BOURGAIN and Haïm BREZIS's endpoint Sobolev inequality [8], [9, Cor. 17]:[12] for every $u \in C_c^\infty(\mathbb{R}^n, \bigwedge^m \mathbb{R}^n)$, if $m \in \{2, \ldots, n-2\}$,

$$\left(\int_{\mathbb{R}^n} |u|^{\frac{n}{n-1}} \right)^{1-\frac{1}{n}} \leq C \int_{\mathbb{R}^n} |du| + |d^* u| \ . \tag{5.58}$$

On the other hand when $m = 1$ and $n \geq 3$, we have $\bigcap_{\xi \in \mathbb{R}^n \setminus \{0\}} A(\xi)[V] = \{0\} \times \bigwedge^0 \mathbb{R}^n$ and when $m = n-1$ and $n \geq 3$, we have $\bigcap_{\xi \in \mathbb{R}^n \setminus \{0\}} A(\xi)[V] = \bigwedge^n \mathbb{R}^n \times \{0\}$, and finally when $m = 1$ and $n = 2$, we have $\bigcap_{\xi \in \mathbb{R}^n \setminus \{0\}} A(\xi)[V] = \bigwedge^2 \mathbb{R}^n \times \bigwedge^0 \mathbb{R}^n$, so that the operator $A(\mathrm{D})$ is not cancelling and the estimate (5.58) does not hold.

Example 5.10 (Symmetric Derivative) The symmetric derivative operator $A(\mathrm{D})u = (\mathrm{D}u + (\mathrm{D}u)^*)/2$ is cancelling if and only if $n \geq 2$. Indeed, for every $w \in \mathbb{R}^n$ and $v \in \mathbb{R}^n$,

$$\langle A(\xi)[v], w \otimes w \rangle = \frac{\langle \xi \otimes v + v \otimes \xi, w \otimes w \rangle}{2} = \langle \xi, w \rangle \langle v, w \rangle \ .$$

Since $n \geq 2$, we can choose $\xi \in \mathbb{R}^n \setminus \{0\}$ such that $\langle \xi, w \rangle = 0$, and thus for every $v \in \mathbb{R}^n$, $\langle A(\xi)[v], w \otimes w \rangle = \{0\}$, and hence

$$\bigcap_{\xi \in \mathbb{R}^n \setminus \{0\}} A(\xi)[\mathbb{R}^n] \subseteq \mathrm{Lin}^2_{\mathrm{sym}}(\mathbb{R}^n, \mathbb{R}) \cap \bigcap_{w \in \mathbb{R}^n} (w \otimes w)^\perp = \{0\} \ .$$

As a consequence of Theorem 5.9, we recover Monty J. STRAUSS's Korn-Sobolev inequality [64]

$$\left(\int_{\mathbb{R}^n} |u|^{\frac{n}{n-1}} \right)^{1-\frac{1}{n}} \leq C \int_{\mathbb{R}^n} |\mathrm{D}_{\mathrm{sym}} u| \ . \tag{5.59}$$

---

[12] Another proof of the endpoint Sobolev estimate for forms (5.58) was given by Loredana Lanzani and Elias M. Stein, and [45].

Example 5.11 (Trace-Free Symmetric Derivative[13]**)** If we consider $A(D)u := D_{\text{sym}}u + \lambda \operatorname{div} u$, then $A(D)$ is cancelling when either $n = 2$ and $\lambda \neq -\frac{1}{2}$, or $n \geq 3$. In particular, in the case $\lambda = -\frac{1}{n}$, $A(D)$ is the trace-free symmetric derivative which is cancelling if and only if $n \geq 3$. Indeed, for every $v, w, \xi \in \mathbb{R}^n$ such that $\langle \xi, w \rangle = 0$, we have $\langle A(\xi)[v], \xi \otimes \xi \rangle = (1+\lambda)|\xi|^2 \langle \xi, v \rangle$ and $\langle A(\xi)[v], w \otimes w \rangle = \lambda |w|^2 \langle \xi, v \rangle$, so that if $|w| = |\xi|$, one has $\langle A(\xi)[v], \lambda \xi \otimes \xi - (1+\lambda)w \otimes w \rangle = 0$ and it follows that

$$\bigcap_{\xi \in \mathbb{R}^n \setminus \{0\}} A(\xi)[\mathbb{R}^n] \subseteq \operatorname{Lin}^2_{\text{sym}}(\mathbb{R}^n, \mathbb{R}) \cap \bigcap_{\substack{\xi, w \in \mathbb{R}^n \setminus \{0\} \\ \langle \xi, w \rangle = 0 \\ |\xi| = |w|}} \{\lambda \xi \otimes \xi - (1+\lambda)w \otimes w\}^\perp = \{0\}.$$

Indeed if $\lambda \neq -\frac{1}{2}$ and $n \geq 2$, then any element in the common image should be a tensor represented in any orthonormal basis as a matrix with zero diagonal; when $\lambda = -\frac{1}{2}$, then the sum of any two distinct elements in the diagonal should be 0, which imply they should all be zero if $n \geq 3$. As a consequence of Theorem 5.9, we get that when either $n = 2$ and $\lambda \neq -\frac{1}{2}$, or $n \geq 3$, for every $u \in C_c^\infty(\mathbb{R}^n, \mathbb{R}^n)$,

$$\left(\int_{\mathbb{R}^n} |u|^{\frac{n}{n-1}} \right)^{1-\frac{1}{n}} \leq C \int_{\mathbb{R}^n} |D_{\text{sym}}u + \lambda \operatorname{div} u|. \tag{5.60}$$

Example 5.12 (Laplacian) The Laplacian $A(D) = \Delta$ is *not* cancelling as an operator from V to V on $\mathbb{R}^n$. Indeed, for every $\xi \in \mathbb{R}^n \setminus \{0\}$, one has $A(\xi)[V] = \{|\xi|^2 v \mid v \in V\} = V$, and thus $\bigcap_{\xi \in \mathbb{R}^n \setminus \{0\}} A(\xi)[V] = V \neq \{0\}$, so that the Laplacian operator is not cancelling.

Example 5.13 (Cauchy-Riemann Operator) The *Cauchy-Riemann operator* $A(D) = \bar{\partial}$, which is a differential operator from $\mathbb{C}$ to $\mathbb{C}$ on $\mathbb{R}^2$ defined for each function $u \in C^\infty(\mathbb{R}^2, \mathbb{C})$ by $A(D)u = (\partial_1 u + i \partial_2 u)/2$ is *not* cancelling. Indeed, for every $\xi = (\xi_1, \xi_2) \in \mathbb{R}^2 \setminus \{0\}$, one has $A(\xi)\mathbb{C} = \mathbb{C}$ and thus $\bigcap_{\xi \in \mathbb{R}^n \setminus \{0\}} A(\xi)[\mathbb{C}] = \mathbb{C} \neq \{0\}$, so that the operator $A(D)$ is not cancelling.

The noncancellation properties of examples 5.12 and 5.13 are particular cases of the general fact that if the differential operator $A(D)$ is injectively elliptic and if $\dim V = \dim E$, then for every $\xi \in \mathbb{R}^n$, one has $\dim A(\xi)[V] = \dim E$ and thus $A(\xi)[V] = E$ and finally $\bigcap_{\xi \in \mathbb{R}^n \setminus \{0\}} A(\xi)[V] = E$.

The cancellation property can be obtained as a consequence of stronger properties such as $\mathbb{C}$-ellipticity [36, Lem. 3.2] or boundary ellipticity [38]. Many of the examples above satisfy a *strong cancellation* property in which the intersection is taken on any two-dimensional subspace [37].

---

[13] The cancellation of the trace-free symmetric derivative if and only if $n \geq 3$ was obtained as a consequence of its $\mathbb{C}$-ellipticity by Dominic Breit et al. [15, Ex. 2.2]; their proof adapts to $\lambda > -\frac{1}{2}$. The $\mathbb{C}$-ellipticity is related to the notion of conformal Killing vectors [27].

5.4 Duality Estimates

Given a cancelling and injectively elliptic homogeneous constant coefficient differential operator $A(\mathrm{D})$, we have constructed in Proposition 5.3 an operator $L(\mathrm{D})$, describing the associated compatibility conditions, such that for every $u \in C_c^\infty(\mathbb{R}^n, V)$, one has $L(\mathrm{D})A(\mathrm{D})u = 0$ and such that, by (5.42),

$$\bigcap_{\xi \in \mathbb{R}^n \setminus \{0\}} \ker L(\xi) = \bigcap_{\xi \in \mathbb{R}^n \setminus \{0\}} A(\xi)[V] = \{0\} \,.$$

Setting $f := A(\mathrm{D})u$, we investigate the estimates in terms of $\|f\|_{L^1(\mathbb{R}^n)}$ that can be obtained taking into account the condition $L(\mathrm{D})f = 0$.

5.4.1 Estimates for Cocancelling Operators

Motivated by the compatibility conditions of Proposition 5.3 and the definition of cancelling operators (Proposition 5.3), we define cocancelling operators:

Definition 5.4 (Cocancelling Operator[14]) Let $n \in \mathbb{N} \setminus \{0\}$ and let E and F be finite-dimensional vector spaces. A homogeneous constant coefficient differential operator $L(\mathrm{D})$ from E to F on $\mathbb{R}^n$ is *cocancelling* whenever

$$\bigcap_{\xi \in \mathbb{R}^n \setminus \{0\}} \ker L(\xi) = \{0\} \,.$$

When $n = 1$, one notes that $L(\mathrm{D})$ is cocancelling if and only if $L(\mathrm{D}) = 0$; hence the notion will only be relevant for $n \ge 2$.

Proposition 5.4 (Characterization of Cocancelling Operators[15]) *Let $n \in \mathbb{N}\setminus\{0\}$, let E and F be finite-dimensional vector spaces, and let $L(\mathrm{D})$ be a homogeneous linear differential operator of order $k \in \mathbb{N} \setminus \{0\}$ on $\mathbb{R}^n$ from E to F. The following are equivalent*

(i) the operator $L(\mathrm{D})$ is cocancelling,
(ii) for every $e \in E$ such that $L(\mathrm{D})(e\delta_0) = 0$ in the sense of distributions, one has $e = 0$,

---

[14] The identification of the cocancellation condition of Definition 5.4 as a necessary and sufficient condition for duality estimates in critical Sobolev spaces is due to the author [74]; similar conditions appears in characterizations of the dimension of measures lying in the kernel of operators [4, 30, 54].

[15] Proposition 5.4 characterizing cocancelling operators is due to the author [74].

(iii) for every $f \in L^1(\mathbb{R}^n, E)$ such that $L(D)f = 0$ in the sense of distributions, one has $\int_{\mathbb{R}^n} f = 0$,
(iv) for every $f \in C_c^\infty(\mathbb{R}^n, E)$ such that $L(D)f = 0$, one has $\int_{\mathbb{R}^n} f = 0$.

Here δ_0 denotes Dirac's measure at 0 on $\mathbb{R}^n$.

Proof of Proposition 5.4 We assume that (i) holds, that is, that the operator $L(D)$ is cocancelling and we consider a vector $e \in E$ such that $L(D)(e\delta_0) = 0$ in the sense of distributions. For every $\varphi \in \mathcal{S}(\mathbb{R}^n, F)$, by definition of the distributional derivative and properties of the Fourier transform $\mathcal{F}\varphi$ of φ, we have

$$\int_{\mathbb{R}^n} \langle (2\pi i)^k L(\xi)[e], \varphi(\xi) \rangle \, d\xi = \langle L(D)(e\delta_0), \mathcal{F}^{-1}\varphi \rangle = 0 .$$

and hence, since $L : \mathbb{R}^n \to \mathrm{Lin}(E, F)$ is continuous, for every $\xi \in \mathbb{R}^n$, we have $L(\xi)[e] = 0$ and thus $e \in \bigcap_{\xi \in \mathbb{R}^n \setminus \{0\}} \ker L(\xi)$. By definition of cocancelling operator (Definition 5.4), we have $e = 0$ and thus (ii) holds.

We assume now that (ii) holds. Given $f \in L^1(\mathbb{R}^n, E)$ such that $L(D)f = 0$ and given $\lambda \in (0, +\infty)$, we define the function $f_\lambda : \mathbb{R}^n \to E$ for each $x \in \mathbb{R}^n$ by $f_\lambda(x) := \frac{1}{\lambda^n} f\left(\frac{x}{\lambda}\right)$. Since $f_\lambda \to \delta_0 \int_{\mathbb{R}^n} f$ in the sense of distributions as $\lambda \to 0$, we have $L(D)f_\lambda \to L(D)(\delta_0 \int_{\mathbb{R}^n} f)$ in the sense of distributions as $\lambda \to 0$. Since $L(D)$ is homogeneous and has constant coefficients, we have for every $\lambda \in (0, +\infty)$, $L(D)f_\lambda = 0$ and hence $L(D)(\delta_0 \int_{\mathbb{R}^n} f) = 0$. By our assumption (ii), we deduce that $\int_{\mathbb{R}^n} f = 0$ and thus (iii) holds.

Since we have $C_c^\infty(\mathbb{R}^n, E) \subset L^1(\mathbb{R}^n, E)$, the assertion (iii) implies immediately the assertion (iv).

Finally, let us assume that (iv) holds. Let $e \in \bigcap_{\xi \in \mathbb{R}^n \setminus \{0\}} \ker L(\xi)$ and let $\psi \in C_c^\infty(\mathbb{R}^n, E)$ such that $\int_{\mathbb{R}^n} \psi = 1$. For every $x \in \mathbb{R}^n$, we have by the Fourier inversion formula,

$$\bigl(L(D)(\psi e)\bigr)(x) = \int_{\mathbb{R}^n} e^{2\pi i \langle \xi, x \rangle} (2\pi i)^k L(\xi)[e] \mathcal{F}\psi(\xi) \, d\xi = 0 .$$

By (iv), we conclude that $e = \int_{\mathbb{R}^n} \psi e = 0$. The operator $L(D)$ is thus cocancelling by definition (Definition 5.4) and (i) holds. □

The cocancelling operator is a necessary and sufficient condition for some duality estimate with critical Sobolev spaces.

Theorem 5.10 (Duality Estimate for Cocancelling Operators[16]**)** *Let $n \in \mathbb{N} \setminus \{0, 1\}$, let V and E be finite-dimensional vector spaces, and let $L(D)$ be a homogeneous constant coefficient differential operator from E to F on $\mathbb{R}^n$. There*

---

[16] Estimates of the form (5.61) originates in an estimate on circulation integrals by Jean Bourgain et al. [10] (see also [68]). The estimate (5.61) was first proved when $L(D)$ is the divergence [8, 9] (see also [69]); it was then extended to the case where $L(D)$ is in some class of second-order operators [70] or higher-operators [9], and finally for the higher-order divergence [72] and then deduced by an algebraic argument for a general cocancelling operator [74]. The approach presented here is a direct proof of the estimate (5.61).

exists a constant $C \in (0, +\infty)$ such that for every $f \in C^\infty(\mathbb{R}^n, E) \cap L^1(\mathbb{R}^n, E)$ that satisfies $L(D)f = 0$ and every $\varphi \in C_c^\infty(\mathbb{R}^n, E)$,

$$\left| \int_{\mathbb{R}^n} \langle f, \varphi \rangle \right| \le C \int_{\mathbb{R}^n} |f| \left(\int_{\mathbb{R}^n} |D\varphi|^n \right)^{\frac{1}{n}} \tag{5.61}$$

if and only if the operator $L(D)$ is cocancelling.

Theorem 5.10 can be seen as a substitute for the missing embedding of the critical Sobolev space $\dot{W}^{1,n}(\mathbb{R}^n, E)$ in $L^\infty(\mathbb{R}^n, E)$, which would be equivalent to having (5.61) without the condition $L(D)f = 0$.

Proof of the Necessity of Cocancellation in Theorem 5.10 We consider a sequence $(\varphi_j)_{j \in \mathbb{N}}$ in $C_c^\infty(\mathbb{R}^n)$, such that

(a) for every $j \in \mathbb{N}$ and $x \in \mathbb{R}^n$, one has $0 \le \varphi_j(x) \le 1$,
(b) for every $x \in \mathbb{R}^n$, the sequence $(\varphi_j(x))_{j \in \mathbb{N}}$ converges to 1,
(c) $\lim_{n \to \infty} \int_{\mathbb{R}^n} |D\varphi_j|^n = 0$.

(One can take for instance $\varphi_j(x) := \phi((\ln|x|)/j)$ with $\phi \in C^\infty(\mathbb{R}, \mathbb{R})$ such that $\phi = 1$ on $(-\infty, 1]$, $0 \le \phi \le 1$ on $[1, 2]$ and $\phi = 0$ on $[2, +\infty)$.) By our assumption (5.61) we have for every $j \in \mathbb{N}$ and $e \in E$

$$\left| \int_{\mathbb{R}^n} \langle f, \varphi_j e \rangle \right| \le C |e| \int_{\mathbb{R}^n} |f| \left(\int_{\mathbb{R}^n} |D\varphi_j|^n \right)^{\frac{1}{n}}. \tag{5.62}$$

By Lebesgue's dominated convergence theorem, we have in view of (a) and (b)

$$\lim_{j \to \infty} \int_{\mathbb{R}^n} \langle f, \varphi_j \rangle = \int_{\mathbb{R}^n} \langle f, e \rangle. \tag{5.63}$$

It follows then from (5.62) and (5.63), taking into account (c), that $\int_{\mathbb{R}^n} f = 0$, and thus by Proposition 5.4, the operator $L(D)$ is cocancelling. □

We review now several examples of cocancelling operators.

Example 5.14 (Divergence) The *divergence* operator $L(D) = \text{div}$ on $\mathbb{R}^n$ is cocancelling when $n \ge 2$. Indeed, for every $\xi \in \mathbb{R}^n$, $\ker L(\xi) = \{\xi\}^\perp$, and thus, since $n \ge 2$, $\bigcap_{\xi \in \mathbb{R}^n \setminus \{0\}} \ker L(\xi) = \{0\}$. As a consequence of Theorem 5.10, one gets that if $f \in L^1(\mathbb{R}^n, \mathbb{R}^n)$ satisfies $\text{div } f = 0$ in the sense of distributions then (5.61) holds.[17] This estimate can be seen, through Stanislav K. SMIRNOV's result on the approximation of divergence-free measures [57], to be equivalent to the estimate on

---

[17] This result was first obtained as a consequence of a stronger estimate by Haïm Brezis and Jean Bourgain [8, 9]; a direct proof of the result was given by the author [69].

circulation integrals of Jean Bourgain et al. [10]:[18] if $\Gamma \subset \mathbb{R}^n$ is a *closed* curve with tangent vector t and length $|\Gamma|$, then for every vector field $\varphi \in C_c^\infty(\mathbb{R}^n, \mathbb{R}^n)$, one has

$$\left| \int_\Gamma \langle \varphi, t \rangle \right| \le C_{22} |\Gamma| \left(\int_{\mathbb{R}^n} |D\varphi|^n \right)^{\frac{1}{n}}. \tag{5.64}$$

Example 5.15 (Exterior Derivative) If $E = \bigwedge^m \mathbb{R}^n$, with $m \in \{0, \ldots, n-1\}$, the *exterior derivative* $L(D) = d$ is cocancelling. Indeed, one has

$$\ker L(\xi) = \left\{ \alpha \in \bigwedge^m \mathbb{R}^n \mid \alpha \wedge \xi = 0 \right\}.$$

Thus if $\alpha \in \ker L(\xi)$, then α has its restriction to $\{\xi\}^\perp$ being equal to 0. If this happens for all $\xi \in \mathbb{R}^n$, then $\alpha = 0$.

Example 5.16 (Saint-Venant Operator) The Saint-Venant differential operator $L(D)$ from $\mathrm{Lin}^2_{\mathrm{sym}}(\mathbb{R}^n, \mathbb{R})$ to $\mathrm{Lin}^4(\mathbb{R}^n, \mathbb{R})$ defined for every $u \in C^\infty(\mathbb{R}^n, \mathrm{Lin}^2_{\mathrm{sym}}(\mathbb{R}^n, \mathbb{R}))$, and $v_1, \ldots, v_4 \in \mathbb{R}^n$ by

$$(L(D)u)[v_1, \ldots, v_4] := \langle v_1 \otimes v_2, D^2 u[v_3, v_4] \rangle + \langle v_3 \otimes v_4, D^2 u[v_1, v_2] \rangle$$
$$- \langle v_1 \otimes v_4, D^2 u[v_2, v_3] \rangle - \langle v_2 \otimes v_3, D^2 u[v_1, v_4] \rangle$$

is cocancelling when $n \ge 2$. This operator corresponds to the Saint-Venant compatibility conditions (5.43) of the symmetric derivative. If for every $\xi \in \mathbb{R}^n \setminus \{0\}$, we have $e \in \ker L(\xi) \subset \mathrm{Lin}^2_{\mathrm{sym}}(\mathbb{R}^n, \mathbb{R})$, then for every $\xi \in \mathbb{R}^n \setminus \{0\}$ and every $v \in \mathbb{R}^n$,

$$(L(\xi)e)[v, v, \xi, \xi] = e(v, v)|\xi|^4 + e(\xi, \xi)\langle \xi, v \rangle^2 - 2e(v, \xi)\langle \xi, v \rangle |\xi|^2 = 0,$$

so that if one has $\langle \xi, v \rangle = 0$, then $e(v, v)|\xi|^4 = 0$ and it follows then that since $n \ge 2$

$$\bigcap_{\xi \in \mathbb{R}^n \setminus \{0\}} \ker L(\xi) \subseteq \bigcap_{\substack{\xi \in \mathbb{R}^n \setminus \{0\} \\ v \in \xi^\perp}} (v \otimes v)^\perp = \{0\}.$$

As a consequence of Theorem 5.10, one gets that if $f \in L^1(\mathbb{R}^n, \mathrm{Lin}^2_{\mathrm{sym}}(\mathbb{R}^n, \mathbb{R}))$ satisfies $L(D)f = 0$, then the estimate (5.61) holds.[19]

---

[18] A direct proof was given by the author [68]; the estimate is equivalent when $n = 2$ to the classical isoperimetric theorem, raising intriguing questions on sharp constants in higher dimensions [18].

[19] The estimate (5.61) under the Saint-Venant compatibility conditions (5.43) was obtained by the author [70].

Example 5.17 (Higher-Order Divergence) The higher-order divergence $L(\mathrm{D}) = \mathrm{div}^k$ is cocancelling on symmetric k-linear forms. Indeed, for every $\xi \in \mathbb{R}^n$ and $\tau \in \mathrm{Lin}_{\mathrm{sym}}^k(\mathbb{R}^n, \mathbb{R})$, one has $L(\xi)[\tau] = \tau[\xi, \ldots, \xi]$ and $\bigcap_{\xi \in \mathbb{R}^n \setminus \{0\}} \ker L(\xi) = \{0\}$.

5.4.2 Estimates Under a Cocancelling Condition

The essential ingredient in the original proof of the estimate in Theorem 5.10 by Jean BOURGAIN and Haïm BREZIS is an approximation result obtained through a Littlewood-Paley decomposition: For every $\varepsilon \in (0, +\infty)$, there exists a constant $M_\varepsilon \in (0, +\infty)$ such that for every function $u \in \dot{W}^{1,n}(\mathbb{R}^n, \mathbb{R})$, one can construct a function $v \in (\dot{W}^{1,n} \cap L^\infty)(\mathbb{R}^n, \mathbb{R})$ satisfying

$$\|\mathrm{D}v\|_{L^n(\mathbb{R}^n)} + \|v\|_{L^\infty(\mathbb{R}^n)} \le M_\varepsilon \|\mathrm{D}u\|_{L^n(\mathbb{R}^n)},$$

and

$$\|\mathrm{D}'u - \mathrm{D}'v\|_{L^n(\mathbb{R}^n)} \le \varepsilon \|\mathrm{D}u\|_{L^n(\mathbb{R}^n)},$$

where D' is the derivative with respect to the $n-1$ first variables; this gives a stronger estimate with $\|f\|_{L^1(\mathbb{R}^n)}$ replaced by the weaker norm $\|f\|_{L^1(\mathbb{R}^n) + \dot{W}^{-1,n/(n-1)}(\mathbb{R}^n)}$ [8, 9].[20]

We present here a proof of Theorem 5.10 based on Morrey's embedding on $\mathbb{R}^{n-1}$ and some appropriate integration by parts.[21]

5.4.2.1 About Hölder-Continuous Functions

The first analytical tool that we will used for the proof of the sufficiency part in Theorem 5.10 will be the Morrey-Sobolev embedding applied on some hyperplanes.

---

[20] This approximation result generalized Jean BOURGAIN and Haïm BREZIS's similar approximation result in their work on the divergence equation [6, 7], and yields some estimates for a large class of cocancelling operators [9]. Algebraic arguments to reach the whole class of cocancelling operators are due to the author [72, 74].

[21] This strategy goes back to the author's elementary proof of an inequality of Jean BOURGAIN, Haïm BREZIS and Petru MIRONESCU on circulation integrals [68], which was successively extended to divergence-free vector fields [69], some class of second order operator [70] and the higher-order divergence [72]. The latter proof together with appropriate algebraic manipulations yielded the general estimate for cocancelling operators [74].

Proposition 5.5 (Morrey–Sobolev Embedding[22]**)** *Let $\ell \in \mathbb{N} \setminus \{0\}$, let V be finite-dimensional vector space and $p \in (\ell, +\infty)$. There exists a constant $C \in (0, +\infty)$ such that for every $\psi \in C_c^\infty(\mathbb{R}^\ell, V)$,*

$$|\psi(y) - \psi(x)| \le C |y-x|^{1-\frac{\ell}{p}} \left(\int_{\mathbb{R}^\ell} |D\psi|^p \right)^{\frac{1}{p}}.$$

The Morrey–Sobolev embedding can be reinterpreted as a bound on the Hölder seminorm $|\varphi|_{C^{0,1-\ell/p}(\mathbb{R}^\ell, V)}$.

Definition 5.5 (Hölder Seminorm) Given $\ell \in \mathbb{N} \setminus \{0\}$, $\alpha \in (0, 1)$ and a finite-dimensional vector space V, we define the *Hölder seminorm* for every function $\psi : \mathbb{R}^\ell \to V$ by

$$|\psi|_{C^{0,\alpha}(\mathbb{R}^\ell)} := \sup \left\{ \frac{|\psi(y) - \psi(x)|}{|y-x|^\alpha} \mid x, y \in \mathbb{R}^\ell \right\}.$$

The next result shows how estimates in terms of the uniform norm of the function and of its derivative can be combined into an estimate in terms of the Hölder seminorm.

Lemma 5.5 (Interpolation to Hölder Spaces) *Let $\ell \in \mathbb{N} \setminus \{0\}$, let V be a linear space, let the mapping $T : C_c^\infty(\mathbb{R}^\ell, V) \to \mathbb{R}$ be linear, and let $\alpha \in (0, 1)$. There exists a constant $C \in (0, +\infty)$ for which if there are constants $A, B \in \mathbb{R}$ such that for every $\psi \in C_c^\infty(\mathbb{R}^\ell, V)$, one has*

$$|\langle T, \psi \rangle| \le A \|\psi\|_{L^\infty(\mathbb{R}^\ell)} \tag{5.65}$$

and

$$|\langle T, \psi \rangle| \le B \|D\psi\|_{L^\infty(\mathbb{R}^\ell)}, \tag{5.66}$$

then for every $\psi \in C_c^\infty(\mathbb{R}^\ell, V)$, we have

$$|\langle T, \psi \rangle| \le C A^{1-\alpha} B^\alpha |\psi|_{C^{0,\alpha}(\mathbb{R}^\ell)}. \tag{5.67}$$

Proof We fix a function $\eta \in C_c^\infty(\mathbb{R}^\ell, \mathbb{R})$ such that $\int_{\mathbb{R}^\ell} \eta = 1$. Given a function $\psi \in C_c^\infty(\mathbb{R}^\ell, \mathbb{R})$, we define for each $\lambda \in (0, +\infty)$, the function $\psi_\lambda : \mathbb{R}^\ell \to V$ by setting for each $x \in \mathbb{R}^\ell$

$$\psi_\lambda(x) := \int_{\mathbb{R}^\ell} \eta(y) \psi(x - \lambda y) \, dy = \int_{\mathbb{R}^\ell} \eta\left(\tfrac{x}{\lambda} - z\right) \psi(\lambda z) \, dz. \tag{5.68}$$

---

[22] For the statement and the proof of the Morrey–Sobolev embedding (Proposition 5.5), see for example [Th. 4.5.11 41], [48, §1.4.5], [1, Lem. 4.28], [50, Th. 3.5.2], [16, Th. 9.12], [77, Lem. 6.4.13], and [32, §5.6.2].

5 Injective Ellipticity, Cancelling Operators and Endpoint Sobolev Inequalities

Since $\int_{\mathbb{R}^\ell} \eta = 1$, we observe that for every $x \in \mathbb{R}^\ell$ we have by (5.68)

$$\psi_\lambda(x) - \psi(x) = \int_{\mathbb{R}^\ell} \eta(y) \big(\psi(x - \lambda y) - \psi(x)\big) \, dy \,, \tag{5.69}$$

and thus by (5.69) and by Definition 5.5,

$$\begin{aligned}|\psi_\lambda(x) - \psi(x)| &\le \int_{\mathbb{R}^\ell} |\eta(y)| \, |\psi(x - \lambda y) - \psi(x)| \, dy \\ &\le |\psi|_{C^{0,\alpha}(\mathbb{R}^\ell)} \lambda^\alpha \int_{\mathbb{R}^\ell} |\eta(y)| \, |y|^\alpha \, dy = C_{23} |\psi|_{C^{0,\alpha}(\mathbb{R}^\ell)} \lambda^\alpha \,.\end{aligned} \tag{5.70}$$

On the other hand, we have for each $x \in \mathbb{R}^\ell$, by (5.68) again

$$\begin{aligned}D\psi_\lambda(x) &= \frac{1}{\lambda} \int_{\mathbb{R}^\ell} D\eta\left(\tfrac{x}{\lambda} - z\right) \psi(\lambda z) \, dz \\ &= \frac{1}{\lambda} \int_{\mathbb{R}^\ell} D\eta(y) \psi(x - \lambda y) \, dy \\ &= \frac{1}{\lambda} \int_{\mathbb{R}^\ell} D\eta(y) \big(\psi(x - \lambda y) - \psi(x)\big) \, dy \,,\end{aligned} \tag{5.71}$$

since $\int_{\mathbb{R}^\ell} D\eta = 0$. Hence, we have by (5.71) and by Definition 5.5

$$\begin{aligned}|D\psi_\lambda(x)| &\le \frac{1}{\lambda} \int_{\mathbb{R}^\ell} |D\eta(y)| \, |\psi(x - \lambda y) - \psi(x)| \, dy \\ &\le \frac{|\psi|_{C^{0,\alpha}(\mathbb{R}^\ell)}}{\lambda^{1-\alpha}} \int_{\mathbb{R}^\ell} |D\eta(y)| \, |y|^\alpha \, dy \\ &= C_{24} \frac{|\psi|_{C^{0,\alpha}(\mathbb{R}^\ell)}}{\lambda^{1-\alpha}} \,.\end{aligned} \tag{5.72}$$

By our assumption (5.65) and (5.66) and by (5.70) and (5.72), we have

$$\begin{aligned}|\langle T, \psi \rangle| &\le |\langle T, \psi - \psi_\lambda \rangle| + |\langle T, \psi_\lambda \rangle| \\ &\le A \|\psi - \psi_\lambda\|_{L^\infty(\mathbb{R}^\ell)} + B \|D\psi_\lambda\|_{L^\infty(\mathbb{R}^\ell)} \\ &\le \left(C_{23} A \lambda^\alpha + \frac{C_{24} B}{\lambda^{1-\alpha}}\right) |\psi|_{C^{0,\alpha}(\mathbb{R}^\ell)} \,.\end{aligned} \tag{5.73}$$

If we choose $\lambda = B/A$ in (5.73), we obtain (5.67) with $C := C_{23} + C_{24}$. □

5.4.2.2 Integrating by Parts

The proof of the sufficiency part of Theorem 5.10 will also rely on an estimate of an integral on a hyperplane.

Lemma 5.6 (Integration by Parts Estimate) *Let $n \in \mathbb{N} \setminus \{0\}$ and let E and F be finite-dimensional vector spaces. If $L(\mathrm{D})$ is a homogenous constant coefficients differential operator of order $k \in \mathbb{N} \setminus \{0\}$ from E to F on $\mathbb{R}^n$, then there exists a constant $C \in (0, +\infty)$ such that for every $f \in (C^\infty \cap L^1)(\mathbb{R}^n, E)$ satisfying $L(\mathrm{D})f = 0$, every affine hyperplane $\Sigma \subset \mathbb{R}^n$ and every $\psi \in C^1(\Sigma, F)$, one has*

$$\left| \int_\Sigma \langle \psi, L(\nu_\Sigma)[f] \rangle \right| \le C \, \|\mathrm{D}\psi\|_{L^\infty(\Sigma)} \int_{\mathbb{R}^n} |f| \, . \tag{5.74}$$

Here ν_Σ denotes the unit normal vector to the hyperplane Σ. We do not assume f to be compactly supported to avoid issues with the rigidity of the condition $L(\mathrm{D})f = 0$.

In order to outline the ideas, we first prove Lemma 5.6 for the divergence operator.

Proof of Lemma 5.6 when $L(\mathrm{D}) = \operatorname{div}$ Without loss of generality we assume that $\Sigma = \mathbb{R}^{n-1} \times \{0\}$ and that $\nu_\Sigma = \nu := (0, \ldots, 0, 1)$. We define the function $\Psi : \mathbb{R}^n_+ \to \mathbb{R}$ where $\mathbb{R}^n_+ := \mathbb{R}^n \times (0, +\infty)$, by setting for each $(x', x_n) \in \mathbb{R}^n_+$, $\Psi(x', x_n) := \psi(x')$. We fix a function $\theta \in C^1([0, +\infty), \mathbb{R})$ such that $\theta = 1$ on $[0, 1]$ and $\theta = 0$ on $[2, +\infty)$, and for every $\lambda > 0$ we define the function $\theta_\lambda \in C^1_c(\mathbb{R}^n_+, \mathbb{R})$ for each $x = (x', x_n) \in \mathbb{R}^n_+$ by $\theta_\lambda(x) := \theta(x_n/\lambda)$. For every $\lambda > 0$, the function $\theta_\lambda \Psi$ is compactly supported in $\overline{\mathbb{R}^n_+}$, and we have thus by the Gauß-Ostrogradski divergence theorem and by the Leibniz rule for the divergence

$$\int_{\mathbb{R}^{n-1} \times \{0\}} \langle \nu, f \rangle \psi = \int_{\mathbb{R}^{n-1} \times \{0\}} \langle \nu, f \Psi \theta_\lambda \rangle = - \int_{\mathbb{R}^n_+} \operatorname{div}(f \Psi \theta_\lambda)$$
$$= - \int_{\mathbb{R}^n_+} (\operatorname{div} f) \Psi \theta_\lambda - \int_{\mathbb{R}^n_+} \langle \mathrm{D}\Psi, f \theta_\lambda \rangle - \int_{\mathbb{R}^n_+} \langle \mathrm{D}\theta_\lambda, f \Psi \rangle \, . \tag{5.75}$$

We deduce then from (5.75), since $\operatorname{div} f = 0$ in $\mathbb{R}^n_+$, that

$$\left| \int_{\mathbb{R}^{n-1} \times \{0\}} \langle \nu, f \rangle \psi \right| \le \left(\|\mathrm{D}\Psi\|_{L^\infty(\mathbb{R}^n_+)} + \frac{C_{25}}{\lambda} \|\Psi\|_{L^\infty(\mathbb{R}^n_+)} \right) \int_{\mathbb{R}^n_+} |f|$$
$$= \left(\|\mathrm{D}\psi\|_{L^\infty(\mathbb{R}^{n-1})} + \frac{C_{25}}{\lambda} \|\psi\|_{L^\infty(\mathbb{R}^{n-1})} \right) \int_{\mathbb{R}^n_+} |f| \, , \tag{5.76}$$

Letting $\lambda \to +\infty$ in (5.76), we obtain (5.74). $\square$

We proceed now to the proof of Lemma 5.6 with a general operator $L(\mathrm{D})$.

Proof of Lemma 5.6 in the General Case Up to a rotation of the Euclidean space $\mathbb{R}^n$, we can assume that $\Sigma = \mathbb{R}^{n-1} \times \{0\}$ and $\nu_\Sigma = \nu := (0, \ldots, 0, 1)$. We fix a function $\eta \in C_c^1(\mathbb{R}^{n-1}, \mathbb{R})$ such that $\int_{\mathbb{R}^{n-1}} \eta = 1$. Given $\psi \in C_c^1(\mathbb{R}^{n-1}, F)$, we define the function $\Psi : \mathbb{R}_+^n \to F$ for each $(x', x_n) \in \mathbb{R}_+^n$ by

$$\Psi(x', x_n) := \frac{1}{x_n^{n-1}} \int_{\mathbb{R}^{n-1}} \psi(y) \eta\left(\frac{x'-y}{x_n}\right) dy = \int_{\mathbb{R}^{n-1}} \psi(x - x_n z) \eta(z) \, dz \,. \tag{5.77}$$

We immediately have from (5.77) that for every $(x', x_n) \in \mathbb{R}_+^n$

$$|\Psi(x', x_n)| \leq C_{26} \|\psi\|_{L^\infty(\mathbb{R}^{n-1})} \,. \tag{5.78}$$

We next observe that for every point $(x', x_n) \in \mathbb{R}^{n-1} \times (0, +\infty) = \mathbb{R}_+^n$ and every vector $v = (v', v_n) \in \mathbb{R}^n$, we have

$$\begin{aligned} D\Psi(x', x_n)[v] &= \int_{\mathbb{R}^{n-1}} D\psi(x' - x_n z)[v' - v_n z] \eta(z) \, dz \\ &= \frac{1}{x_n^{n-1}} \int_{\mathbb{R}^{n-1}} D\psi(y)[H(\tfrac{x'-y}{x_n})] \, dy \,, \end{aligned} \tag{5.79}$$

where the function $H \in C_c^\infty(\mathbb{R}^{n-1}, \mathrm{Lin}(\mathbb{R}^n, \mathbb{R}^{n-1}))$ is defined for every $z \in \mathbb{R}^{n-1}$ and $v = (v', v_n) \in \mathbb{R}^n$ by $H(z)[v] := (v' - v_n z) \eta(z)$. Hence we deduce from (5.79) that for every $j \in \mathbb{N} \setminus \{0\}$ and $x \in \mathbb{R}_+^n$

$$\begin{aligned} |D^j \Psi(x', x_n)| &\leq \frac{1}{x_n^{n+j-2}} \int_{B_{x_n}(x)} |D\psi(y)| |D^{j-1} H(\tfrac{x'-y}{x_n})| \, dy \\ &\leq \frac{C_{27} \|D\psi\|_{L^\infty(\mathbb{R}^{n-1})}}{x_n^{j-1}} \,. \end{aligned} \tag{5.80}$$

Moreover, for every point $y \in \mathbb{R}^{n-1}$, we have

$$\begin{aligned} \lim_{(x', x_n) \to (y, 0)} \Psi(x', x_n) &= \lim_{(x', x_n) \to (y, 0)} \int_{\mathbb{R}^{n-1}} \psi(x' - x_n z) \eta(x) \, dz \\ &= \int_{\mathbb{R}^{n-1}} \psi(y) \eta(z) \, dz = \psi(y) \,, \end{aligned}$$

by Lebesgue's dominated convergence theorem, since the function ψ is continuous and since $\int_{\mathbb{R}^{n-1}} \eta = 1$.

We fix a function $\theta \in C^k((0, +\infty), \mathbb{R})$ such that $\theta = 1$ on $(0, 1]$ and $\theta = 0$ on $[2, +\infty)$ and we define for each $\lambda \in (0, +\infty)$ the function $\theta_\lambda^k : \mathbb{R}_+^n \to \mathbb{R}$ for every $(x', x_n) \in \mathbb{R}^{n-1} \times (0, +\infty)$ by

$$\theta_\lambda^k(x', x_n) := \frac{x_n^{k-1}}{(k-1)!} \theta(x_n/\lambda). \qquad (5.81)$$

By k successive integration by parts we have, since the function $\theta_\lambda^k \Psi$ is compactly supported in $\bar{\mathbb{R}}_+^n$,

$$\int_{\mathbb{R}_+^n} \langle \partial_n^k L(\nu)[f], \theta_\lambda^k \Psi \rangle - (-1)^k \langle L(\nu)[f], \partial_n^k(\theta_\lambda^k \Psi) \rangle$$
$$= -\sum_{j=0}^{k-1}(-1)^j \int_{\mathbb{R}^{n-1} \times \{0\}} \langle \partial_n^{k-1-j} L(\nu)[f], \partial_n^j(\theta_\lambda^k \Psi) \rangle. \qquad (5.82)$$

The general Leibniz rule yields for each $j \in \{0, \ldots, k\}$ that

$$\partial_n^j(\theta_\lambda^k \Psi) = \sum_{i=0}^{j} \binom{j}{i}(\partial_n^i \theta_\lambda^k)(\partial_n^{j-i}\Psi). \qquad (5.83)$$

Since $\partial_n^i \theta_\lambda^k = 0$ on $\mathbb{R}^{n-1} \times \{0\}$ if $i \in \{0, \ldots, k-2\}$ and $\partial_n^{k-1}\theta_\lambda^k = 1$ on $\mathbb{R}^{n-1} \times \{0\}$, we have by (5.83) $\partial_n^j(\theta_\lambda^k \Psi) = 0$ on $\mathbb{R}^{n-1} \times \{0\}$ when $j \in \{1, \ldots, k-2\}$ whereas $\partial_n^{k-1}(\theta_\lambda^k \Psi) = \psi$ on $\mathbb{R}^{n-1} \times \{0\}$. Hence, the right-hand side in (5.82) reduces to

$$\sum_{j=0}^{k-1}(-1)^j \int_{\mathbb{R}^{n-1} \times \{0\}} \langle \partial_n^{k-j} L(\nu)[f], \partial_n^j(\theta_\lambda^k \Psi) \rangle = (-1)^{k-1} \int_{\mathbb{R}^{n-1} \times \{0\}} \langle L(\nu)[f], \psi \rangle, \qquad (5.84)$$

and hence by the identity (5.82), we have

$$\int_{\mathbb{R}^{n-1} \times \{0\}} \langle L(\nu)[f], \psi \rangle = \int_{\mathbb{R}_+^n} (-1)^k \langle \partial_n^k L(\nu)[f], \theta_\lambda^k \Psi \rangle - \langle L(\nu)[f], \partial_n^k(\theta_\lambda^k \Psi) \rangle. \qquad (5.85)$$

We observe that for every $j \in \{0, \ldots, k-1\}$ and every $(x', x_n) \in \mathbb{R}_+^n$, in view of (5.81) and of the general Leibniz rule

$$\partial_n^j \theta_\lambda^k(x', x_n) = \sum_{i=0}^{j} \binom{j}{i} \frac{x_n^{k-1-(j-i)} \theta^{(i)}(x_n/\lambda)}{(k-1-(j-i))!\lambda^i},$$

5 Injective Ellipticity, Cancelling Operators and Endpoint Sobolev Inequalities

so that, since for every $i \in \mathbb{N}$ and $t \in (0, +\infty)$, $|\theta^{(i)}(t)| \le C_{28}/t^i$,

$$|\partial_n^j \theta_\lambda^k(x', x_n)| \le C_{29} \, x_n^{k-1-j} \tag{5.86}$$

whereas

$$\partial_n^k \theta_\lambda^k(x', x_n) = \sum_{i=1}^{k} \binom{k}{i} \frac{x_n^{i-1} \theta^{(i)}(x_n/\lambda)}{(i-1)! \lambda^i} \,,$$

and thus since for every $i \in \mathbb{N} \setminus \{0\}$ and $t \in (0, +\infty)$, $|\theta^{(i)}(t)| \le C_{30}/t^{i-1}$,

$$|\partial_n^k \theta_\lambda^k(x', x_n)| \le \frac{C_{31}}{\lambda} \,. \tag{5.87}$$

For every $j \in \{0, \ldots, k-1\}$, we have by (5.80) and (5.86)

$$|\partial_n^j \theta_\lambda^k(x', x_n) \partial_n^{k-j} \Psi(x', x_n)| \le C_{32} \|D\psi\|_{L^\infty(\mathbb{R}^{n-1})} \,, \tag{5.88}$$

and by (5.78) and (5.87)

$$|\partial_n^k \theta_\lambda^k(x', x_n) \Psi(x', x_n)| \le C_{33} \frac{\|\psi\|_{L^\infty(\mathbb{R}^{n-1})}}{\lambda} \,, \tag{5.89}$$

and hence by (5.83), (5.88) and (5.89)

$$|\partial_n^k(\theta_\lambda^k \Psi)| \le C_{34} \Big(\|D\psi\|_{L^\infty(\mathbb{R}^{n-1})} + \frac{\|\psi\|_{L^\infty(\mathbb{R}^{n-1})}}{\lambda} \Big) \,, \tag{5.90}$$

so that

$$\Big| \int_{\mathbb{R}_+^n} \langle L(\nu)[f], \partial_n^k(\theta_\lambda^k \Psi) \rangle \Big|$$
$$\le C_{35} \Big(\|D\psi\|_{L^\infty(\mathbb{R}^{n-1})} + \frac{\|\psi\|_{L^\infty(\mathbb{R}^{n-1})}}{\lambda} \Big) \int_{\mathbb{R}_+^n} |L(\nu)[f]| \,. \tag{5.91}$$

Finally, we write for every $\xi = (\xi', \xi_n) \in \mathbb{R}^n$

$$L(\xi) = \sum_{j=0}^{k} \xi_n^{k-j} L_j(\xi') \,, \tag{5.92}$$

where for every $j \in \{0, \ldots, k\}$, $L_j(D')$ is a homogeneous linear differential operator on $\mathbb{R}^{n-1}$ of degree j from E to F. Since $L(D)f = 0$ by assumption and since $L(\nu) = L_0(\xi)$, we have by (5.92)

$$\int_{\mathbb{R}^n_+} \langle \partial_n^k L(\nu)[f], \theta_\lambda^k \Psi \rangle = -\sum_{j=1}^{k} \int_{\mathbb{R}^n_+} \langle \partial_n^{k-j} L_j(D')[f], \theta_\lambda^k \Psi \rangle . \tag{5.93}$$

By integration by parts, we compute for every $j \in \{1, \ldots, k\}$,

$$\int_{\mathbb{R}^n_+} \langle \partial_n^{k-j} L_j(D')[f], \theta_\lambda^k \Psi \rangle = (-1)^k \int_{\mathbb{R}^n_+} \langle f, L_j^*(D') \partial_n^{k-j} (\theta_\lambda^k \Psi) \rangle , \tag{5.94}$$

where $L_j(\xi')^* \in \mathrm{Lin}(E, F)$ is the adjoint to $L_j(\xi')$. We compute, for each $j \in \{1, \ldots, k\}$, by the general Leibniz rule again

$$L_j(D')^* \partial_n^{k-j}(\theta_\lambda^k \Psi) = \partial_n^{k-j}(\theta_\lambda^k L_j(D')^* \Psi)$$
$$= \sum_{i=0}^{k-j} \binom{k-j}{i} (\partial_n^i \theta_\lambda^k)(\partial_n^{k-j-i} L_j(D')^* \Psi) , \tag{5.95}$$

so that if $j \in \{1, \ldots, k\}$, we have by (5.80), (5.86) and (5.95)

$$|L_j(D')^* \partial_n^{k-j}(\theta_\lambda^k \Psi)| \le C_{36} \|D\psi\|_{L^\infty(\mathbb{R}^{n-1})} , \tag{5.96}$$

and thus by (5.93), (5.94) and (5.96),

$$\left| \int_{\mathbb{R}^n_+} \langle \partial_n^k L(\nu)[f], \theta_\lambda^k \Psi \rangle \right| \le C_{37} \|D\psi\|_{L^\infty(\mathbb{R}^{n-1})} \int_{\mathbb{R}^n_+} |f| . \tag{5.97}$$

In order to conclude we combine the identity (5.85) and the inequalities (5.91) and (5.97) to get

$$\left| \int_{\mathbb{R}^{n-1} \times \{0\}} \langle L(\nu)[f], \psi \rangle \right| \le C_{38} \left(\|D\psi\|_{L^\infty(\mathbb{R}^{n-1})} + \frac{\|\psi\|_{L^\infty(\mathbb{R}^{n-1})}}{\lambda} \right) \int_{\mathbb{R}^n_+} |f| ; \tag{5.98}$$

letting $\lambda \to +\infty$ we obtain (5.74). $\square$

5.4.2.3 Completing the Proof of the Duality Estimate

Our main step towards the proof of Theorem 5.10 is the following estimate on a component of f.

Lemma 5.7 (Unidirectional Cocancelling Estimate) *Let $n \in \mathbb{N} \setminus \{0\}$ and let E and F be finite-dimensional vector spaces. If $L(D)$ is a homogenous constant coefficients differential operator from E to F on $\mathbb{R}^n$ of order $k \in \mathbb{N} \setminus \{0\}$, then there exists a constant $C \in (0, +\infty)$ such that for every $f \in C^\infty(\mathbb{R}^n, E) \cap L^1(\mathbb{R}^n, E)$ satisfying $L(D) f = 0$, for every $\varphi \in C^\infty(\mathbb{R}^n, F)$ and for every $\xi \in \mathbb{R}^n$,*

$$\left| \int_{\mathbb{R}^n} \langle \varphi, L(\xi)[f] \rangle \right| \le C |\xi| \int_{\mathbb{R}^n} |f| \left(\int_{\mathbb{R}^n} |D\varphi|^n \right)^{\frac{1}{n}}. \tag{5.99}$$

Proof Without loss of generality, we assume that $\xi = (0, \dots, 0, 1)$. By Fubini's theorem, we have

$$\int_{\mathbb{R}^n} \langle L(\xi)[f], \varphi \rangle = \int_{\mathbb{R}} \int_{\mathbb{R}^{n-1}} \langle L(\xi)[f(x', x_n)], \varphi(x', x_n) \rangle \, dx' \, dx_n. \tag{5.100}$$

We observe that for every $x_n \in \mathbb{R}$ and every $\psi \in C_c^1(\mathbb{R}^{n-1}, F)$, we have immediately

$$\left| \int_{\mathbb{R}^{n-1}} \langle L(\xi)[f(x', x_n)], \psi(x', x_n) \rangle \, dx' \right| \le C_{39} \|\psi\|_{L^\infty(\mathbb{R}^{n-1})} \int_{\mathbb{R}^{n-1}} |f(\cdot, x_n)|, \tag{5.101}$$

and, by Lemma 5.6 we have

$$\left| \int_{\mathbb{R}^{n-1}} \langle L(\xi)[f(x', x_n)], \psi(x', x_n) \rangle \, dx' \right| \le C_{40} \|D\psi\|_{L^\infty(\mathbb{R}^{n-1})} \int_{\mathbb{R}^n} |f|. \tag{5.102}$$

By the interpolation into Hölder spaces (Lemma 5.5) with $\alpha = \frac{1}{n}$ and then by the Morrey-Sobolev embedding (Proposition 5.5), we deduce from (5.101) and (5.102) that

$$\left| \int_{\mathbb{R}^{n-1}} \langle L(\xi)[f(x', x_n)], \psi(x', x_n) \rangle \, dx' \right|$$
$$\le C_{41} |\psi|_{C^{0,1/n}(\mathbb{R}^{n-1})} \left(\int_{\mathbb{R}^{n-1}} |f(\cdot, x_n)| \right)^{1-\frac{1}{n}} \left(\int_{\mathbb{R}^n} |f| \right)^{\frac{1}{n}} \tag{5.103}$$
$$\le C_{42} \left(\int_{\mathbb{R}^{n-1}} |D\psi|^n \right)^{\frac{1}{n}} \left(\int_{\mathbb{R}^{n-1}} |f(\cdot, x_n)| \right)^{1-\frac{1}{n}} \left(\int_{\mathbb{R}^n} |f| \right)^{\frac{1}{n}}.$$

We now deduce from (5.100) and (5.103) and by Hölder's inequality that

$$\left| \int_{\mathbb{R}^n} \langle L(\xi)[f], \varphi \rangle \right|$$
$$\leq C_{43} \int_{\mathbb{R}} \left(\int_{\mathbb{R}^{n-1}} |D\varphi(\cdot, x_n)|^n \right)^{\frac{1}{n}} \left(\int_{\mathbb{R}^{n-1}} |f(\cdot, x_n)| \right)^{1-\frac{1}{n}} \left(\int_{\mathbb{R}^n} |f| \right)^{\frac{1}{n}} dx_n$$
$$\leq C_{43} \left(\int_{\mathbb{R}} \left(\int_{\mathbb{R}^{n-1}} |D\varphi(\cdot, x_n)|^n \right) dx_n \right)^{\frac{1}{n}} \left(\int_{\mathbb{R}} \left(\int_{\mathbb{R}^{n-1}} |f(\cdot, x_n)| \right) dx_n \right)^{1-\frac{1}{n}}$$
$$\left(\int_{\mathbb{R}^n} |f| \right)^{\frac{1}{n}},$$
(5.104)

which proves (5.99). □

Our last step to prove Theorem 5.10 is the following algebraic lemma.

Lemma 5.8 (Intersecting Kernel Quotient) *Let $n \in \mathbb{N} \setminus \{0\}$, let E, F and W be finite-dimensional vector spaces, let $L(D)$ be a homogeneous constant coefficient differential operator from E to F on $\mathbb{R}^n$, and let $Q \in \mathrm{Lin}(E, W)$. If*

$$\bigcap_{\xi \in \mathbb{R}^n \setminus \{0\}} \ker L(\xi) \subseteq \ker Q \,,$$

then there exist $m \in \mathbb{N}$, $\xi_1, \ldots, \xi_m \in \mathbb{R}^n \setminus \{0\}$, and $Q_1, \ldots, Q_m \in \mathrm{Lin}(F, W)$ such that

$$Q = \sum_{j=1}^{m} Q_j \circ L(\xi_j) \,. \qquad (5.105)$$

Proof Since the space E has finite dimension, there exists a nonnegative integer $m \in \mathbb{N}$ and vectors $\xi_1, \ldots, \xi_m \in \mathbb{R}^n \setminus \{0\}$ such that

$$\bigcap_{\xi \in \mathbb{R}^n \setminus \{0\}} \ker L(\xi) = \bigcap_{1 \leq i \leq m} \ker L(\xi_i) \,.$$

If we define $R \in \mathrm{Lin}(E, F^m)$ for each $e \in E$ by $R(e) := (L(\xi_1)[e], \ldots, L(\xi_m)[e]) \in F^m$, we have $\ker R \subseteq \ker Q$ and there exists thus $Q_1, \ldots, Q_m \in \mathrm{Lin}(F, W)$ such that (5.105) holds. □

Proof of the Sufficiency Part of Theorem 5.10 By Lemma 5.8 with Q being taken to be the identity id_E, there exists $m \in \mathbb{N}$, $\xi_1, \ldots, \xi_m \in \mathbb{R}^n \setminus \{0\}$ and $Q_1, \ldots, Q_m \in \mathrm{Lin}(F, E)$ such that $\mathrm{id}_E = \sum_{j=1}^{m} Q_j \circ L(\xi_j)$. We have then,

$$\int_{\mathbb{R}^n} \langle f, \varphi \rangle = \sum_{j=1}^{m} \int_{\mathbb{R}^n} \langle L(\xi_j)[f], Q_j^*[\varphi] \rangle,$$

and the conclusion (5.61) follows then from Lemma 5.7. □

Remark 5.3 The estimate in Theorem 5.10 can be proved by replacing the assumption that $L(\mathrm{D})$ is cocancelling by the assumption that $\varphi(x) \in \left(\bigcap_{\xi \in \mathbb{R}^n \setminus \{0\}} \ker L(\xi) \right)^\perp$. In order to prove this one follows the proof above, taking $P : E \to E$ to be a projection on the space $\left(\bigcap_{\xi \in \mathbb{R}^n \setminus \{0\}} \ker L(\xi) \right)^\perp$ so that $\ker P^* = P[E]^\perp = \bigcap_{\xi \in \mathbb{R}^n \setminus \{0\}} \ker L(\xi)$ and one can then write $P^* = \sum_{j=1}^{m} Q_j \circ L(\xi_j)$ for some $Q_1, \ldots, Q_m \in \mathrm{Lin}(F, E)$ thanks to Lemma 5.8.

Although the proof of Theorem 5.10 given here relies heavily on the Euclidean structure of the underlying space $\mathbb{R}^n$ through the application of the Fubini theorem in (5.100) and (5.104) and the decomposition of the vector space (5.105), suitable integral-geometric formulae allow to obtain similar results for the hyperbolic plane [21] and symmetric spaces of noncompact type [22].

5.4.3 Back to Cancelling Operators

We apply now Theorem 5.10 to get the endpoint Sobolev estimate for cancelling operators.

Lemma 5.9 (Singular Integral Representation of Order -1) *Under the assumptions of Proposition 5.2 and with $G_A : \mathbb{R}^n \setminus \{0\} \to \mathrm{Lin}(E, V)$ given by the same proposition, for every $u \in C_c^\infty(\mathbb{R}^n, V)$, if $A(\mathrm{D})u = \mathrm{div}\, F$, with $F \in C^\infty(\mathbb{R}^n, \mathrm{Lin}(\mathbb{R}^n, E))$ and if $\int_{\mathbb{R}^n} |F(x)|/(1 + |x|^n) \, \mathrm{d}x < \infty$, then*

$$\mathrm{D}^{k-1} u(x) = \int_{\mathbb{S}^{n-1}} \mathrm{D}^{k-1} G_A(z)[F(x)[z]] \mathrm{d}z + \lim_{\varepsilon \to 0} \int_{\mathbb{R}^n \setminus B_\varepsilon(0)} \mathrm{D}^k G_A(h)[F(x-h)] \mathrm{d}h. \tag{5.106}$$

Proof The first $k - 1$ derivatives of G_A are locally integrable in view of Proposition 5.2 (ii) and thus differentiating $k - 1$ times (5.15), we get that (5.47) holds for every $x \in \mathbb{R}^n$. We fix a function $\eta \in C_c^\infty(\mathbb{R}^n, \mathbb{R})$ such that $\eta = 1$ on $B_{1/2}(0)$ and $\eta = 0$ on $\mathbb{R}^n \setminus B_1(0)$. If $R > 0$ is large enough so that $\mathrm{supp}\, u \subseteq B_R(0)$, we deduce from (5.47) that for every $\varepsilon \in (0, +\infty)$,

$$D^k u(x) = \int_{\mathbb{R}^n} \eta\left(\tfrac{|x-y|}{R}\right)\left(1 - \eta\left(\tfrac{|x-y|}{\varepsilon}\right)\right) D^{k-1} G_A(x-y) [\operatorname{div}(F - F(x))(y)] \, dy$$
$$+ \int_{\mathbb{R}^n} \eta\left(\tfrac{|h|}{\varepsilon}\right) D^{k-1} G_A(h) [A(D) u(x-h)] \, dh \, .$$
(5.107)

Integrating by parts in the first integral on the right-hand side of (5.107), we get

$$\int_{\mathbb{R}^n} \eta\left(\tfrac{|x-y|}{R}\right)\left(1 - \eta\left(\tfrac{|x-y|}{\varepsilon}\right)\right) D^{k-1} G_A(x-y) [\operatorname{div}(F - F(x))(y)] \, dy$$
$$= \int_{\mathbb{R}^n} \eta\left(\tfrac{|x-y|}{R}\right)\left(1 - \eta\left(\tfrac{|x-y|}{\varepsilon}\right)\right) D^k G_A(x-y) [(F - F(x))(y)] \, dy$$
$$- \int_{\mathbb{R}^n} \left(\tfrac{1}{R} \eta'\left(\tfrac{|x-y|}{R}\right) + \tfrac{1}{\varepsilon} \eta'\left(\tfrac{|x-y|}{\varepsilon}\right)\right) D^{k-1} G_A(x-y) \left[(F(y) - F(x))[\tfrac{x-y}{|x-y|}]\right] dy \, .$$
(5.108)

Applying Proposition 5.2 (iii), letting $\varepsilon \to 0$ and $R \to \infty$, we reach the conclusion. □

Proof of Sufficiency of Injective Ellipticity and Cancellation in Theorem 5.9 As the operator $A(D)$ is injectively elliptic, by Proposition 5.3, there exists a homogeneous constant coefficient differential operator from E to F on $\mathbb{R}^n$ such that for every $\xi \in \mathbb{R}^n \setminus \{0\}$, $\ker L(\xi) = A(\xi)[V]$ and thus by Definition 5.3 and Definition 5.4, the operator $L(D)$ is cocancelling. By Theorem 5.10 and the representation of bounded linear functionals on Sobolev spaces, there exists $F \in L^{\frac{n}{n-1}}(\mathbb{R}^n, E \otimes \mathbb{R}^n)$ such that $\operatorname{div} F = A(D) u$ in the sense of distributions and

$$\left(\int_{\mathbb{R}^n} |F|^{\frac{n}{n-1}} \right)^{1-\frac{1}{n}} \le C_{44} \int_{\mathbb{R}^n} |A(D) u| \, .$$

Letting $\eta \in C_c^\infty(\mathbb{R}^n, \mathbb{R})$ such that $\int_{\mathbb{R}^n} \eta = 1$ and setting for $\varepsilon \in (0, +\infty)$ and $x \in \mathbb{R}^n$, $\eta_\varepsilon(x) := \eta(x/\varepsilon)/\varepsilon^n$, we have $A(D)(\eta_\varepsilon * u) = \operatorname{div}(\eta_\varepsilon * F)$, we get by Theorem 5.4 and Lemma 5.9,

$$\left(\int_{\mathbb{R}^n} |D^{k-1}(\eta_\varepsilon * u)|^{\frac{n}{n-1}} \right)^{1-\frac{1}{n}} \le C_{45} \left(\int_{\mathbb{R}^n} |F|^{\frac{n}{n-1}} \right)^{1-\frac{1}{n}} \le C_{46} \int_{\mathbb{R}^n} |A(D)[u]| \, ,$$
(5.109)

letting $\varepsilon \to 0$, we get (5.56) when $\ell = k - 1$; the case $\ell \in \{1, \ldots, k-2\}$ then follows from the classical Sobolev embedding theorem (Theorem 5.7). □

5.4.4 Characterization of Cancelling Operators

As a byproduct of the characterization of cocancelling operators (Proposition 5.4), we can characterize cancelling operators under the assumption of injective ellipticity.

Proposition 5.6 (Characterization of Cancelling Operators) *Let $n \in \mathbb{N} \setminus \{0\}$, let V and E be finite-dimensional vector spaces, and let $A(\mathrm{D})$ be a homogeneous differential operator of order $k \in \mathbb{N} \setminus \{0\}$ on $\mathbb{R}^n$ from V to E. If the operator $A(\mathrm{D})$ is injectively elliptic, then the following are equivalent*

(i) the operator $A(\mathrm{D})$ is cancelling,
(ii) for every $u \in L^1_{\mathrm{loc}}(\mathbb{R}^n, V)$ such that $A(\mathrm{D})u \in L^1(\mathbb{R}^n, E)$, one has

$$\int_{\mathbb{R}^n} A(\mathrm{D})u = 0, \tag{5.110}$$

(iii) for every $u \in C^\infty(\mathbb{R}^n, V)$ such that $\mathrm{supp}\, A(\mathrm{D})u$ is compact, (5.110) holds,
(iv) for every distribution $u \in C_c^\infty(\mathbb{R}^n, V)^*$ and every $e \in E$ such that $A(\mathrm{D})u = e\delta_0$, one has $e = 0$.

It is crucial that no decay assumption is imposed on u in (iii). Indeed, if for every $j \in \{0, \ldots, k-1\}$, we assume that $\lim_{|x| \to \infty} |\mathrm{D}^j u(x)| |x|^{n-j} = 0$, then one has $\int_{\mathbb{R}^n} A(\mathrm{D})u = 0$, for any constant coefficient differential operator $A(\mathrm{D})$ of order $k \in \mathbb{N} \setminus \{0\}$.

Proof of Proposition 5.6 We first note that since $A(\mathrm{D})$ is injectively elliptic, proposition 5.3 applies and yields a homogeneous differential operator $L(\mathrm{D})$ on $\mathbb{R}^n$ from E to F such that for every $\xi \in \mathbb{R}^n \setminus \{0\}$, we have $\ker L(\xi) = A(\xi)[V]$ and $L(\mathrm{D})$ is cocancelling.

We first prove that (i) implies (ii). We assume that $u \in L^1_{\mathrm{loc}}(\mathbb{R}^n, E)$ and $A(\mathrm{D})u \in L^1(\mathbb{R}^n, E)$. By construction of $L(\mathrm{D})$, we have $L(\mathrm{D})(A(\mathrm{D})u) = 0$. Since $A(\mathrm{D})$ is cancelling, the operator $L(\mathrm{D})$ is cocancelling, (5.110) holds in view of proposition 5.4 (iii).

It is clear that (ii) implies (iii). Assume now that assertion (iii) holds and that $A(\mathrm{D})u = e\delta_0$. If $\eta \in C_c^\infty(\mathbb{R}^n, \mathbb{R})$ and $\int_{\mathbb{R}^n} \eta = 1$, then $A(\mathrm{D})(\eta * u) = e\eta$, and thus by assertion (iii), $e = \int_{\mathbb{R}^n} A(\mathrm{D})(\eta * u) = 0$ so that (iv) holds.

Finally we assume that (iv) holds and that $e \in \bigcap_{\xi \in \mathbb{R}^n \setminus \{0\}} A(\xi)[V]$. Since the operator $A(\mathrm{D})$ is injectively elliptic, then by Proposition 5.2 (v) $A(\mathrm{D})[G_A[e]] = e\delta_0$, where G_A is the representation kernel given by Proposition 5.2, and thus by (iv) we have $e = 0$ so that assertion (i) holds. □

5.5 Related Questions and Variants

5.5.1 Cocancelling Estimates and BMO

The cocancelling duality estimates of Proposition 5.4 can be seen as a replacement for the failure of the endpoint Sobolev embedding of $\dot{W}^{1,n}(\mathbb{R}^n, \mathbb{R})$ into $L^\infty(\mathbb{R}^n, \mathbb{R})$. On the other hand $\dot{W}^{1,n}(\mathbb{R}^n, \mathbb{R})$ is known to be embedded into the space of functions of bounded mean oscillation $\mathrm{BMO}(\mathbb{R}^n, \mathbb{R})$. It turns out that cocancelling estimates capture in general a stronger property than vanishing mean oscillation. In order to state this precisely, we define the following semi-norm associated to an operator.

Definition 5.6 (Cocancellation Spaces[23]) Let $n \in \mathbb{N} \setminus \{0\}$, let E and F be finite-dimensional vector spaces, and let $L(\mathrm{D})$ be a homogeneous constant coefficient differential operator from E to F on $\mathbb{R}^n$. We define for every $\varphi \in C_c^\infty(\mathbb{R}^n, \mathbb{R})$, the semi-norm

$$\|\varphi\|_{D_L(\mathbb{R}^n)} := \sup\left\{\left|\int_{\mathbb{R}^n} \varphi f\right| \,\Big|\, f \in C^\infty(\mathbb{R}^n, E),\ L(\mathrm{D})f = 0 \text{ and } \int_{\mathbb{R}^n} |f| \le 1\right\}.$$

This semi-norm turns out to be stronger than the bounded mean oscillation seminorm.

Proposition 5.7 (Cocancelling Estimates Imply Bounded Mean Oscillation[24])
Let $n \in \mathbb{N} \setminus \{0\}$, let E and F be finite-dimensional vector spaces, and let $L(\mathrm{D})$ be a homogeneous constant coefficient differential operator from E to F on $\mathbb{R}^n$. If there exists a finite-dimensional vector space V and a homogeneous constant coefficients differential operator $A(\mathrm{D})$ from V to E on $\mathbb{R}^n$ such that for every $\xi \in \mathbb{R}^n \setminus \{0\}$, one has $A(\xi) \ne 0$ and $L(\xi) \circ A(\xi) = 0$, then there exists a constant $C \in (0, +\infty)$ such that for every $\varphi \in C_c^\infty(\mathbb{R}^n, \mathbb{R})$, one has

$$\|\varphi\|_{\mathrm{BMO}(\mathbb{R}^n)} \le C \|\varphi\|_{D_L(\mathbb{R}^n)}.$$

The assumption of Proposition 5.7 is always satisfied when the operator $L(\mathrm{D})$ is given by Proposition 5.3 with $A(\mathrm{D})$. The assumption is also satisfied when $L(\xi)$ is surjective for every $\xi \in \mathbb{R}^n \setminus \{0\}$: indeed, similarly to (5.44) in the proof of Proposition 5.3 one can define then the operator $A(\mathrm{D})$ by setting for each $\xi \in \mathbb{R}^n \setminus \{0\}$

$$A(\xi) := \det\left(L(\xi) \circ L(\xi)^*\right) \left(\mathrm{id}_E - L(\xi)^* \circ (L(\xi) \circ L(\xi)^*)^{-1} \circ L(\xi)\right). \tag{5.111}$$

---

[23] These space generalize the spaces $\mathrm{D}_k(\mathbb{R}^n)$ defined in [71] which correspond to the case where $L(\mathrm{D})$ is the exterior differential acting on $C_c^\infty(\mathbb{R}^n, \bigwedge^k \mathbb{R}^n)$.

[24] Proposition 5.7 generalizes the author's result when the operator $L(\mathrm{D})$ is the exterior differential d [71].

Proof of Proposition 5.7 By assumption, for every $\xi \in \mathbb{R}^n \setminus \{0\}$, we have $A(\xi) \neq 0$ and thus $\operatorname{tr}(A(\xi) \circ A(\xi)^*) > 0$. By the theory of multipliers on the real Hardy space $\mathcal{H}^1(\mathbb{R}^n)$,[25] given $g \in C^\infty(\mathbb{R}^n, \mathbb{R}) \cap \mathcal{H}^1(\mathbb{R}^n, \mathbb{R})$, we can define $f \in C^\infty(\mathbb{R}^n, E \otimes E) \cap \mathcal{H}^1(\mathbb{R}^n, E \otimes E)$ by the condition that for every $\xi \in \mathbb{R}^n \setminus \{0\}$,

$$\mathcal{F}f(\xi) := \frac{A(\xi) \circ A(\xi)^*}{\operatorname{tr}(A(\xi) \circ A(\xi)^*)} \mathcal{F}g(\xi) \, ;$$

the function f also satisfies the estimate

$$\|f\|_{L^1(\mathbb{R}^n, E \otimes E)} \leq C_{47} \|f\|_{\mathcal{H}^1(\mathbb{R}^n, E \otimes E)}$$
$$\leq C_{48} \|g\|_{\mathcal{H}^1(\mathbb{R}^n, \mathbb{R})} \, .$$

By construction, we have $\operatorname{tr} f = g$ on $\mathbb{R}^n$ and $L(D)f = 0$ on $\mathbb{R}^n$ and hence

$$\left| \int_{\mathbb{R}^n} g\varphi \right| = \left| \operatorname{tr}\left(\int_{\mathbb{R}^n} f\varphi \right) \right|$$
$$\leq \left| \int_{\mathbb{R}^n} f\varphi \right|$$
$$\leq C_{49} \|\varphi\|_{D_L(\mathbb{R}^n)} \int_{\mathbb{R}^n} |f|$$
$$\leq C_{50} \|\varphi\|_{D_L(\mathbb{R}^n)} \|g\|_{\mathcal{H}^1(\mathbb{R}^n, \mathbb{R})} \, ,$$

and we conclude by the characterization of bounded mean oscillation (BMO) functions by duality with the Hardy space $\mathcal{H}^1(\mathbb{R}^n)$.[26] □

Remark 5.4 In the particular case where $n \geq 2$ and $L(D) = \operatorname{div}$, we can take $V := \mathbb{R}^n \otimes \mathbb{R}^n$ and $A(\xi)[v] := v \cdot \xi - v^* \cdot \xi$ in Proposition 5.7. Hence we obtain in the proof $A(D)^* \circ A(D)(u) = \Delta u - \nabla \nabla \cdot u$, and thus $f = (1 - \mathcal{R} \otimes \mathcal{R})g$, where $\mathcal{R}$ is the vector Riesz transform operator.

5.5.2 Fractional Estimates

Theorem 5.9 can be generalized to Sobolev embeddings into fractional Sobolev spaces $W^{k-1+s, p}(\mathbb{R}^n, V)$ with $s \in (0, 1)$.

---

[25] For the theory of multipliers on the Hardy space $\mathcal{H}^1(\mathbb{R}^n)$, see for example [63]*Ch. 3 §3 and Ch. I §6.2.1.
[26] For the duality between BMO($\mathbb{R}^n$) and the real Hardy space $\mathcal{H}^1(\mathbb{R}^n)$, see for instance [63, Ch. IV §1.2], [33], and [34].

Theorem 5.11 (Fractional Embedding for Cancelling Operators[27]**)** *Let $n \in \mathbb{N} \setminus \{0\}$, let V and E be finite-dimensional vector spaces, let $A(\mathrm{D})$ be a homogeneous constant coefficient differential operator of order $k \in \mathbb{N} \setminus \{0\}$ from V to E on $\mathbb{R}^n$. If $A(\mathrm{D})$ is injectively elliptic, and if $s \in (0, 1)$ and $p \in [1, +\infty)$ satisfy $\frac{1}{p} = 1 - \frac{1-s}{n}$, then there exists a constant $C \in (0, +\infty)$ such that for every $u \in C_c^\infty(\mathbb{R}^n, V)$,*

$$\left(\int_{\mathbb{R}^n} \int_{\mathbb{R}^n} \frac{|\mathrm{D}^{k-1} u(y) - \mathrm{D}^{k-1} u(x)|^p}{|y-x|^{n+sp}} \, \mathrm{d}y \, \mathrm{d}x \right)^{\frac{1}{p}} \leq C \int_{\mathbb{R}^n} |A(\mathrm{D})[u]|,$$

if and only if the operator $A(\mathrm{D})$ is cancelling.

In particular, the endpoint fractional Sobolev inequality

$$\left(\int_{\mathbb{R}^n} \int_{\mathbb{R}^n} \frac{|u(y) - u(x)|^p}{|y-x|^{n+sp}} \, \mathrm{d}y \, \mathrm{d}x \right)^{\frac{1}{p}} \leq C \int_{\mathbb{R}^n} |\mathrm{D}u|, \tag{5.112}$$

with $\frac{1}{p} = 1 - \frac{1-s}{n}$, holds for every function $u \in C_c^\infty(\mathbb{R}, \mathbb{R})$ if and only if the condition $n \geq 2$ holds.[28]

The main new tool we need to prove Theorem 5.11 is a fractional counterpart of Theorem 5.10.

Theorem 5.12 (Fractional Cocancelling Estimate[29]**)** *Let $n \in \mathbb{N} \setminus \{0, 1\}$, let V and E be finite-dimensional vector spaces, let $L(\mathrm{D})$ be a homogeneous constant coefficient differential operator from E to F on $\mathbb{R}^n$, and let $p \in (n, +\infty)$. There exists a constant $C \in (0, +\infty)$ such that for every $f \in C^\infty(\mathbb{R}^n, E) \cap L^1(\mathbb{R}^n, E)$ that satisfies $L(\mathrm{D}) f = 0$ in the sense of distributions and for every $\varphi \in C_c^\infty(\mathbb{R}^n, E)$, one has*

$$\left| \int_{\mathbb{R}^n} \langle f, \varphi \rangle \right| \leq C \int_{\mathbb{R}^n} |f| \left(\int_{\mathbb{R}^n} \int_{\mathbb{R}^n} \frac{|\varphi(y) - \varphi(x)|^p}{|y-x|^{2n}} \, \mathrm{d}y \, \mathrm{d}x \right)^{\frac{1}{p}} \tag{5.113}$$

if and only if the operator $L(\mathrm{D})$ is cocancelling.

---

[27] Theorem 5.11 is due to the author [73, Th. 8.1]; particular cases where obtained in [49, 73].

[28] The inequality (5.112) can also be obtained as a consequence of the inequality $\|u\|_{B_1^{s,p}(\mathbb{R}^n)} \leq C \|\mathrm{D}u\|_{L^1(\mathbb{R}^n)}$ of Solonnikov [60, Th. 2] and [44, Th. 4] or as a consequence of the interpolation estimate

$$\|u\|_{B_1^{s,p}(\mathbb{R}^n)} \leq C \|\mathrm{D}u\|_{L^1(\mathbb{R}^n)}^s \|u\|_{L^{\frac{d}{d-1}}(\mathbb{R}^n)}^{1-s}$$

by Albert Cohen et al. [24, Th. 1.4] (see also Jean Bourgain et al. [10, Lem. D.2]) together with the standard embeddings between Besov spaces and the identification between Besov spaces and fractional Sobolev-Slobodeckiĭ spaces [67, 2.3.2(5), 2.3.5(3) and 2.5.7(9)]. A counterexample for $n = 1$ is obtained by regularizing a characteristic function [55].

[29] Theorem 5.12 is due to the author [74], following several particular cases statements [8, Rem. 2], [9, Rem. 11], [69, Rem. 2], [49], and [73]; in some cases a strong version à la Bourgain-Brezis could be obtained relying on the constructions of Pierre BOUSQUET, Petru MIRONESCU, Emmanuel RUSS, WANG Yi and YUNG Po-Lam [13, 14].

5 Injective Ellipticity, Cancelling Operators and Endpoint Sobolev Inequalities

The proof of Theorem 5.12 follows the proof of Theorem 5.10. This approach is possible first because of the following fractional counterpart of the Morrey-Sobolev embedding of Proposition 5.5.

Proposition 5.8 (Fractional Sobolev-Morrey Embedding[30]) *Let $\ell \in \mathbb{N} \setminus \{0\}$, let V be a finite-dimensional vector space, let $s \in (0, 1)$ and let $p \in (\ell/s, +\infty)$. There exists a constant $C \in (0, +\infty)$ such that for every $\varphi \in C_c^\infty(\mathbb{R}^\ell, V)$, one has*

$$|\varphi(y) - \varphi(x)| \le C|y-x|^{1-\frac{\ell}{p}} \left(\int_{\mathbb{R}^\ell} \int_{\mathbb{R}^\ell} \frac{|\varphi(y) - \varphi(x)|^p}{|y-x|^{\ell+sp}} \, dy \, dx \right)^{\frac{1}{p}}.$$

The second ingredient that make the generalization possible is the following counterpart of Fubini's theorem for fractional Sobolev spaces.

Proposition 5.9 (Fractional Fubini Theorem) *Let $n \in \mathbb{N} \setminus \{0\}$, let $\ell \in \{1, \ldots, n-1\}$, let $s \in (0, 1)$ and let $p \in [1, +\infty)$. There exists a constant $C \in (0, +\infty)$ such that for every Borel-measurable function $\varphi : \mathbb{R}^n \to \mathbb{R}$ and every $t \in \mathbb{R}^{n-\ell}$ one has*

$$\int_{\mathbb{R}^\ell} \int_{\mathbb{R}^\ell} \frac{|\varphi(y,t) - \varphi(x,t)|^p}{|y-x|^{\ell+sp}} \, dy \, dx \le C \int_{\mathbb{R}^\ell} \int_{\mathbb{R}^n} \frac{|\varphi(z) - \varphi(x,t)|^p}{|z-(x,t)|^{n+sp}} \, dz \, dx. \tag{5.114}$$

Proposition 5.9 and Fubini's theorem imply then that we have the following fractional counterpart of Fubini's theorem

$$\int_{\mathbb{R}^{n-\ell}} \int_{\mathbb{R}^\ell} \int_{\mathbb{R}^\ell} \frac{|\varphi(y,t) - \varphi(x,t)|^p}{|y-x|^{\ell+sp}} \, dy \, dx \, dt \le C \int_{\mathbb{R}^n} \int_{\mathbb{R}^n} \frac{|\varphi(y) - \varphi(x)|^p}{|y-x|^{n+sp}} \, dy \, dx.$$

Proof of Proposition 5.9 We have for every $x, y \in \mathbb{R}^\ell$,

$$|\varphi(x,t) - \varphi(y,t)|^p$$
$$\le 2^{p-1} \left(\fint_{B_{\frac{|y-x|}{2}}(\frac{x+y}{2}, t)} |\varphi - \varphi(x,t)|^p + \fint_{B_{\frac{|y-x|}{2}}(\frac{x+y}{2}, t)} |\varphi - \varphi(y,t)|^p \right). \tag{5.115}$$

By monotonicity of the integral, we have for every $x \in \mathbb{R}^\ell$,

$$\int_{\mathbb{R}^\ell} \fint_{B_{\frac{|y-x|}{2}}(\frac{x+y}{2}, t)} \frac{|\varphi(z) - \varphi(x,t)|^p}{|y-x|^{n+sp}} \, dz \, dy$$
$$\le C_{51} \int_{\mathbb{R}^\ell} \int_{B_{|y-x|}(x,t)} \frac{|\varphi(z) - \varphi(x,t)|^p}{|y-x|^{n+\ell+sp}} \, dz \, dy \tag{5.116}$$

---

[30] For statements and proofs of the fractional Sobolev-Morrey embedding of Proposition 5.8, see [31, §8], [65, Lem. 11], [67, §2.7.1], and [46, Th. 14.17].

and thus exchanging the integrals we deduce from (5.116) that

$$\int_{\mathbb{R}^\ell} \fint_{B_{\frac{|y-x|}{2}}(\frac{x+y}{2},t)} \frac{|\varphi(z) - \varphi(x,t)|^p}{|y-x|^{n+sp}} \, dz \, dy$$
$$\leq C_{51} \int_{\mathbb{R}^n} \int_{\mathbb{R}^\ell \setminus B_{|z-(x,t)|}(x)} \frac{|\varphi(z) - \varphi(x,t)|^p}{|y-x|^{n+\ell+sp}} \, dy \, dz \qquad (5.117)$$
$$= C_{52} \int_{\mathbb{R}^n} \frac{|\varphi(z) - \varphi(x,t)|^p}{|z-(x,t)|^{n+sp}} \, dz \, .$$

Combining the inequalities (5.115) and (5.117), we have

$$\int_{\mathbb{R}^\ell} \int_{\mathbb{R}^\ell} \frac{|\varphi(y,t) - \varphi(x,t)|^p}{|y-x|^{\ell+sp}} \, dy \, dx$$
$$\leq 2^{p-1} C_{52} \Big(\int_{\mathbb{R}^\ell} \int_{\mathbb{R}^n} \frac{|\varphi(z) - \varphi(x,t)|^p}{|z-(x,t)|^{n+sp}} \, dz \, dx + \int_{\mathbb{R}^\ell} \int_{\mathbb{R}^n} \frac{|\varphi(z) - \varphi(y,t)|^p}{|z-(y,t)|^{n+sp}} \, dz \, dy \Big),$$

and the conclusion (5.114) follows. □

5.5.3 Hardy Estimates

The cancellation condition is also a necessary and sufficient condition for a family of endpoint Hardy inequalities.

Theorem 5.13 (Hardy Estimate for Cancelling Operators[31]**)** *Let $n \in \mathbb{N} \setminus \{0\}$, let V and E be finite-dimensional vector spaces, and let $A(D)$ be a homogeneous constant coefficient differential operator of order $k \in \mathbb{N} \setminus \{0\}$ from V to E on $\mathbb{R}^n$. If $A(D)$ is injectively elliptic and if $\ell \in \mathbb{N} \setminus \{0\}$ satisfies $0 < k - \ell < n$, then there exists a constant $C \in (0, +\infty)$ such that for every $u \in C_c^\infty(\mathbb{R}^n, V)$,*

$$\int_{\mathbb{R}^n} \frac{|D^\ell u(x)|}{|x|^{k-\ell}} \, dx \leq C \int_{\mathbb{R}^n} |A(D)[u]| \, , \qquad (5.118)$$

if and only if the operator $A(D)$ is cancelling.

Our main tool in the proof of Theorem 5.13 will be the following duality estimate.

---

[31] The Hardy estimate of Theorem 5.13 is due to Vladimir Gilelevich Maz'ya [47] for the operator $(\Delta, \nabla \text{div})$. Our proof follows the proof by Pierre Bousquet and Petru Mironescu [11] for the latter operator and its adaptations to cancelling operators by Pierre BOUSQUET and the author [12].

Proposition 5.10 (Weighted Duality Estimate[32]**)** *Let $n \in \mathbb{N} \setminus \{0, 1\}$, let E and F be finite-dimensional vector spaces, let $L(\mathrm{D})$ be a homogeneous constant coefficient differential operator of order ℓ from E to F on $\mathbb{R}^n$. There exists a constant $C \in (0, +\infty)$ such that for every $f \in C^\infty(\mathbb{R}^n, E) \cap L^1(\mathbb{R}^n, E)$ that satisfies $L(\mathrm{D})f = 0$ in $\mathbb{R}^n$ in the sense of distributions and every $\varphi \in C_c^1(\mathbb{R}^n, E)$ such that for every $x \in \mathbb{R}^n$,*

$$\varphi(x) \in \Bigl(\bigcap_{\xi \in \mathbb{R}^n \setminus \{0\}} \ker L(\xi) \Bigr)^\perp, \tag{5.119}$$

one has

$$\Bigl| \int_{\mathbb{R}^n} \langle f, \varphi \rangle \Bigr| \le \sum_{j=1}^{\ell} C \int_{\mathbb{R}^n} |f(x)| |x|^j |D^j \varphi(x)| \, \mathrm{d}x . \tag{5.120}$$

Proof We define the mapping $T : E \to E$ to be the orthogonal projection on the space $(\bigcap_{\xi \in \mathbb{R}^n \setminus \{0\}} \ker L(\xi))^\perp$. This implies that $\ker T = T[E]^\perp = \bigcap_{\xi \in \mathbb{R}^n \setminus \{0\}} \ker L(\xi)$ and thus that, by Lemma 5.8, there exists $m \in \mathbb{N}$, vectors $\xi_1, \ldots, \xi_m \in \mathbb{R}^n \setminus \{0\}$ and $Q_1, \ldots, Q_m \in \mathrm{Lin}(F, E)$ such that $T = \sum_{i=1}^m Q_i \circ L(\xi_i)$. If we define the polynomial $P : \mathbb{R}^n \to \mathrm{Lin}(E, F)$ for every $x \in \mathbb{R}^n$ by

$$P(x) := \sum_{i=1}^m \frac{\langle \xi_i, x \rangle^m Q_i^*}{m!},$$

we have then $T = L(\mathrm{D})^*[P]$, and thus in view of (5.119)

$$\int_{\mathbb{R}^n} \langle f, \varphi \rangle = \int_{\mathbb{R}^n} \langle f, T[\varphi] \rangle = \int_{\mathbb{R}^n} \langle f, L(\mathrm{D})^*[P]\varphi \rangle.$$

Since $L(\mathrm{D})f = 0$ and by integration by parts, we deduce that

$$\int_{\mathbb{R}^n} \langle f, L(\mathrm{D})^*[P]\varphi \rangle = \int_{\mathbb{R}^n} \langle f, L(\mathrm{D})^*[P]\varphi \rangle - \int_{\mathbb{R}^n} \langle L(\mathrm{D})f, P\varphi \rangle$$

$$= \int_{\mathbb{R}^n} \langle f, L(\mathrm{D})^*[P]\varphi - L(\mathrm{D})^*[P\varphi] \rangle ,$$

and the conclusion (5.120) then follows. □

---

[32] Proposition 5.10 is due to Pierre Bousquet and to the author [12] under the additional assumption that $L(\mathrm{D})$ is cocancelling, and in the general case to Bogdan Raiţă [53].

Proof of Theorem 5.13 Let $G_A : \mathbb{R}^n \setminus \{0\} \to \mathrm{Lin}(V, E)$ be the representation kernel given by Proposition 5.2. Fixing a function $\varrho \in C_c^\infty(\mathbb{R}^n, \mathbb{R})$ such that $\varrho = 1$ on $B_{1/4}(0)$ and $\varrho = 0$ on $\mathbb{R}^n \setminus B_{1/2}(0)$, we define the kernels $H_A : \mathbb{R}^n \times \mathbb{R}^n \to \mathrm{Lin}(V, E)$ and $K_A : \mathbb{R}^n \times \mathbb{R}^n \to \mathrm{Lin}(V, E)$ for $x, y \in (\mathbb{R}^n \setminus \{0\}) \times \mathbb{R}^n$ with $x \neq y$ by

$$H_A(x, y) := \varrho\Big(\frac{y}{|x|}\Big) \mathrm{D}^\ell G_A(x)$$

and

$$K_A(x, y) := \mathrm{D}^\ell G_A(x - y) - \varrho\Big(\frac{y}{|x|}\Big) \mathrm{D}^\ell G_A(x),$$

so that $\mathrm{D}^\ell G_A(x - y) = H_A(x, y) + K_A(x, y)$. Letting $L(\mathrm{D})$ be the differential operator of order m given by Proposition 5.3, $L(\mathrm{D})$ is cocancelling and $L(\mathrm{D})A(\mathrm{D})u = 0$ in $\mathbb{R}^n$. By Proposition 5.10 and by the homogeneity of $\mathrm{D}_A^\ell G$ following from the condition that $\ell > k - n$ (Proposition 5.2 (ii)), we have for every $x \in \mathbb{R}^n \setminus \{0\}$,

$$\begin{aligned}
\Big|\int_{\mathbb{R}^n} H_A(x, y)[A(\mathrm{D})u(y)] \, dy\Big| & \\
&\leq C_{53} \sum_{j=1}^m \int_{B_{|x|/2}(0)} |A(\mathrm{D})u(y)||y|^j |\mathrm{D}_y^j H(x, y)| \, dy \\
&\leq C_{54} \sum_{j=1}^m \int_{B_{|x|/2}(0)} \frac{|A(\mathrm{D})u(y)||y|^j}{|x|^{n-k+\ell+j}} \, dy \\
&\leq C_{55} \int_{B_{|x|/2}(0)} \frac{|A(\mathrm{D})u(y)||y|}{|x|^{n-k+\ell+1}} \, dy,
\end{aligned} \quad (5.121)$$

and thus

$$\begin{aligned}
\int_{\mathbb{R}^n} \Big|\int_{\mathbb{R}^n} H_A(x, y)[A(\mathrm{D})u(y)] \, dy\Big| & \frac{dx}{|x|^{k-\ell}} \\
&\leq C_{56} \int_{\mathbb{R}^n} \int_{B_{|x|/2}(0)} \frac{|y||A(\mathrm{D})u(y)|}{|x|^{n+1}} \, dy \, dx \\
&= C_{56} \int_{\mathbb{R}^n} \int_{\mathbb{R}^n \setminus B_{2|y|}(0)} \frac{|y||A(\mathrm{D})u(y)|}{|x|^{n+1}} \, dx \, dy \leq C_{57} \int_{\mathbb{R}^n} |A(\mathrm{D})u|.
\end{aligned} \quad (5.122)$$

5 Injective Ellipticity, Cancelling Operators and Endpoint Sobolev Inequalities

Next we have

$$\int_{\mathbb{R}^n} \left| \int_{\mathbb{R}^n} K_A(x,y)[A(\mathrm{D})u(y)] \, \mathrm{d}y \right| \frac{\mathrm{d}x}{|x|^{k-\ell}}$$
$$\leq \int_{\mathbb{R}^n} \int_{\mathbb{R}^n} \frac{|K_A(x,y)|}{|x|^{k-\ell}} \, \mathrm{d}x \, |A(\mathrm{D})u(y)| \, \mathrm{d}y \, . \tag{5.123}$$

Again by homogeneity of $\mathrm{D}^\ell G_A$, we have if $|x| < 2|y|$

$$|K_A(x,y)| \leq \frac{C_{58}}{|x-y|^{n-k+\ell}},$$

and if $|x| \geq 2|y|$

$$|K_A(x,y)| \leq \frac{C_{59}|y|}{|x|^{n-k+\ell+1}},$$

so that, since $k - \ell < n$,

$$\int_{\mathbb{R}^n} \frac{|K_A(x,y)|}{|x|^{k-\ell}} \, \mathrm{d}x$$
$$\leq C_{60} \Big(\int_{B_{2|y|}(0)} \frac{\mathrm{d}x}{|x-y|^{n-k+\ell}|x|^{k-\ell}} + \int_{\mathbb{R}^n \setminus B_{2|y|}(0)} \frac{|y|}{|x|^{n+1}} \, \mathrm{d}x \Big) \leq C_{61} \, . \tag{5.124}$$

We conclude from (5.123) and (5.124) that

$$\int_{\mathbb{R}^n} \left| \int_{\mathbb{R}^n} K_A(x,y)[A(\mathrm{D})u(y)] \, \mathrm{d}y \right| \frac{\mathrm{d}x}{|x|^{k-\ell}} \leq C_{62} \int_{\mathbb{R}^n} |A(\mathrm{D})u| \, . \tag{5.125}$$

The estimate (5.118) then follows from the inequalities (5.122) and (5.125).

For the necessity of the cancellation, taking a vector $e \in \bigcap_{\xi \in \mathbb{R}^n \setminus \{0\}} A(\xi)[V]$ and letting the sequence $(u_j)_{j \in \mathbb{N}}$ be given by Lemma 5.4 for this vector e, we have by (5.118) and by Fatou's lemma

$$\int_{B_1(0)} \frac{|\mathrm{D}^\ell G_A(x)[e]|}{|x|^{k-\ell}} \, \mathrm{d}x \leq \liminf_{j \to \infty} \int_{\mathbb{R}^n} \frac{|\mathrm{D}^\ell u_j(x)|}{|x|^{k-\ell}} \, \mathrm{d}x$$
$$\leq \lim_{j \to \infty} C_{63} \int_{\mathbb{R}^n} |A(\mathrm{D})u_j|$$
$$\leq C_{64}|e| \, .$$

Since $\ell > n - k$, $\mathrm{D}^\ell G_A[e] : \mathbb{R}^n \setminus \{0\} \to V$ is homogeneous of degree $n - (k - \ell)$ (by Proposition 5.2 (ii)); this implies that $G_A[e] = 0$ on $\mathbb{R}^n$ and thus $e = 0$. □

5.5.4 Uniform Estimate and Weakly Cancelling Operators

By Theorem 5.13 with $\ell = k - (n-1) > 0$ and the Sobolev representation formula, one has when $A(D)$ is injectively elliptic and cancelling that for every $x \in \mathbb{R}^n$

$$|D^{k-n}u(x)| \le C_{65} \int_{\mathbb{R}^n} \frac{|D^{k-n-1}u(y)|}{|x-y|^{n-1}}\, dx \le C_{66} \int_{\mathbb{R}^n} |A(D)[u]|\,. \tag{5.126}$$

It turns out however that the cancellation condition is not necessary for such an L^∞ estimate and can be replaced, as showed by Bogdan RAIȚĂ, by a weaker condition.

Theorem 5.14 (L^∞ **Estimate for Cancelling Operators**[33]) *Let $n \in \mathbb{N} \setminus \{0\}$, let V and E be finite-dimensional vector spaces, and let $A(D)$ be a homogeneous constant coefficient differential operator of order $k \in \mathbb{N} \setminus \{0\}$ from V to E on $\mathbb{R}^n$. If $A(D)$ is injectively elliptic and if $k \ge n$, then there exists a constant $C \in (0, +\infty)$ such that for every $u \in C_c^\infty(\mathbb{R}^n, V)$ and every $x \in \mathbb{R}^n$*

$$|D^{k-n}u(x)| \le C \int_{\mathbb{R}^n} |A(D)[u]|\,, \tag{5.127}$$

if and only if for every $e \in \bigcap_{\xi \in \mathbb{R}^n \setminus \{0\}} A(\xi)[V]$, one has

$$\int_{\mathbb{S}^{n-1}} \xi^{\otimes k-n} A(\xi)^{-1}[e]\, d\xi = 0\,. \tag{5.128}$$

A vector differential operator $A(D)$ satisfying the condition (5.128) is known as a *weakly cancelling* operator [53].

Proof of Theorem 5.14 We first assume that the condition (5.128) holds. We let $G_A : \mathbb{R}^n \setminus \{0\} \to \mathrm{Lin}(V, E)$ be the representation kernel given by Proposition 5.2. Differentiating $k - n$ times the identity (5.15), we get for every $x \in \mathbb{R}^n$

$$\begin{aligned} D^{k-n}u(x) &= \int_{\mathbb{R}^n} D^{k-n} G_A(x-y)[A(D)u(y)]\, dy \\ &= \int_{\mathbb{R}^n} D^{k-n} G_A(h)[A(D)u(x-h)]\, dh\,. \end{aligned} \tag{5.129}$$

With $P_A : \mathbb{R}^n \to \mathrm{Lin}(V, E)$ being the homogeneous polynomial of degree $k - n$ given by Proposition 5.2 5.2, we observe that $D^{k-n} P_A : \mathbb{R}^n \to$

---

[33] Theorem 5.14 is due to Bogdan Raiță [53]. The stronger cancellation condition was proved to be sufficient by Pierre BOUSQUET and the author [12], as a consequence of the Hardy inequality of Theorem 5.13.

5 Injective Ellipticity, Cancelling Operators and Endpoint Sobolev Inequalities

$\mathrm{Lin}_{\mathrm{sym}}^{k-n}(\mathbb{R}^n, \mathrm{Lin}(E, V))$ is a constant polynomial that we identify to some $P_A \in \mathrm{Lin}(V, E)$ and we have for every $h \in \mathbb{R}^n \setminus \{0\}$,

$$\mathrm{D}^{k-n} G_A(h) = \mathrm{D}^{k-n} G_A(h/|h|) + \mathrm{D}^{k-n} P_A \ln|h| . \tag{5.130}$$

Since $\mathrm{D}^{k-n} G_A$ is bounded on the unit sphere of $\mathbb{R}^n$, we first have immediately

$$\left| \int_{\mathbb{R}^n} \mathrm{D}^{k-n} G_A(h/|h|)[A(\mathrm{D})u(x-h)] \, \mathrm{d}y \right| \le C_{67} \int_{\mathbb{R}^n} |A(\mathrm{D})u| . \tag{5.131}$$

Next, we take a function $\theta \in C_c^\infty(\mathbb{R}, \mathbb{R})$ satisfying $\theta(t) = t$ when $|t| \le 1/2$ and $\theta(t) = 0$ when $|t| \ge 1$. For every $\lambda \in (0, +\infty)$ and every $\mu \in \mathrm{Lin}_{\mathrm{sym}}^{k-n}(\mathbb{R}^n, V)$, we define the function $\varphi_{\lambda,\mu} \in C_c^\infty(\mathbb{R}^n, E)$ by the condition that for every $h \in \mathbb{R}^n \setminus \{0\}$ and every $e \in E$, we have

$$\langle \varphi_{\lambda,\mu}(h), e \rangle = \langle \mathrm{D}^{k-n} P_A[e], \mu \rangle \frac{\theta(\lambda \ln|h|)}{\lambda} . \tag{5.132}$$

Next, by Proposition 5.2 (vi) and by our assumption (5.128), for every vector $e \in \bigcap_{\xi \in \mathbb{R}^n \setminus \{0\}} A(\xi)[V]$, we have $\mathrm{D}^{k-n} P_A[e] = 0$. Hence, for every $h \in \mathbb{R}^n$ and every $\lambda \in (0, +\infty)$, we have

$$\varphi_{\lambda,\mu}(h) \in \left(\bigcap_{\xi \in \mathbb{R}^n \setminus \{0\}} A(\xi)[V] \right)^\perp \tag{5.133}$$

in view of (5.132). Moreover, for every $j \in \mathbb{N} \setminus \{0\}$ and $x \in \mathbb{R}^n \setminus \{0\}$, we have as a consequence of (5.132)

$$|\mathrm{D}^j \varphi_{\lambda,\mu}(x)| \le C_{68}(1 + \lambda^{j-1})|\mu|/|x|^j . \tag{5.134}$$

In view of (5.133), Proposition 5.10 is applicable and yields for every $\lambda \in (0, 1)$,

$$\left| \int_{\mathbb{R}^n} \langle A(\mathrm{D})u(x-h), \varphi_{\lambda,\mu}(h) \rangle \, \mathrm{d}h \right| \le C_{69} |\mu| \int_{\mathbb{R}^n} |A(\mathrm{D})u| . \tag{5.135}$$

Since on the other hand, we have by (5.132)

$$\int_{\mathbb{R}^n} \langle \mathrm{D}^{k-n} P_A[A(\mathrm{D})u(x-h)], \mu \rangle \ln|h| \, \mathrm{d}h$$

$$= \lim_{\lambda \to 0} \int_{\mathbb{R}^n} \langle \mathrm{D}^{k-n} P_A[A(\mathrm{D})u(x-h)], \mu \rangle \frac{\theta(\lambda \ln|h|)}{\lambda} \, \mathrm{d}h$$

$$= \lim_{\lambda \to 0} \int_{\mathbb{R}^n} \langle \varphi_{\lambda,\mu}(h), A(\mathrm{D})u(x-h) \rangle \, \mathrm{d}h ,$$

$$\tag{5.136}$$

we reach the inequality (5.127) through the identities (5.129), (5.130) and (5.136), and the estimates (5.131) and (5.135).

For the necessity of the condition (5.128), taking a vector $e \in \bigcap_{\xi \in \mathbb{R}^n \setminus \{0\}} A(\xi)[V]$ and letting the sequence $(u_j)_{j \in \mathbb{N}}$ be given by Lemma 5.4 for this vector e, we have by (5.127) and by Fatou's lemma

$$\limsup_{j \to \infty} \sup_{x \in \mathbb{R}^n} |u_j(x)| \le \lim_{j \to \infty} C_{70} \int_{\mathbb{R}^n} |A(D) u_j| \le C_{71} |e| . \tag{5.137}$$

On the other hand by Proposition 5.2 (ii) and (vi), we have for every $e \in E$ and $x \in \mathbb{R}^n \setminus \{0\}$

$$\begin{aligned} D^{k-n} G_A(x)[e] &= D^{k-n} G_A(x/|x|)[e] + \ln|x| \, P_A(x)[e] \\ &= D^{k-n} G_A(x/|x|)[e] + \ln|x| \int_{\mathbb{S}^{n-1}} \frac{\xi^{\otimes k-n} A(\xi)^{-1}[e]}{(2\pi i)^n} \, d\xi . \end{aligned} \tag{5.138}$$

Since $u_j \to G_A[e]$ on $B_1(0)$, we conclude by (5.137) and (5.138) that the condition (5.128) holds. □

References

1. Adams, R.A., Fournier, J.J.F.: Sobolev Spaces, 2nd edn. Pure and Applied Mathematics, vol. 140. Elsevier/Academic Press, Amsterdam (2003)
2. Agmon, S.: The L_p approach to the Dirichlet problem. I: Regularity theorems. Ann. Scuola Norm. Sup. Pisa **13**, 405–448 (1959)
3. Agmon, S.: Lectures on Elliptic Boundary Value Problems. Van Nostrand Mathematical Studies, vol. 2, Van Nostrand, Princeton (1965)
4. Arroyo-Rabasa, A., De Philippis, G., Hirsch, J., Rindler, F.: Dimensional estimates and rectifiability for measures satisfying linear PDE constraints. Geom. Funct. Anal. **29**(3), 639–658 (2019). https://doi.org/10.1007/s00039-019-00497-1
5. Booß-Bavnbek, B., Wojciechowski, K.P.: Elliptic boundary problems for Dirac operators. In: Mathematics: Theory & Applications. Birkhäuser, Boston (1993). https://doi.org/10.1007/978-1-4612-0337-7
6. Bourgain, J., Brezis, H.: Sur l'équation div $u = f$. C. R. Math. Acad. Sci. Paris **334**(11), 973–976 (2002). https://doi.org/10.1016/S1631-073X(02)02344-0
7. Bourgain, J., Brezis, H.: On the equation div $Y = f$ and application to control of phases. J. Am. Math. Soc. **16**(2), 393–426 (2003). https://doi.org/10.1090/S0894-0347-02-00411-3
8. Bourgain, J., Brezis, H.: New estimates for the Laplacian, the div-curl, and related Hodge systems. C. R. Math. Acad. Sci. Paris **338**(7), 539–543 (2004). https://doi.org/10.1016/j.crma.2003.12.031
9. Bourgain, J., Brezis, H.: New estimates for elliptic equations and Hodge type systems. J. Eur. Math. Soc. (JEMS) **9**(2), 277–315 (2007). https://doi.org/10.4171/JEMS/80
10. Bourgain, J., Brezis, H., Mironescu, P.: $H^{1/2}$ maps with values into the circle: minimal connections, lifting, and the Ginzburg-Landau equation. Publ. Math. Inst. Hautes Études Sci. **99**, 1–115 (2004). https://doi.org/10.1007/s10240-004-0019-5

11. Bousquet, P., Mironescu, P.: An elementary proof of an inequality of Maz'ya involving L^1 vector fields. In: Nonlinear Elliptic Partial Differential Equations. Contemporary Mathematics, vol. 540, pp. 59–63. American Mathematics Society, Providence (2011). https://doi.org/10.1090/conm/540/10659
12. Bousquet, P., Van Schaftingen, J.: Hardy-Sobolev inequalities for vector fields and canceling differential operators. Ind. Univ. Math. J. **63**(5), 1419–1445 (2014). https://doi.org/10.1512/iumj.2014.63.5395
13. Bousquet, P., Mironescu, P., Russ, E.: A limiting case for the divergence equation. Math. Z. **274**(1–2), 427–460 (2013). https://doi.org/10.1007/s00209-012-1077-x
14. Bousquet, P., Russ, E., Wang, Y., Yung, P.-L.: Approximation in higher-order Sobolev spaces and Hodge systems, J. Funct. Anal. **276**(5), 1430–1478 (2019). https://doi.org/10.1016/j.jfa.2018.08.003
15. Breit, D., Diening, L., Gmeineder, F.: On the trace operator for functions of bounded A-variation. Anal. PDE **13**(2), 559–594 (2020). https://doi.org/10.2140/apde.2020.13.559
16. Brezis, H.: Functional Analysis, Sobolev Spaces and Partial Differential Equations. Universitext. Springer, New York (2011). https://doi.org/10.1007/978-0-387-70914-7
17. Brezis, H., Van Schaftingen, J.: Boundary estimates for elliptic systems with L^1-data. Calc. Var. Partial Differ. Equ. **30**(3), 369–388 (2007). https://doi.org/10.1007/s00526-007-0094-9
18. Brezis, H., Van Schaftingen, J.:Circulation integrals and critical Sobolev spaces: problems of optimal constants. In: Perspectives in Partial Differential Equations, Harmonic Analysis and Applications. Proceedings of Symposia in Pure Mathematics, vol. 79, pp. 33–47. American Mathematical Society, Providence (2008). https://doi.org/10.1090/pspum/079/2500488
19. Calderon, A.P., Zygmund, A.: On the existence of certain singular integrals. Acta Math. **88**, 85–139 (1952). https://doi.org/10.1007/BF02392130
20. Cantor, M.: Elliptic operators and the decomposition of tensor fields. Bull. Amer. Math. Soc. **5**(3), 235–262 (1981). https://doi.org/10.1090/S0273-0979-1981-14934-X
21. Chanillo, S., Van Schaftingen, J., Yung, P.L.: Variations on a proof of a borderline Bourgain-Brezis Sobolev embedding theorem. Chin. Ann. Math. Ser. B **38**(1), 235–252 (2017). https://doi.org/10.1007/s11401-016-1069-y
22. Chanillo, S., Van Schaftingen, J., Yung, P.L.: Bourgain-Brezis inequalities on symmetric spaces of non-compact type. J. Funct. Anal. **273**(4), 1504–1547 (2017). https://doi.org/10.1016/j.jfa.2017.05.005
23. Ciarlet, P.G.: Linear and Nonlinear Functional Analysis with Applications. Society for Industrial and Applied Mathematics, Philadelphia (2013)
24. Cohen, A., Dahmen, W., Daubechies, I., DeVore, R.: Harmonic analysis of the space BV. Rev. Mat. Iberoamericana **19**(1), 235–263 (2003). https://doi.org/10.4171/RMI/345
25. Conti, S., Faraco, D., Maggi, F.: A new approach to counterexamples to L^1 estimates: Korn's inequality, geometric rigidity, and regularity for gradients of separately convex functions. Arch. Ration. Mech. Anal. **175**(2), 287–300 (2005). https://doi.org/10.1007/s00205-004-0350-5
26. Dacorogna, B., Gangbo, W., Kneuss, O.: Symplectic factorization, Darboux theorem and ellipticity. Ann. Inst. H. Poincaré Anal. Non Linéaire **35**(2), 327–356 (2018). https://doi.org/10.1016/j.anihpc.2017.04.005
27. Dain, S.: Generalized Korn's inequality and conformal Killing vectors. Calc. Var. Partial Differ. Equ. **25**(4), 535–540 (2006). https://doi.org/10.1007/s00526-005-0371-4
28. de Leeuw, K., Mirkil, H.: Majorations dans L_∞ des opérateurs différentiels à coefficients constants. C. R. Acad. Sci. Paris **254**, 2286–2288 (1962)
29. de Leeuw, K., Mirkil, H.: A priori estimates for differential operators in L_∞ norm. Ill. J. Math. **8**, 112–124 (1964). https://doi.org/10.1215/ijm/1256067459
30. De Philippis, G., Rindler, F.: On the structure of $\mathcal{A}$-free measures and applications. Ann. Math. **184**(3), 1017–1039 (2016). https://doi.org/10.4007/annals.2016.184.3.10
31. Di Nezza, E., Palatucci, G., Valdinoci, E.: Hitchhiker's guide to the fractional Sobolev spaces. Bull. Sci. Math. **136**(5), 521–573 (2012). https://doi.org/10.1016/j.bulsci.2011.12.004
32. Evans, L.C.: Partial Differential Equations. Graduate Studies in Mathematics, 2nd edn., vol. 19. American Mathematical Society, Providence (2010). https://doi.org/10.1090/gsm/019

33. Fefferman, C.: Characterizations of bounded mean oscillation. Bull. Am. Math. Soc. **77**, 587–588 (1971). https://doi.org/10.1090/S0002-9904-1971-12763-5
34. Fefferman, C., Stein, E.M.: H^p spaces of several variables. Acta Math. **129**(3–4), 137–193 (1972). https://doi.org/10.1007/BF02392215
35. Gagliardo, E.: Proprietà di alcune classi di funzioni in più variabili. Ricerche Mat. **7**, 102–137 (1958)
36. Gmeineder, F., Raiță, B.: Embeddings for $\mathbb{A}$-weakly differentiable functions on domains. J. Funct. Anal. **277**(12), 108278 (2019). https://doi.org/10.1016/j.jfa.2019.108278
37. Gmeineder, F., Raiță, B., Van Schaftingen, J.: On limiting trace inequalities for vectorial differential operators. Ind. Univ. Math. J. **70**(5), 2133–2176 (2021). https://doi.org/10.1512/iumj.2021.70.8682
38. Gmeineder, F., Raiță, B., Van Schaftingen, J.: Boundary ellipticity and limiting L^1-estimates on halfspaces (2022). arXiv:2211.08167
39. Grubb, G.: Pseudo-differential boundary problems in L_p spaces. Commun. Partial Differ. Equ. **15**(3), 289–340 (1990). https://doi.org/10.1080/03605309908820688
40. Hörmander, L.: Differentiability properties of solutions of systems of differential equations. Ark. Mat. **3**, 527–535 (1958). https://doi.org/10.1007/BF02589514
41. Hörmander, L.: The Analysis of linear partial differential operators. In: I: Distribution Theory and Fourier Analysis. Grundlehren der Mathematischen Wissenschaften, 2nd edn., vol. 256. Springer, Berlin (1990). https://doi.org/10.1007/978-3-642-61497-2
42. Kirchheim, B., Kristensen, J.: Automatic convexity of rank-1 convex functions. C. R. Math. Acad. Sci. Paris **349**(7–8), 407–409 (2011). https://doi.org/10.1016/j.crma.2011.03.013
43. Kirchheim, B., Kristensen, J.: On rank one convex functions that are homogeneous of degree one. Arch. Ration. Mech. Anal. **221**(1), 527–558 (2016). https://doi.org/10.1007/s00205-016-0967-1
44. Kolyada, V.I.: On the embedding of Sobolev spaces (in Russian). Mat. Zametki **54**(3), 48–71 (1993); English transl., Math. Notes **54** (1993), no. 3–4, 908–922 (1994)
45. Lanzani, L., Stein, E.M.: A note on div curl inequalities. Math. Res. Lett. **12**(1), 57–61 (2005) . https://doi.org/10.4310/MRL.2005.v12.n1.a6
46. Leoni, G.: A First Course in Sobolev Spaces. Graduate Studies in Mathematics, 2nd edn., vol. 181. American Mathematical Society, Providence (2017)
47. Maz'ya, V.: Estimates for differential operators of vector analysis involving L^1-norm. J. Eur. Math. Soc. **12**(1), 221–240 (2010). https://doi.org/10.4171/JEMS/195
48. Maz'ya, V.: Sobolev Spaces with Applications to Elliptic Partial Differential Equations. Grundlehren der Mathematischen Wissenschaften, 2nd edn., vol. 342. Springer, Heidelberg (2011). https://doi.org/10.1007/978-3-642-15564-2
49. Mitrea, I., Mitrea, M.: A remark on the regularity of the div-curl system. Proc. Am. Math. Soc. **137**(5), 1729–1733 (2009). https://doi.org/10.1090/S0002-9939-08-09671-8
50. Morrey, C.B. Jr.: Multiple Integrals in the Calculus of Variations. Die Grundlehren der mathematischen Wissenschaften, Band 130. Springer, New York (1966). https://doi.org/10.1007/978-3-540-69952-1
51. Nirenberg, L.: On elliptic partial differential equations. Ann. Scuola Norm. Sup. Pisa **13**, 115–162 (1959)
52. Ornstein, D.: A non-equality for differential operators in the L^1 norm. Arch. Rational Mech. Anal. **11**, 40–49 (1962). https://doi.org/10.1007/BF00253928
53. Raiță, B.: Critical L^p-differentiability of $BV^{\mathbb{A}}$-maps and canceling operators. Trans. Am. Math. Soc. **372**(10), 7297–7326 (2019). https://doi.org/10.1090/tran/7878
54. Roginskaya, M., Wojciechowski, M.: Singularity of vector valued measures in terms of Fourier transform. J. Fourier Anal. Appl. **12**(2), 213–223. https://doi.org/10.1007/s00041-005-5030-9
55. Schmitt, B.J., Winkler, M.: On embeddings between BV and $\dot{W}^{s,p}$, Preprint no. 6, Lehrstuhl I für Mathematik, RWTH Aachen (2000)
56. Schulze, B.W.: Adjungierte elliptischer Randwert-Probleme und Anwendungen auf über- und unterbestimmte Systeme. Math. Nachr. **89**, 225–245 (1979). https://doi.org/10.1002/mana.19790890120

57. Smirnov, S.K.: Decomposition of solenoidal vector charges into elementary solenoids, and the structure of normal one-dimensional flows (in Russian). Algebra i Analiz **5**)(4), 206–238 (1993); English transl., St. Petersburg Math. J. **5** (1994), no. 4, 841–867
58. Sobolev, S.: Sur un théorème d'analyse fonctionnelle (Russian with French Summary). Rec. Math. Moscou n. Ser. **4**, 471–497 (1938)
59. Solonnikov, V. A.: Overdetermined elliptic boundary value problems (in Russian). Zap. Naučn. Sem. Leningrad. Otdel. Mat. Inst. Steklov. **21**, 112–158 (1971)
60. Solonnikov, V.A.: Inequalities for functions of the classes $\dot{W}_p^m(\mathbb{R}^n)$. Zapiski Nauchnykh Seminarov Leningradskogo Otdeleniya Matematicheskogo Instituta im (in Russian). V. A. Steklova Akademii Nauk SSSR **27**, 194–210 (1972); English transl., J. Sov. Math. **3** (1975), 549–564
61. Spencer, D.C.: Overdetermined systems of linear partial differential equations. Bull. Am. Math. Soc. **75**, 179–239 (1969). https://doi.org/10.1090/S0002-9904-1969-12129-4
62. Stein, E.M.: Singular Integrals and Differentiability Properties of Functions. Princeton Mathematical Series, vol. 30. Princeton University Press, Princeton (1970)
63. Stein, E.M.: Harmonic Analysis: Real-Variable Methods, Orthogonality, and Oscillatory Integrals. With the Assistance of Timothy S. Murphy, Princeton Mathematical Series, vol. 43. Princeton University Press, Princeton (1993)
64. Strauss, M.J.: Variations of Korn's and Sobolev's Equalities. Partial Differential Equations (Univ. California, Berkeley, Calif., 1971). Proceedings of Symposia in Pure Mathematics, vol. XXIII, pp. 207–214. American Mathematical Society, Providence (1973)
65. Taibleson, M.H.: On the theory of Lipschitz spaces of distributions on Euclidean n-space. I. Principal properties. J. Math. Mech. **13**, 407–479 (1964)
66. Timoshenko, S., Goodier, J.N.: Theory of Elasticity, 2nd edn. McGraw-Hill, New York (1951)
67. Triebel, H.: Theory of Function Spaces. Monographs in Mathematics, vol. 78. Birkhäuser, Basel (1983). https://doi.org/10.1007/978-3-0346-0416-1
68. Van Schaftingen, J.: A simple proof of an inequality of Bourgain, Brezis and Mironescu. C. R. Math. Acad. Sci. Paris **338**(1), 23–26 (2004). https://doi.org/10.1016/j.crma.2003.10.036
69. Van Schaftingen, J.: Estimates for L^1-vector fields. C. R. Math. Acad. Sci. Paris **339**(3), 181–186 (2004). https://doi.org/10.1016/j.crma.2004.05.013
70. Van Schaftingen, J.: Estimates for L^1 vector fields with a second order condition. Acad. Roy. Belg. Bull. Cl. Sci. **15**(1–6), 103–112 (2004)
71. Van Schaftingen, J.: Function spaces between BMO and critical Sobolev spaces. J. Funct. Anal. **236**(2), 490–516 (2006). https://doi.org/10.1016/j.jfa.2006.03.011
72. Van Schaftingen, J.: Estimates for L^1 vector fields under higher-order differential conditions. J. Eur. Math. Soc. **10**(4), 867–882 (2008). https://doi.org/10.4171/JEMS/133
73. Van Schaftingen, J.: Limiting fractional and Lorentz space estimates of differential forms. Proc. Am. Math. Soc. **138**(1), 235–240 (2010). https://doi.org/10.1090/S0002-9939-09-10005-9
74. Van Schaftingen, J.: Limiting Sobolev inequalities for vector fields and canceling linear differential operators. J. Eur. Math. Soc. **15**(3), 877–921 (2013). https://doi.org/10.4171/JEMS/380
75. Van Schaftingen, J.: Limiting Bourgain-Brezis estimates for systems of linear differential equations: theme and variations. J. Fixed Point Theory Appl. **15**(2), 273–297 (2014). https://doi.org/10.1007/s11784-014-0177-0
76. Van Schaftingen, J.: Limiting Sobolev estimates for vector fields and cancelling differential operators. In: Lukeš, J., Mihula, Z., Pick, L., Turčinová, H. (eds.) Function Spaces and Applications XII (Pazeky nad Jizerou 2023), pp. 135–152. MatfyzPress, Charles University, Prague (2023)
77. Willem, M.: Functional Analysis: Fundamentals and Applications. Cornerstones, Birkhäuser/Springer, New York (2013). https://doi.org/10.1007/978-1-4614-7004-5

LECTURE NOTES IN MATHEMATICS

Editors in Chief: J.-M. Morel, B. Teissier;

Editorial Policy

1. Lecture Notes aim to report new developments in all areas of mathematics and their applications – quickly, informally and at a high level. Mathematical texts analysing new developments in modelling and numerical simulation are welcome.

    Manuscripts should be reasonably self-contained and rounded off. Thus they may, and often will, present not only results of the author but also related work by other people. They may be based on specialised lecture courses. Furthermore, the manuscripts should provide sufficient motivation, examples and applications. This clearly distinguishes Lecture Notes from journal articles or technical reports which normally are very concise. Articles intended for a journal but too long to be accepted by most journals, usually do not have this "lecture notes" character. For similar reasons it is unusual for doctoral theses to be accepted for the Lecture Notes series, though habilitation theses may be appropriate.

2. Besides monographs, multi-author manuscripts resulting from SUMMER SCHOOLS or similar INTENSIVE COURSES are welcome, provided their objective was held to present an active mathematical topic to an audience at the beginning or intermediate graduate level (a list of participants should be provided).

    The resulting manuscript should not be just a collection of course notes, but should require advance planning and coordination among the main lecturers. The subject matter should dictate the structure of the book. This structure should be motivated and explained in a scientific introduction, and the notation, references, index and formulation of results should be, if possible, unified by the editors. Each contribution should have an abstract and an introduction referring to the other contributions. In other words, more preparatory work must go into a multi-authored volume than simply assembling a disparate collection of papers, communicated at the event.

3. Manuscripts should be submitted either online at www.editorialmanager.com/lnm to Springer's mathematics editorial in Heidelberg, or electronically to one of the series editors. Authors should be aware that incomplete or insufficiently close-to-final manuscripts almost always result in longer refereeing times and nevertheless unclear referees' recommendations, making further refereeing of a final draft necessary. The strict minimum amount of material that will be considered should include a detailed outline describing the planned contents of each chapter, a bibliography and several sample chapters. Parallel submission of a manuscript to another publisher while under consideration for LNM is not acceptable and can lead to rejection.

4. In general, **monographs** will be sent out to at least 2 external referees for evaluation.

    A final decision to publish can be made only on the basis of the complete manuscript, however a refereeing process leading to a preliminary decision can be based on a pre-final or incomplete manuscript.

    Volume Editors of **multi-author works** are expected to arrange for the refereeing, to the usual scientific standards, of the individual contributions. If the resulting reports can be forwarded to the LNM Editorial Board, this is very helpful. If no reports are forwarded or if other questions remain unclear in respect of homogeneity etc, the series editors may wish to consult external referees for an overall evaluation of the volume.

5. Manuscripts should in general be submitted in English. Final manuscripts should contain at least 100 pages of mathematical text and should always include

    - a table of contents;
    - an informative introduction, with adequate motivation and perhaps some historical remarks: it should be accessible to a reader not intimately familiar with the topic treated;
    - a subject index: as a rule this is genuinely helpful for the reader.
    - For evaluation purposes, manuscripts should be submitted as pdf files.

6. Careful preparation of the manuscripts will help keep production time short besides ensuring satisfactory appearance of the finished book in print and online. After acceptance of the manuscript authors will be asked to prepare the final LaTeX source files (see LaTeX templates online: https://www.springer.com/gb/authors-editors/book-authors-editors/manuscriptpreparation/5636) plus the corresponding pdf- or zipped ps-file. The LaTeX source files are essential for producing the full-text online version of the book, see http://link.springer.com/bookseries/304 for the existing online volumes of LNM). The technical production of a Lecture Notes volume takes approximately 12 weeks. Additional instructions, if necessary, are available on request from lnm@springer.com.

7. Authors receive a total of 30 free copies of their volume and free access to their book on SpringerLink, but no royalties. They are entitled to a discount of 33.3 % on the price of Springer books purchased for their personal use, if ordering directly from Springer.

8. Commitment to publish is made by a *Publishing Agreement*; contributing authors of multiauthor books are requested to sign a *Consent to Publish form*. Springer-Verlag registers the copyright for each volume. Authors are free to reuse material contained in their LNM volumes in later publications: a brief written (or e-mail) request for formal permission is sufficient.

Addresses:
Professor Jean-Michel Morel, CMLA, École Normale Supérieure de Cachan, France
E-mail: moreljeanmichel@gmail.com

Professor Bernard Teissier, Equipe Géométrie et Dynamique,
Institut de Mathématiques de Jussieu – Paris Rive Gauche, Paris, France
E-mail: bernard.teissier@imj-prg.fr

Springer: Ute McCrory, Mathematics, Heidelberg, Germany,
E-mail: lnm@springer.com

SPRINGER NATURE

GPSR Compliance

The European Union's (EU) General Product Safety Regulation (GPSR) is a set of rules that requires consumer products to be safe and our obligations to ensure this.

If you have any concerns about our products, you can contact us on ProductSafety@springernature.com

In case Publisher is established outside the EU, the EU authorized representative is:

Springer Nature Customer Service Center GmbH
Europaplatz 3
69115 Heidelberg, Germany

The manufacturer's authorised representative in the EU is Springer Nature Customer Service Centre GmbH, Europaplatz 3, 69115 Heidelberg, Germany. If you have any concerns regarding our products, please contact ProductSafety@springernature.com

Printed and bound by CPI Group (UK) Ltd, Croydon, CR0 4YY
26/03/2026
02078933-0016